VICTORIAN
INTERDISCIPLINARITY
AND THE SCIENCES

SCIENCE AND CULTURE IN THE NINETEENTH CENTURY
Bernard Lightman, Editor

VICTORIAN
INTERDISCIPLINARITY
AND THE SCIENCES

Rethinking *the* Specialization Thesis

EDITED BY BERNARD LIGHTMAN
AND EFRAM SERA-SHRIAR

University of Pittsburgh Press

Published by the University of Pittsburgh Press, Pittsburgh, Pa., 15260
Copyright © 2024, University of Pittsburgh Press
All rights reserved
Manufactured in the United States of America
Printed on acid-free paper
10 9 8 7 6 5 4 3 2 1

Cataloging-in-Publication data is available from the Library of Congress

ISBN 13: 978-0-8229-4814-8
ISBN 10: 0-8229-4814-1

Cover art: "Microscopio de tres cuerpos para las observaciones simultáneas." *El mundo físi-co: gravedad, gravitación, luz, calor, electricidad, magnetismo, etc.* by A. Guillemin. Barcelona Montaner y Simón, 1882.
Cover design: Melissa Dias-Mandoly

To our good friend, Rosemary Mitchell.
Her brilliance and warmth will always be remembered.
May she rest in peace.

CONTENTS

Acknowledgments

This project emerged through a collaboration between the Centre for Nineteenth-Century Studies at Durham University and the Leeds Centre for Victorian Studies at Leeds Trinity University. It all began when Bennett Zon approached Rosemary Mitchell about the possibility of cohosting a series of workshops on Victorian studies and interdisciplinarity. Soon we were also brought into this program of activity, and the result of this labor forms the foundation of this collection. We are thankful to both Bennett and Rosemary for initiating these conversations and supporting us in bringing these papers to print. We would also like to thank the contributors for all their hard work in writing their respective chapters, especially during the challenges of a global health pandemic. Thank you to Abby Collier and the University of Pittsburgh Press, as well as the anonymous reviewers for their collective support and constructive feedback throughout the review process. Finally, we are grateful to our families, who have been cheering us on throughout the period of compiling this collection.

VICTORIAN
INTERDISCIPLINARITY
AND THE SCIENCES

Introduction

RETHINKING DISCIPLINARY SPECIALISM IN VICTORIAN SCIENCES

BERNARD LIGHTMAN AND EFRAM SERA-SHRIAR

In 1847 the British physician James Cowles Prichard (1786–1848) delivered his presidential address before members of the Ethnological Society of London (f. 1843). For over three decades he had been a figurehead for the emerging research field, and his book *Researches into the Physical History of Mankind*, which was first published in 1813, was a formative text that defined both the theoretical foundation of early ethnological research and its methodological framework.[1] Rather than using the occasion to summarize the discipline's major achievements during the previous year, Prichard chose instead to discuss a more pressing issue—the position of ethnological research at the British Association for the Advancement of Science (BAAS; f. 1831).[2] During the 1840s ethnology remained a subsection within the Department of Natural History at BAAS meetings, and for many early ethnologists, such as Prichard, this organizational structure was problematic. It was believed that so long as ethnology remained subordinate to the larger research field of natural history, its progress and findings would be stunted. For ethnology to mature and be an important scientific specialism in its own right, it needed to be independent and recognized as a distinct discipline. Ethnologists wanted to have ownership over their research program, and to achieve this goal, it was essential for its practitioners to differentiate their research from that of other disciplines. Prichard therefore focused the rest of his presidential address on the relationship between ethnology and other branches of knowledge.

As he recounted in his address, ethnology may have begun as a form of natural history, but over the previous thirty years it had grown into something different. Ethnology, according to Prichard, was a historical pursuit that focused entirely on humanity's place in nature, while natural history was a contemporary study that explored the whole of the organic world. As he explained, it was "distinct from natural history, inasmuch as the object of its investigations is not *what is*, but *what has been*."[3] He continued to expound on how ethnology borrowed from many other "departments of knowledge," including linguistics, geography, archaeology, literature, anatomy, and history, to name a few examples. It used data from these other fields to trace the historical development of human groups over multiple generations.[4] Ethnology's engagement with these other departments of knowledge was, to use a somewhat anachronistic framing, "interdisciplinary" in nature.

What is striking about Prichard's address before members of the Ethnological Society of London in 1847 is how it raises questions concerning the conventional historiographical narrative about nineteenth-century disciplinary specialization. Prichard's discussion of a specific Victorian scientific discipline, ethnology, is but one of many examples that suggest the need to reassess the "specialization thesis"—the idea that nineteenth-century science fragmented into separate and specialized forms of knowledge, which led to the creation of modern disciplines. While it is certainly the case that new disciplines emerged during the nineteenth century, the intellectual landscape was far muddier, and in many cases these new forms of specialist knowledge continued to cross disciplinary boundaries while integrating ideas from other disciplines. As the Prichard example shows, his attempt to highlight ethnology's supposed expert knowledge was somewhat tenuous. Instead of showcasing the distinctive nature of ethnological research, he actually showed that the boundaries between ethnology and other newly formed disciplines were quite nebulous. The conception of ethnological expertise was actually intended to serve the rhetorical purpose of solidifying the place of the discipline within the larger scientific community. It was part of a vocational strategy for Prichard and other ethnological researchers to strengthen their authority within a dramatically transforming intellectual culture in Victorian Britain.

Prichard's story is by no means unique, even later in the century when it might be claimed that specialization had increased in importance. If we fast-forward to the mid-1870s, we see a similar rhetorical argument in the writings of the scientific naturalist Thomas Henry Huxley (1825–1895). In his famous address "On the Study of Biology," which he delivered in 1876 in connection with the special loan collection of scientific apparatuses to the South Kensington Museum, he framed biology as a kind of umbrella science that covered "all the phenomena which are exhibited by living things." Biology, according to Huxley, necessarily included humans, and therefore it had authority over other disciplines, including psychology, political science, and economics. If humans, as "living things," were responsible for its production, biologists could claim ownership of it.[5] It was therefore, like ethnology, a highly interdisciplinary research field.

Huxley's broad definition of biology was significant and part of a rhetorical strategy, one that was often used by the scientific naturalists as part of their attempts to gain cultural hegemony within Victorian Britain. The construction of disciplinary parameters during the middle of the nineteenth century was as much an exercise in expanding the boundaries of human knowledge as it was a performance in cultural politics.[6] Getting the balance right between interdisciplinary breadth and disciplinary focus was a difficult task. This volume follows this historical narrative by exploring the history of Victorian interdisciplinarity in the sciences through a series of interconnected case studies. The term *interdisciplinarity*, as we are using it here, can be defined as the process of bringing together two or more areas of knowledge into a single research field. This volume will expose the tension between the rhetorical push for disciplinary specialization and the actual practice of interdisciplinary integration during the nineteenth century.

Using the words *interdisciplinary* and *interdisciplinarity* when discussing Victorian science may strike some scholars as controversial and anachronistic, as they were not terms that were used commonly in the nineteenth century. We have even gone so far as to use the term *interdisciplinarity* in the title for the entire volume. It is nearly impossible to define a term such as *interdisciplinarity* precisely because of its messy epistemic ambition, which runs counter to the very reasons for disciplinary boundary making in the first place. And yet there remains the important task of historicizing the term's origin because of its continued impact on, and legacy within, academic fields today. Understanding the roots of this intellectual epistemic shift in knowledge production allows us to better appreciate the making of modern science in the broadest sense. Because the Victorian period is traditionally seen as the moment in which these very ideas of specialization formed, it is an ideal starting point for tracing the history of interdisciplinarity.

In early twentieth-century editions of James Murray's *A New En-*

glish Dictionary on Historical Principles, there is no entry for *interdiscipli-narity*. More recent sources confirm that this term, as well as *interdisci-plinary*, were rarely used by historical actors of the nineteenth century. When searched, the digital database C19: Nineteenth Century Index lists no book titles including either of these terms. Google Ngrams for "interdisciplinary" and "interdisciplinarity," which display graphs show-ing how often phrases or words have appeared in a corpus of books over time, confirm that these terms were used during the nineteenth century, though not commonly. The big jump in their usage occurs in the 1970s. So why do we persist in applying recently evolved terms to a historical period when they rarely appeared in print?

While we acknowledge the difficulties that can arise in using terms such as *interdisciplinary* and *interdisciplinarity*, it can be argued that there are several reasons for using a modern label. First, what we mean by interdisciplinarity is a fairly accurate description of how the relationship among scientific disciplines were conceived of for much of the nine-teenth century. What we now refer to as physics and astronomy, for example, were a part of what historical actors termed *natural philoso-phy*. Scientists routinely moved between physics and astronomy when they discussed issues in natural philosophy. Second, the twentieth-century understanding of interdisciplinarity did not arise ex nihilo. It had its roots in nineteenth-century currents of thought. While the dis-ciplinary landscape of the nineteenth century was transformed by the disaggregation of natural philosophy and natural history, the two pri-mary bodies of knowledge of the early and mid-century, the process of disaggregation was messy and lengthy.

Although new disciplines were formed, often they were conditioned by the disciplines that they were to be distinguished from. The chapters in this volume deal with this complicated process of disaggregation in different ways. Some of the new disciplines can be conceived of as being composed of hybrid fields. Or they can be seen as meta-disciplines, that is, a branch of knowledge designed to discipline other disciplines. Some of the chapters even go so far as to use the term *interdisciplinary* to refer to the new sciences, as the lines of demarcation dividing them from each other were porous and blurry. However, others note that the term is anachronistic and resist using it, or adopt fuzzier expressions like "disci-plinary transgression" or "cross-disciplinary work." Rather than viewing the volume as working toward a firm consensus, it should be understood as being more of a dialogue about the viability of interdisciplinarity as a category applicable to the nineteenth century. This dialogue has generat-ed novel insights into the complex evolution of scientific disciplines. The term *interdisciplinary*, while admittedly retrospective, nevertheless offers scholars a heuristic tool that helps us to identify and group together a set of intellectual practices characteristic of the nineteenth century.

SPECIALIZATION AND PROFESSIONALIZATION IN THE HISTORY OF VICTORIAN SCIENCE

To fully understand the history of interdisciplinarity, one should begin by critically reflecting on the "specialization thesis" that is dominant in the historiography on Victorian science. It is this thesis, after all, that has obscured our historical perspectives relating to disciplinary forma-tion that emerged during the nineteenth century. Previous scholars have presented a much more straightforward narrative about the process of specialization. At the beginning of the nineteenth century what we call natural science was divided into two bodies of knowledge, natural phi-losophy and natural history. By the end of the century natural philoso-phy and natural history were no longer commonly used designations for the organization of knowledge. The terms more generally used to delin-eate scientific disciplines were the ones we are now more familiar with, such as *biology, geology, physics, astronomy, chemistry,* and *anthropology.* As the historian of science Jan Golinski has observed, the period from 1780 to 1850 was a time in which "new scientific disciplines such as geolo-gy, biology, and physiology were founded and existing ones (especially physics and chemistry) dramatically reconfigured. Remarkable changes in conceptual content and practice occurred in institutional settings that were themselves being transformed."[7]

The question is how to interpret this reconfiguration of the scien-tific disciplines during the nineteenth century. According to what we have referred to as the specialization thesis, British science began to be transformed by the formation of specialist societies beginning primarily in the 1820s, and by the adoption of disciplinary sections within the BAAS. In this era of the "gentlemen of science," the phrase used by Jack Morrell and Arnold Thackray to describe the dominant group of scien-tists in the first half of the century, embracing natural theology went hand in hand with specialization.[8] Some gentlemen of science, such as William Whewell (1794–1866) and John Herschel (1792–1871), had res-ervations about increasing specialization. Whewell famously coined the term *scientist* at an early meeting of the BAAS in order to counter what he saw as the fragmentation of science.[9] Herschel felt overwhelmed by the huge amount of information contained in the papers given at BAAS meetings. Like Whewell a polymath at heart, Herschel regretted that specialization was a necessity due to the growth of scientific knowledge. "Such is science now-a-days," he wrote to Whewell in 1835, "no man can now hope to know more than one part of one science."[10] Yet despite their reservations, historians subsequently adopted the specialization thesis and treated specialization as an almost unstoppable force throughout the rest of the nineteenth century, as scientific institutions developed and scientific knowledge increased significantly over time.

Scholars have often dealt with disciplinary development in the Victorian period by focusing on one specific discipline. Take, for example, George Stocking's *Victorian Anthropology* (1987) or William Coleman's *Biology in the Nineteenth Century* (1971).[11] Inevitably this approach maintains the specialization thesis. Our aim is to break out of disciplinary silos—to challenge the older scholarship that chopped up the history of science into the history of separate disciplines. The most extensive scholarly treatments of specialization across the sciences can be found in older works like Morrell and Thackray's *Gentlemen of Science* (1981) and Colin Russell's *Science and Social Change in Britain and Europe 1700–1900* (1983). Morrell and Thackray have discussed how the establishment in 1831 in the BAAS of subcommittees on mathematical and physical science, chemistry, mineralogy, geology and geography, zoology and botany, and the mechanical arts eventually led by 1836 to the formation of seven sections with their own presidents, vice presidents, and secretaries. They emphasize how the sections of the BAAS provided "a context in which the devotees of different disciplines could fashion a sense of common identity." They also argue that the development of sections within the BAAS was linked to the development of specialist societies. Not only did the sections work in league with existing societies, according to Morrell and Thackray, they could also lead to the formation of new disciplinary societies. Sectional activities were "symbiotic" with the work of the national societies.[12]

Where Morrell and Thackray focus on the disciplinary sections in the BAAS as an engine of specialization, Russell pays more attention to the development of specialist societies. In a chapter titled "The Rise of the Specialist," Russell asserts that "the new specialist consciousness was institutionalized in a relatively large number of societies which sprang up in the early years of the nineteenth century." The Geological Society of London, founded in 1807, was the first of these London learned societies, followed by the Astronomical (1820), Meteorological (1823), Zoological (1826), Geographical (1830), Entomological (1833), Botanical (1836), Microscopical (1839), Pharmaceutical (1841), and Chemical (1841). In focusing on the creation of scientific societies, Russell, like Morrell and Thackray, sees specialization largely in institutional, rather than intellectual, terms. He starts the chapter discussing the founding of specialized societies in the cities and in the provinces, and then moves to the founding of professorships as well as museums in specific disciplines as further indications of the rise of the specialist.[13] In addition to Morrell and Thackray's and Russell's overviews of specialization writ large in the first half of the nineteenth century, there are also accounts of the way specific disciplines were becoming more narrowly defined.[14] However, how specialists conceived of the boundaries between their discipline and others is rarely discussed.[15]

Past studies have tended to link specialization with professionalization. Russell's chapter on the rise of the specialist is followed by a chapter titled "The Road to Professionalization," implying that the specialized societies came first, inevitably followed by professionalization. In other words, for Russell the road to professionalization passes through specialization. Frank Turner, in his classic article on the professional dimension of the Victorian conflict between science and religion, quotes Bernard Barber on the major features associated with nascent professionalism. One of the features is the control of behavior through voluntary associations organized and operated by the work of specialists.[16] Turner's article treats the scientific naturalists—figures like Huxley, John Tyndall (1820–1893), and Herbert Spencer (1820–1903)—as a group that sought to professionalize science, producing the conflict between science and religion as a by-product of their efforts. In the past this led historians to think of the scientific naturalists as important agents of specialization in the second half of the nineteenth century. More recently, in her book on the X Club, a central node of scientific naturalism, Ruth Barton has contended that its members wanted "more narrowly focused specialist journals."[17]

It would, however, be a mistake to take Barton's comment about their desire for specialist journals as an indication that she is in agreement with the older scholarship's emphasis on an intimate connection between professionalization and specialization. She points out that the push for specialist publishing matches only one of the three characteristics of professionalization. The X Clubbers, Barton maintains, were not concerned with specialist education and formal qualifications.[18] Barton, in fact, has for some time been one of the leading exponents of a reevaluation of how historians of science should think about professionalization in general. More than twenty years ago she convincingly argued that historians had neglected the importance of amateur members of the X Club, such as John Lubbock (1834–1913).[19] Like Barton, Adrian Desmond, Paul White, Theodore Porter, and Jim Endersby have questioned the idea that contemporary notions of professionalization are applicable to the second half of the nineteenth century. The newer scholarship has complicated the historian of science's understanding of the meaning and nature of the process of professionalization in this period.[20]

The reassessment of professionalization has not led to a reexamination of specialization, although they were frequently linked in the past. We now have a more sophisticated grasp of professionalization thanks to Barton and others, but we do not have anything similar when it comes to appreciating specialization in the latter half of the nineteenth century and the role that it played in the thinking of the scientific naturalists. Even in his brilliant piece on Joseph Dalton Hooker (1817–1911) as scientific naturalist, Endersby has little to say about specialization. In his "Odd Man Out: Was Joseph Dalton Hooker an Evolutionary Naturalist?," Endersby aims to revise our understanding of scientific naturalism by questioning the role of the three "isms"—professionalism, secularism, and naturalism—in Hooker's conception of himself as a scientist.[21] However, he never broaches the subject of specialism. In this volume we are proposing to remedy the situation by focusing on the way specialization functioned in conjunction with an interdisciplinary ethos, not only within the thought of scientific naturalists, but throughout different fields of science.[22] During the nineteenth century, at the same time that new, fluid, and malleable disciplines were created, the desire to unify science continued to inform a broader vision of what constituted knowledge.[23]

VICTORIAN INTERDISCIPLINARITY THROUGH CASE STUDIES

Interdisciplinarity during the Victorian period took on many forms, much like it still does today. There was no singular model of practice, and it is for this reason that a series of interconnected case studies that explore the history of Victorian interdisciplinarity is so useful. It allows us to consider what the historian George Stocking has described as the "multiple contextualizations" in which these ideas and activities occurred.[24] Such a historiographical approach offers an important pathway for cross-comparing how a diverse group of historical actors engaged with their areas of study through both competing and complementary interdisciplinary frameworks. By bringing these diverse perspectives together, it is then possible to highlight the complexities and subtleties that existed within Victorian science, thereby helping us to critically reevaluate the specialization thesis.

The volume is divided into five thematic sections, with each focusing on a different form of Victorian interdisciplinarity. Section one, "Between Disciplines," opens with a chapter by Bernard Lightman, who considers how a diverse group of scientific naturalists, usually seen as those scientific researchers most interested in professionalization and specialization, unexpectedly developed a rather interdisciplinary framework for investigating the natural world in a Science Primer series that was published by Macmillan during the second half of the nineteenth century. Through an examination of the various volumes in the series, which included contributions by key figures such as Thomas Henry Huxley, H. E. Roscoe (1833–1915), and Balfour Stewart (1828–1887), Lightman exposes an example of cross-disciplinarity that was occurring within the supposed disciplinary specialization of the period.

Exploring this theme of cross-disciplinary dialogues further, chapter 2 by Geoffrey Cantor considers how investigations into electrochemistry did not fit unambiguously into a single scientific discipline, but instead straddled two main fields of study. As Cantor explains, electricity typically falls within physics (broadly construed), while chemical action, a key part of electrochemical research, is typically within the remit of chemistry. Both disciplines are concerned with the phenomena of electrochemistry, albeit in significantly different ways, therefore allowing us to examine a fascinating cross-disciplinary dialogue that emerged through this research program. Cantor anchors his investigation in a close study of Michael Faraday's electrochemical work.

In the second thematic section, "Synthesizers," Janet Browne begins chapter 3 by positioning Charles Darwin (1809–1882) as one of the great synthetic writers of the nineteenth century. Coming out of a classic naturalist tradition, typically associated with the first half of the nineteenth century, Darwin's research program covered a range of different fields, including geology, paleontology, zoology, and botany, to highlight a few examples. As Browne argues, this broad scope provides the basis for thinking of Darwin as an interdisciplinary scholar. Browne traces Darwin's interdisciplinary research activities through a close reading of several of Darwin's works on natural selection, thus exposing his interdisciplinary and synthetic approach to understanding the natural world.

In chapter 4 Ian Hesketh provides a similar analysis of another significant synthetic figure from the first half of the nineteenth century, Henry Thomas Buckle (1821–1862). Although Buckle is typically seen as one of the progenitors of a scientific discipline of history, Hesketh provides a different picture, one that shows how Buckle's historical framework was far more interdisciplinary in breadth than scholars have typically recognized. When Buckle attempted to explain the development of human civilization as resulting from natural laws and processes, he drew heavily on other specialist fields of knowledge, including philosophy, politics, classics, and statistics. Thus, much like Darwin's synthetic approach in the *Origin of Species* (1859), Buckle's main book, *History of Civilization in England* (1857), harmonized the theories and methods of multiple disciplines.

The third thematic section "Practices and Displays," provides a different perspective on Victorian interdisciplinarity. Through a detailed study of the International Health Exhibition of 1884, Elsa Richardson in chapter 5 traces the way interdisciplinary knowledge about health and wellness was physically and spatially exhibited throughout the event. The chapter therefore fosters important discussions about Victorian interdisciplinary displays and workspaces, because the International Health Exhibition was an extremely active space of interdisciplinarity, where different forms of knowledge and expertise intersected in fascinating ways. Laboratories sat alongside working dairies, while drainpipes and cooking appliances jostled against ice cream stalls and Japanese restaurants. The design of the exhibition also blurred distinctions among disciplinary fields. Thus medicine and engineering, sanitary science and elementary education, agriculture and manufacturing all had potentially equal roles to play in shaping Victorian understandings of health and wellness.

In chapter 6 Iwan Morus explores Victorian interdisciplinarity through a study of physics. While at first one might think that such an examination would reinforce the specialization thesis, what emerges through Morus's discussion is a rather complex narrative that shows the diversity of the discipline of physics during the nineteenth century. Morus's discussion underscores the tension between the rhetorical push for disciplinary specialization and the actual practice of interdisciplinary integration during this period. Through a series of snapshots of the field of physics across the Victorian era, Morus explores some of the ways that different conceptualizations of physics practice emerged within the discipline, thus exposing how interdisciplinary the research programs of physicists were at the time. Focusing particularly on public performances of physical experiments, led by leading figures such as William Robert Grove (1811–1896) and John Tyndall, Morus, much like Richardson in the preceding chapter, demonstrates how practices of display and spatiality intersected with Victorian interdisciplinary paradigms within the sciences, creating new forms of knowledge.

The fourth thematic section, "Reluctant Collaborations," provides yet another perspective to our narrative of Victorian interdisciplinarity in the sciences. Unlike previous sections, we see instances not of harmony and synthetization, but of disunity and contestation. For example, chapter 7 by Chris Manias examines competing discourses of disciplinary theories and practices within Victorian anthropology and paleontology. With the supposed advance of increased specialization and professionalization within these two disciplines, clear divisions began to emerge as "branches" of distinct knowledge. And yet, despite the growth of these divergent disciplinary perspectives, a rhetoric persisted that argued for more dialogue among these new disciplinary specialisms. Through his detailed and thoughtful examination of anthropology and paleontology,

Manias shows how some forms of interdisciplinary knowledge during the Victorian period were contentious and resistant to intellectual and practical collaboration.

In chapter 8 we see another example of a disunified and competing interdisciplinary research program, with Nanna Katrine Lüders Kaalund's examination of Arctic science during the second half of the nineteenth century. There was no consensus on how to undertake scientific research in the Arctic during the nineteenth century, and as Kaalund shows in her chapter, questions about theory and practice were widely debated by researchers interested in the icy north. Moreover, because Arctic science was still in a nascent stage, it borrowed theories, practices, and topics from a range of disciplines, further exposing the interdisciplinary character of the research field. However, the uncertainty of the Arctic generated other problems, and much of the scientific work conducted in this harsh environment was opportunistic, which therefore resulted in a vibrant interdisciplinary research program during Arctic expeditions. These activities, more often than not, however, were conducted through reluctant collaboration. Kaalund explores all these issues through both a detailed examination of questionnaires given to travelers visiting the Arctic during the second half of the nineteenth century, and with a critical analysis of the findings generated by the British Arctic Expedition of 1875, led by George Strong Nares (1831–1915).

The final thematic section focuses on "Hybrid Fields," and chapter 9 by Efram Sera-Shriar explores the emergence of the interdisciplinary field of psychical research during the late Victorian era. As Sera-Shriar argues, one of the more heated cultural debates of the late Victorian period was determining whether or not spirits and psychic forces were real. It was a topic that attracted researchers from all corners of the scholarly world. Yet, despite numerous attempts by believers and skeptics alike to resolve the matter once and for all, no single discipline seemed able to offer a definitive conclusion on whether there was any real weight to the possibility of the genuine existence of supernormal forces. Any verdict regarding the veracity of spiritualism or telepathy was informed by an interdisciplinary approach, and spirit investigations and psychical research more broadly were a kind of hybridized field. Sera-Shriar explores these issues through a close historical reconstruction of the supposed exposure of the medium Henry Slade (1835–1905) as a fraud by the biologist and scientific naturalist E. Ray Lankester (1847–1929) in 1876.

In chapter 10 James Stark and Richard Bellis explore how far-reaching the process of disciplinary specialization in the sciences during the nineteenth century impacted heterogeneous fields such as nutrition. Drawing on nineteenth-century medical and scientific texts, as well as

Victorian manuscript recipes, Stark and Bellis argue that the heterogeneity of nutrition and its relationships with the practice of preparing food and medicaments in the home, demonstrates two key points. First, investigators studying human nutrition drew on many different research specialties, and second, the example of nutrition provides a model for thinking about activities that functioned beyond the supposed boundaries of a nascent disciplinary framework, and that continues to dominate historiographical understandings of the structure of Victorian science today. Thus nutrition, much like spirit investigations, represents an important case study for understanding the interdisciplinary hybridization of some nineteenth-century research programs.

In the Afterword, Bennett Zon approaches the topic of Victorian interdisciplinarity as an improvised form of conversation. Zon contends that modern conceptions of disciplinary identities are formed both spatially and temporally. The problem, then, is a cognitive one, where people struggle to comprehend the "becoming of things" and instead position themselves outside of these processes as a way of coming to terms with them intellectually. Zon goes on to explore the ontology of interdisciplinarity through the lens of what the psychologist John Shotter calls "withness-thinking," which flips the cognitive problem of engaging these processes from the outside to a model that is centered on responsive thinking from within.[25] Drawing together key themes from the various chapters in the collection, what emerges in Zon's discussion is a new framework for understanding the processes of constructing disciplinary, and by extension interdisciplinary, identities, or what he terms "Victorian withness."

Moving the Discussion Forward

The specialization thesis has in the past been an integral part of how historians have described the changing disciplinary map of nineteenth-century British science. The lines demarcating one discipline from the other supposedly became more and more rigid, until by the end of century they were almost set in stone. This picture of the disciplinary map frames the process of specialization as if it were the product of a necessary law of nature. By contrast, this collection has encouraged historians to recognize the more fluid "nature" of Victorian science in at least three ways. First, it points to how scientific spaces were sites in which a variety of disciplines were brought together. We see this in Richardson's International Health Exhibition, an interdisciplinary space where different registers of knowledge and diverse forms of expertise circulated. Here visitors encountered information drawn from biology, chemistry, bacteriology, epidemiology, and engineering. We also catch a glimpse of other interdisciplinary spaces in Manias's discussion of the Manchester Museum, in some of the experimental sites of physics examined by Morus, and in the séance room explored by Sera-Shriar.

More often the chapters in this collection focus on scientific figures whose writings and research drew on a multitude of disciplines. This is the second way in which the collection moves the discussion forward. Cantor's analysis of Faraday as a powerful symbol of unity in nature who explored the quantitative interrelation between electrical and chemical forces, in addition to the connections among magnetism, heat, light, and gravity, remind us of the similar aims of important early and mid-century figures like John Herschel, William Whewell, and Mary Somerville (1780–1872). But the volume also investigates the cross-disciplinary exploits of significant late-century scientists, such as Darwin, Grove, and Huxley. Browne argues that Darwin saw himself as a naturalist who ranged across a number of different fields, including geology, paleontology, zoology, and botany. By crossing disciplines Darwin aimed to produce a synthesis that interconnected the study of living things through the application of one crucial insight—the theory of evolution by natural selection—to many fields. Yet, Browne maintains, Darwin also actively contributed to the consolidation of specialized domains such as zoology and geology. Strikingly, in his chapter Morus also emphasizes how Victorian physicists tended toward disciplinary transgression—that Grove, James Clerk Maxwell (1831–1879), and Oliver Joseph Lodge (1851–1940) had more of an interest in overstepping boundaries rather than policing them. Where Darwin attempted to perpetuate the tradition of the naturalist, Morus's physicists desired to continue the tradition of the natural philosopher. The desire to move across disciplines is evident in both the life and the physical sciences.

It is also remarkable that the group of scientists from the latter half of the century most identified with the push for specialization, the scientific naturalists, are presented in a different light in a number of chapters in this volume. As Hesketh demonstrates, Buckle's attempt to save both history and science from becoming too specialized inspired scientific naturalists in the middle of the 1860s after first being rejected by them. The scientific naturalists who were involved in producing Macmillan's Science Primers during the 1870s, which appeared to chop up science into its constituent disciplines, actually embraced a vision of unity in their volumes. Lightman shows how each of the authors of the primers insisted that the discipline being covered did not stand on its own but was part of an interconnected whole, unified by a common method, a shared purpose, or overlapping subjects. If scientific naturalism was the main force behind the production of the primers, then we need to reevaluate the customary picture of Huxley, Hooker, and the others as being champions of narrow specialization.

The third way that this volume moves our understanding of disciplinary specialization forward is its handling of the disciplines themselves. Many of the chapters discuss how even after the modern disciplines began to emerge during the nineteenth century and boundaries were being drawn to demarcate these disciplines from each other, the traffic between them continued to be heavy. No doubt the boundaries were continuously contested and therefore remained fuzzy, as in the case of electrochemistry and physics in the second half of the century. Add to the mix the development of nascent disciplines that drew on the more established disciplines, such as nutrition science, the study of human origins, Arctic science, or even spiritualism. The latter, as Sera-Shriar shows, was at the center of heated controversies in which both sides drew on an array of disciplines and expertise, including physics, psychology, chemistry, folklore, and anthropology. Though less controversial, as Stark and Bellis demonstrate, the nutrition sciences drew on physiology, chemistry, and physics as practitioners in the field struggled to establish a coherent disciplinary framework. In her account of the formation of a discipline focused on Arctic research, Kaalund insists that it was inherently interdisciplinary since it drew on multiple specialisms that required a division of labor between specialists and generalists. Manias also claims that the new study of human antiquity required collaborations among those working in geology, archaeology, and paleontology. These collaborations could conceal a strategy of annexing multiple fields under the umbrella of a single analytical framework, as in the case of James Hunt's (1833–1869) use of anthropology.

The more complicated story of the formation of disciplines told in this volume, which does not align with the specialization thesis, serves as a reminder that disciplines cannot be seen as unambiguous mirrors of the organization of nature into discrete bodies of knowledge. Disciplines are not actual groupings of natural phenomena. In his *Origin of Species*, Darwin asserted that in the future species would have to be treated in the same manner as naturalists treated genera: mere artificial combinations made for convenience. "This may not be a cheering prospect," Darwin declared, "but we shall at least be free from the vain search for the undiscovered and undiscoverable essence of the term species."[26] Like species, disciplines are not fixed. Unlike species, they are not formed through some law of evolution or specialization. Disciplines are contingent and in flux, since they are made by human beings attempting to find order in nature. This is what makes them such fascinating subjects for historical analysis.

I

Between Disciplines

Chapter 1

INTERDISCIPLINARITY IN MACMILLAN'S SCIENCE PRIMERS

BERNARD LIGHTMAN

In an anonymous review of Mary Somerville's (1780–1872) *On the Connexion of the Physical Sciences* (1834), the master of Trinity College, Cambridge, William Whewell (1794–1866), warned of the dangerous "tendency of the sciences" toward "separation and dismemberment." Science was like "a great empire falling to pieces" as it was "endlessly subdivided, and subdivisions insulated." As a result, Whewell cautioned his readers, science was losing "all traces of unity." Whewell praised Somerville for attempting to "remove the evil" by "showing how detached branches have, in the history of science, united by the discovery of general principles."[1] In her book Somerville observed that the progress of modern science had tended to "simplify the laws of nature, and to unite detached branches by general principles." Magnetism, electricity, light, and heat were "so connected" that they would "ultimately be referred to some one power of a higher order, in conformity with the general economy of the system of the world, where the most varied and complicated effects are produced by a small number of universal laws." Somerville maintained that these natural laws, such as gravitation, could only have been "selected by Divine Wisdom out of an infinity of others, as being the most simple, and that which gives the greatest stability to the celestial motion."[2]

Although Whewell was warning against the fragmentation of British science in the early nineteenth century, scholars have tended to view this period as one in which scientists moved relatively effortlessly across the disciplines that we are familiar with today. The very term *scientist* was coined by Whewell during a meeting of the newly founded British Association for the Advancement of Science, and it was intended to fulfill the need for a term to refer collectively to all seekers of knowledge of the material world. Whewell recalled in his review of Somerville's book that the members of the association all felt the need for a proper title that symbolized the unity of science. Whewell himself was celebrated as a polymath—an individual with a wide range of learning. An Anglican priest, philosopher, theologian, and historian of science, as a scientist he wrote on such topics as mechanics, mineralogy, geology, and astronomy. Whewell was not that unusual. He lived in a time when science was organized into two main disciplines: natural history and natural philosophy. Those working within natural history were accustomed to moving across geology, zoology, and botany, while natural philosophers often studied astronomy, physics, and chemistry together.

The second half of the nineteenth century, by contrast, is often seen as a period in which Whewell's fears came to pass. The bodies of knowledge within natural history and natural philosophy were disaggregated into the autonomous disciplines familiar to current scientists. Important scientific figures of this era are often depicted as being bent on professionalizing science, which involved an emphasis on the exclusion of amateurs, laboratory training, the secularization of scientific thinking and terminology, and, most importantly for us, the aggressive pursuit of specialization.[3]

Scholars often point to "Darwin's Bulldog," Thomas Henry Huxley (1825–1895), to demonstrate the move toward specialization in the latter half of the century. Throughout the 1860s and 1870s Huxley was pushing for scientists to drop natural history and take up what he called "biology." Huxley's historical account of the origins of the term *biology* in his essay "On the Study of Biology" (1876) was totally in line with this goal. Dramatically proclaiming the end of natural history, Huxley concluded that it had been a victim of its own success. The "marvelous progress" of the subjects that were an integral part of natural history, such as physical geography, geology, mineralogy, the history of plants, and the history of animals, at the "latter end of the last and the beginning of the present century," led "thinking men" to realize that "very heterogeneous constituents" had been included "under this title of 'Natural History.'" Geology and mineralogy, for example, were "in many respects widely different from botany and zoology." It was possible to "obtain an extensive knowledge of the structure and functions of plants and animals, without having need to enter upon the study of geology or mineralogy and *vice versa*." Moreover, "as knowledge advanced," it was realized that botany and zoology were very closely allied, since they both dealt with living beings. They could therefore be united "into one whole" and dealt with "as one discipline." According to Huxley "all clear thinkers and lovers of consistent nomenclature" now used the term *biology* to denote the "whole of the sciences which deal with living things, whether they be animals or whether they be plants." Only the muddled thinkers and lovers of inconsistent nomenclature retained "the old confusing name of 'Natural History,'" which had "conveyed so many meanings."[4] Huxley's account of the origins of the specialized discipline of biology banished natural history to the dustbin of history. His history of the rise of biology was part of a larger vocational strategy for all the sciences.

So, we have an age of polymaths in the first half of the century and an age of narrow professionals bent on specializing the sciences in the second half. Of course, this is a simplistic picture, and scholars have for several decades worked on complicating it. We have had important critiques of the concept of "professionalization" from Adrian Desmond, Ruth Barton, and Jim Endersby, to name but a few, but we have paid less attention to the process of specialization.[5] Here I want to explore the nature of interdisciplinary thinking in science in the second half of the nineteenth century. I will examine the group of scientists usually seen as the most interested in professionalization and specialization, the scientific naturalists, the group of intellectuals to which Huxley belonged. I will argue that there is an unexpected degree of interdisciplinarity in scientific naturalism that we frequently ignore due to our overexaggerated sensitivity to their push for disciplinary autonomy. I will illustrate this point by examining the Science Primer series published by Macmillan. Since Huxley was one of the editors of the series, it is usually seen, in its entirety, as pushing the scientific naturalists' agenda. Looking at the Science Primers would not seem to be a promising topic to tackle for a study of interdisciplinarity in science. However, rather than viewing the scientific naturalists as narrow specialists, I will depict them as scientists with broad, interdisciplinary sensibilities.

SCIENTIFIC NATURALISM AND THE X CLUB

The link between scientific naturalism and the Science Primer series was observed by Katy Ring in her PhD dissertation "The Populariza-tion of Elementary Science through Popular Science Books" (1988). She asserted that Macmillan recognized in the late 1860s that the lucrative future of publishing in elementary science lay with the textbook because educational reform was in the air. Ring then quoted from Morgan's *The House of Macmillan* to provide evidence that Macmillan's plan for the Science Primers was based on enlisting the help of well-known scientific practitioners. As Macmillan, through his interest in science, was already friendly with members of the 'X-Club," he decided to go, "not to the ordinary teacher but to the recognized masters in each branch of science," in order to recruit authors for an elementary science series. So it was that T. H. Huxley, H. E. Roscoe [1833–1915], and Balfour Stewart [1828–1887] came to edit and contribute volumes to 'Macmillan's Science Primers.'"[6] Ring's assertions need to be unpacked before we can get to the primers themselves. Putting aside for the moment the fact that Roscoe and Stewart were not members of the X Club, we are still left with the questions: What was the X Club? What was its relation to scientific naturalism?

Founded in 1864, the X Club was a private, informal society where the members could exchange ideas on literature, politics, and science over dinner. For twenty years the members of the club met once a month from October to June. The X Club wielded tremendous power in the scientific world. The formation of the club allowed its members to pursue a number of common objectives, the foremost among them to turn science into a professional meritocratic, publicly respected, and state-endowed activity. The members included men who were to become prominent in their respective fields of research, including mathematicians William Spottiswoode (1825–1883) and Thomas Archer Hirst (1830–1892), physicist John Tyndall (1820–1893), botanist Joseph Dalton Hooker (1817–1911), chemist Edward Frankland (1825–1899), archaeologist and entomologist John Lubbock (1834–1913), philosopher of evolution Herbert Spencer (1820–1903), and zoologist and paleontologist George Busk (1807–1886). The X Club wielded tremendous power in the scientific world.[7]

The club formed the nucleus of a larger group of figures referred to by scholars as "scientific naturalists." For the late, great Frank Turner, an intellectual historian, scientific naturalists were those men who put forward new interpretations of nature, society, and humanity derived from the theories, methods, and categories of empirical science. Scientific naturalists were naturalistic in the sense that they ruled out recourse to causes not present in empirically observed nature. They were therefore hostile to traditional natural theology, and they were scientific in that they interpreted nature in accordance with three major mid-century scientific theories: the atomic theory of matter, the conservation of energy, and evolution. According to Turner, the larger group of scientific naturalists included mathematician William Kingdon Clifford (1845–1879), the founder of eugenics Francis Galton (1822–1911), statistician Karl Pearson (1857–1936), anthropologist Edward Burnett Tylor (1832–1917), biologist E. Ray Lankester (1847–1929), doctor Henry Maudsley (1835–1918), and a group of journalists, editors, and writers, such as Leslie Stephen (1832–1904), G. H. Lewes (1817–1878), John Morley (1838–1923), Grant Allen (1848–1899), and Edward Clodd (1840–1930).[8] Turner did not add Charles Darwin's (1809–1882) name to this list, but I think there are good arguments for including him as well. Recent work on scientific naturalism has offered new perspectives on this important group by focusing on issues of community, identity, and continuity. Retaining Turner's term *scientific naturalism*, scholars nevertheless now recognize that this category is much more fluid and mutable than hitherto acknowledged. This allows us to see more diversity among scientific naturalists and to realize that some were not averse to forming alliances with other groups.[9] Keeping the scholarship on the X Club and scientific naturalism in mind, let us turn to the Science Primers themselves.

The Science Primers and Science Textbooks

In 1872 the journal *Nature* published a review of the first two volumes of the Science Primers, which were on chemistry and physics. The reviewer treated them as science books designed for "systematic elementary teaching." "Scientific class-books hitherto have been either too difficult or too easy," the reviewer asserted. But in Macmillan's Science Primers "both extremes are avoided," as they led the beginner into more sophisticated realms of knowledge. "Any man," the reviewer believed, "ignorant even of the first principles of chemistry and physics, yet fairly dexterous and intelligent, who will patiently master the books, and try each experiment for himself, is in a position to transmit their contents successfully and clearly." He recommended the volumes for use both in the middle and the higher schools. But they were also of use outside the classroom. "The officer may lecture to the soldiers of his regiment," he wrote, "the clergyman to the artisans of his parish, the national schoolmaster to the children of his school." The reviewer lavished praise upon the primers, predicting that they would usher in a new era in education. "We tender them our hearty thanks," he declared, "for work which marks a stage in the advance of scientific education."[10]

Whether or not the Science Primers had a crucial impact on science education, they have not received a lot of scholarly attention. Science textbooks in general, until recently, were neglected sources. In her introduction to the *Isis* Focus section on "Textbooks in the Sciences," Margaret Vicedo asks why this is the case.[11] She argues that textbooks have been seen as passive receptacles of the results of scientific creativity and research. In the old view, they showcase accumulated knowledge but do not contribute to scientific development. Their primary purpose is to initiate the student in "the well-established views and practices of specific scientific communities." She points to how Thomas Kuhn treated textbooks in his essay "The Function of Dogma in Scientific Research," where he presented them as repositories of exemplars from the reigning paradigm within a field. Their main role, according to him, is to "indoctrinate" students into the wisdom of the elders.[12]

But Vicedo discusses how in the previous decade a small number of scholars had developed a more sophisticated approach to the analysis of textbooks: exploring their role in pedagogical and training practices; understanding what they reveal about the epistemological concerns of a field; investigating how they have been used in priority disputes; looking at how the production and commercialization of textbooks are affected by economic and social forces; and, most relevant for our purposes, seeing how they attempted to create new disciplines. The articles in the Focus section go on to examine new ways to treat textbooks, though Vicedo believes that one salient theme connects them: "the ability of textbooks to break free from the constraints normally imposed by their genre."[13]

I want to treat the Science Primers as attempts to create or redefine the scientific disciplines. They were produced at a particularly important point in time, just after the Education Act of 1870 passed as a response to the perceived need to educate workers. In those areas of the country where churches and other voluntary organizations had not provided enough schools, the act decreed that locally elected school boards were given the power to establish schools and to levy a rate to pay for them. In addition to their importance for the history of education, however, the Science Primers also were published at a crucial moment in the development of the scientific disciplines. It is no coincidence that Huxley's essay on biology was published in 1874. During the 1870s the organization of scientific knowledge into natural history and natural philosophy was widely discussed. The Science Primers were part of that discussion, as they represented an attempt to set in stone the reconfiguration of the disciplines that Huxley and his allies were interested in establishing in the more specialized landscape of the 1870s. Huxley's time in the early 1870s was dominated by the issue of science education and the disciplines. He was a member of the Devonshire Commission, a Royal Commission on Science Instruction and the Advancement of Science that met from 1870 to 1874. In 1871 he was busy setting up his new state-of-the-art biology laboratory in the new Science Schools built at South Kensington. For Huxley, the Science Primers were part of the activity involved in reforming British science.

The Science Primers also fit into Macmillan's reading of the market for science books. Between 1860 and 1872 Alexander Macmillan (1818–1896) had established Macmillan and Company as one of the leading publishing houses in London. The Macmillan family had strong connections to liberal Anglican F. D. Maurice (1805–1872) and the Christian Socialists, and was receptive to liberal ideas. Macmillan attracted prominent literary figures to social gatherings in London where the leading topics of the day were discussed. In addition to Huxley, Alfred Tennyson (1809–1892), Herbert Spencer, Francis Turner Palgrave (1824–1897), Coventry Patmore (1823–1896), Charles Kingsley (1819–1875), Thomas Hughes (1822–1896), and Maurice frequently attended. Over the course of the 1860s and 1870s, Macmillan became more interested in science and began to produce more publications in the area, including elementary science books. As Ring observes, even before the Education Act, Macmillan "recognized in the later 1860's that the lucrative future of publishing in elementary science lay with the text-book because educational reform was in the air."[14] It was Macmillan who had founded the illustrated scientific weekly *Nature* in 1869. Whereas in the early 1860s scientific books constituted one in every ten published by Macmillan, by the latter half of the 1870s this had risen to almost one in every four. By this time Macmillan had become a leading publisher of scientific books for the general reader, along with H. S. King, Longman, and John Murray.[15]

Friendly with Huxley and other scientists, Macmillan recruited Huxley and two new faculty members of Owen's College, Balfour Stewart, a physicist, and Henry Roscoe, a chemist, to act as joint editors, as well as to contribute to the series. Contentious subject matters were ruled out. Anticipating a large institutional demand as a result of the Education Act, each volume of the series was to have a large initial run of ten thousand copies, and each copy was to be priced at one shilling. Sales for the first three years were good, but by 1878 there was a noticeable decline, in part because the market was saturated with a glut of elementary science books designed for class use. The Society for the Promotion of Christian Knowledge (SPCK) had begun their Manuals of Elementary Science in 1873. They also sold for a shilling per volume. In 1875 Chambers brought out their series Chambers Elementary Science Manual, selling them at a similar price. Other presses set up their own series later in the decade.[16] That did not stop Macmillan from continuing to publish volumes in the Science Primer series. By 1880 ten volumes had appeared, covering astronomy, botany, chemistry, geology, logic, physical geography, physics, physiology, political economy, and the introductory volume by Huxley.[17]

HUXLEY AND THE PRIMERS

As one of the editors of the Science Primers, Huxley played a key role in the conceptualization of the book series. Throughout the 1870s he was corresponding with Macmillan about who should write the books and what topics should be covered. His suggestions were not always taken up. In 1877 he wrote to Macmillan that the mathematician W. K. Clifford should be recruited to write a geometry primer, but a book on that topic never appeared in the series.[18] On August 18, 1872, he wrote to Macmillan, "there is nothing I should like better than to get a book by Tylor into our Science Series." Recognizing that some would think it "rather a strong measure, if we included a book with the title 'Man and Civilization' in our series," Huxley recommended that the anthropologist be invited to write a book on "elementary Natural History" that focused on humans. The book would "tell of his ways and doings just as if he were a Bee—undoubtedly this is a matter of science and is indeed the scientific basis of historical study and we have a right to do it. Q.E.D." Huxley proposed the title *The Natural History of Man*, hoping that by including the familiar term *natural history* it would render the potentially explosive subject more palatable.[19] Tylor's *Primitive Culture*, in which he attempted to naturalize anthropology, had just been published a year earlier.[20]

Although a primer on anthropology did not become a part of the series, Huxley's justification for including one reveals his general plan for the project. He told Macmillan, "As I claim every thing knowable as Science (unknowable is left for theology) I am not troubled with any classificatory scruples."[21] Huxley borrowed Herbert Spencer's distinction between the knowable and the unknowable from the latter's book *First Principles* (1862), the first volume of the System of Synthetic Philosophy. This primer series was to deal with everything that came under Spencer's heading of "the knowable," which included both the natural sciences and the human sciences. Huxley's lack of concern for "classificatory scruples" indicated that he saw these bodies of knowledge as connected in a way that arbitrary divisions between them were not important. Although each book covered a different discipline, the primer series was held together by an underlying interdisciplinary principle.

Huxley's views on how the textbook series should be conceptualized as a whole are embedded in the introductory primer that he eventually wrote in 1880. The basic plan for his contribution was formulated in 1871. Writing to Roscoe about his chemical primer in 1871, which he praised as "admirable," Huxley sent a sketch for his own introductory primer. "When it touches upon chemical matters," Huxley wrote, "it would deal with them in a more rudimentary fashion than yours does, and only prepare the minds of the fledglings for you."[22] Later, in 1879 after he had finished writing his introductory primer, Huxley explained to Roscoe how it would prepare students and other readers for the primers that went into more detail on each of the disciplines: "The idea is to develop Science out of common observation, and to lead up to Physics, Chemistry, Biology, and Psychology."[23] It was wildly successful. Macmillan wrote to Huxley on April 23, 1880, that eight hundred copies of the introductory Science Primer had been sold the previous week bringing the total sales to over nineteen thousand copies.[24]

Huxley divides his primer into three sections: "Nature and Science," "Material Objects," and "Immaterial Objects." The first section is intended to introduce his readers to basic concepts at the heart of science. Our senses give us information about nature, which is composed of things or objects. There is an order in nature, and nothing happens by accident. Science *is* knowledge of the laws of nature, which are known through observation, experiment, and reasoning. But Huxley makes it clear that "no line can be drawn between common knowledge of things and scientific knowledge," as "all accurate knowledge is *Science*." The method of observation and experiment that scientists have used to uncover grand secrets of nature is "identically the same as that which is employed by every one, every day of his life, but refined and rendered precise." Science, therefore, is nothing but "perfected common sense." Huxley then refers the reader to the primer on logic should they wish to learn more about the process of reasoning.[25]

Huxley then moves on to material objects, asserting that science can be divided into two sections that correspond with the two parts of nature, material and immaterial objects. In this section of the book he continually uses water in his examples, since it is the most common of common natural objects that are material, and everybody comes into contact with it almost daily. "So," Huxley concludes, "we may as well make a beginning of science by studying water."[26] Like all matter, water occupies space, offers resistance, has weight, and can transfer motion which it has acquired. "It is therefore a form of Matter," Huxley declares.[27] All matter is probably composed of either molecules or atoms, and this unifies the scientific disciplines that study it. Huxley spends a large chunk of the second section of the book demonstrating how water possesses all of the characteristics of matter. When he gets to the point about how water can be broken up into its elements, he recommends that the reader turn to the primer on chemistry. For once it is shown that water can combine with another body to create something different from either, we are "led to the science of *chemistry*, which tells us exactly how bodies combine, what comes of their combination, and how compounds may be separated into their constituents."[28]

Huxley subdivided "Material Objects" into two parts, mineral and living bodies. After bringing readers to the portal through which they could enter the science of chemistry, he now shows them the door to biology. Here again water plays an important role. All animals and plants are composed of the same substance, a compound of water, proteids, fat, amyloids, and mineral matter, which he refers to as protoplasm.[29] This was a term that he had introduced in the late 1860s that led to charges that he was a materialist. The term was used here to emphasize the unity of living bodies that allows them to be analyzed as a whole. Since Huxley is not overly concerned with "classificatory scruples," only at the end of the section on living bodies does he explain how their study is divided up into smaller disciplines that can be explored more deeply. "The further study of living bodies leads to the province of *Biology*, of which there are two great divisions—*Botany*, which deals with plants, and *Zoology*, which treats of animals," he explains. Each of these divisions has its own subdivisions, morphology and physiology.[30] But Huxley wants the reader to recognize the unity of nature rather than dwelling on how science chops it up into disciplines. There is an interdisciplinary vision grounding the primer series. On the same page he points out what mineral and living matter share in common. "The elements of living matter are identical with those of mineral bodies;" he insists, "and the fundamental laws of nature and motion apply as well to living matter as to mineral matter; but every living body is, as it were, a complicated piece of mechanism which 'goes,' or lives, only under certain conditions."[31]

The final section of the book, "Immaterial Objects," brings the human sciences into the picture. But this section is relatively brief compared to the first two. He begins by reviewing how he has divided up mineral and living bodies. Both come under the heading of material objects. They are "either not living, that is to say, mineral bodies, or they are living bodies. Everything which occupies space, offers resistance, has weight, and transfers motion, belongs to one or other of these two great provinces of nature."[32] Though they are separate provinces, they are part of one country of science. Then, for the first time, Huxley outlines in detail how each province contains smaller regions of territory. Mineral bodies are studied through the sciences of astronomy, mineralogy, physics, and chemistry, while knowledge of living bodies can be found in biology, subdivided into zoology and botany.[33] However "natural knowledge is not exhausted by this catalogue of its topics."[34] At the beginning of the primer Huxley had distinguished between material objects and sensations. He reminds his readers that "sensations, emotions, and thoughts, thus constitute a peculiar group of natural phenomena, which are termed mental."[35] Like other natural phenomena, they can be studied scientifically. The "province" of the science of psychology is the order in which mental phenomena succeed one another, "and of the relation of cause and effect which obtain between them and material phenomena."[36] Huxley concludes this chapter, and the book, with a statement on what properly belongs within the purview of the scientist. "All the phenomena of nature are either material or immaterial, physical or mental;" he asserts, "and there is no science, except such as consists in the knowledge of one or other of these groups of natural objects, and of the relations which obtain between them."[37] Huxley excluded from science everything that fell outside of the world of nature, including theological topics. This made scientific naturalism constitutive of natural science. Every schoolchild that read this introductory primer would be trained to reject the very premises of theologies of nature, and to see the scientific disciplines as connected through their focus on natural objects with shared physical characteristics. For Huxley, a secular, interdisciplinary vision lay at the heart of the Science Primers.

HUXLEY'S COEDITORS

Huxley's introductory volume to the series was actually published in 1880, after many of the other Science Primers had appeared. Even though Huxley was one of the editors for the series, he was unable to find the time to complete his contribution since he was juggling multiple projects and frequently dealing with exhaustion. His fellow editors, Roscoe and Stewart, kept prodding him. Their primers had appeared, as agreed, in 1872. As the other primers were being published throughout the rest of the 1870s, his coeditors became exasperated with Huxley. Roscoe wrote to him on January 24, 1875, that the sales of the primers were splendid. Stewart's book on physics had increased in sales from five thousand in the previous financial year to over seven thousand during the previous six months. "Now we really want your Introductory! And I think you will not find it difficult to put the thing together," Roscoe asserted. When the idea for the series had been in its infancy, and authors had not yet been recruited, writing an introduction would have been based on anticipating the final shape of each volume. Many of the volumes had come out by then, so Huxley had an exact notion of the scope and content of the series. All he had to do was "give a sketch of the different special divisions of physical science beginning . . . from the simplest concepts and ending in life phenomena."[38]

To motivate Huxley, Roscoe pointed out that other publishers, including the religious firms, were stealing the idea behind Macmillan's

Science Primer series, and that to compete they desperately needed the introductory volume that tied everything together. He told Huxley that "your Introductory is now necessary for us to meet the coming struggle with the Christian Knowledge Society and other sinners who are pirating our ideas!" The public saw the Christian Knowledge Society publications as copies of their series, but, Roscoe wrote, "our plan differs from all the wretched imitations in making each book an experimental inquiry." Hoping to rouse Huxley to action, he wrote that the Christian Knowledge Society publications were "in reality whitened sepulchers for they are hideously bad (at least Chemistry is) inside!" Roscoe believed that Huxley's introductory volume was crucial, as it would give their series a distinctive identity.[39] Huxley's inability to fulfill his commitments to the series became more and more distressing as the years rolled by. In 1877 he admitted to a correspondent that the introductory primer "weighs heavily on my mind."[40] Finally, in the summer of 1879 he quickly wrote the introduction and sent the "long-promised Primer" to Roscoe and Stewart for their feedback.[41] Huxley's views on the relationship among the disciplines in his primer, inflected with his scientific naturalism, could not have supplied a concrete template for the previous primers.

The other primers were in alignment with scientific naturalism insofar that none of them presented a theology of nature and none of them used religious language. But not all of the authors were hardcore scientific naturalists; at least one was hostile to it. However, all of the books in the series presented a vision of how the body of knowledge under examination was related to other scientific disciplines. The vision of unity in the Science Primers was created, at first, without the aid of Huxley's introduction. The most common strategy for indicating the interdisciplinary nature of science relied on a system of references to the other volumes in the series, which pointed the reader to these connections. On October 19, 1871, Roscoe wrote to Huxley that he and Stewart believed that although the Science Primers should be self-contained, "frequent references should be made from one to the other of the Primers."[42] These references implied that one primer could not be read independently of the others, as key points in each discipline could not be explained without consulting sections in other primers. No scientific discipline, therefore, stood on its own. Each was interwoven with the others.

The primers on chemistry and physics were the first two published in 1872. Authored by Henry Roscoe and Balfour Stewart, respectively, these volumes, not Huxley's, provided a model for the subsequent primer authors to follow. Roscoe, a dissenter, was trained at University College London, where in 1853 he took his degree of bachelor of arts, with honors in chemistry. He continued his chemical education in Germany with the distinguished chemist Robert Bunsen at Heidelberg, receiving a PhD from that institution. In 1856 Roscoe was selected to be professor of chemistry at Owens College of Manchester, where he built Manchester into a leading chemistry center in Britain. In 1866 Roscoe established Science Lectures for the People and brought important scientists, such as Huxley, William Benjamin Carpenter (1813–1885), John Tyndall, and William Huggins (1824–1910), to the city.[43] An energetic, German-trained chemist with extensive laboratory experience, Roscoe was a natural choice as one of Huxley's coeditors of the Science Primer series.

Macmillan's selection of Balfour Stewart, the physicist and meteorologist, as the third coeditor is somewhat puzzling. Trained in Scotland at the University of Edinburgh under James David Forbes (1809–1868), he joined the staff of Kew Observatory as assistant observer in 1856 after giving up work in a Leith merchant's business. In 1859 he was appointed director of the Kew Observatory, where he devoted himself to meteorology, especially terrestrial magnetism. In 1870 he was appointed professor of natural philosophy in Owens College, where he established a new physics laboratory.[44] As Roscoe's colleague at the college and a physicist who appreciated laboratory research, Stewart may have seemed like a good choice as the third coeditor of the primers. But whereas Roscoe would have been sympathetic to Huxley's scientific naturalism, Stewart was not. Later, in 1875 Stewart wrote the anonymously published *Unseen Universe* with the physicist Peter Tait (1831–1901). It was a powerful defense of the harmonious relationship between science and religion. Stewart is included within the group labeled as the North British Physicists by Crosbie Smith, a group that often clashed with the scientific naturalists over the need for a religious framework for science.[45]

However, if there were any tensions between Stewart and those sympathetic to scientific naturalism, it was not evident in his primer on physics. Both his volume and Roscoe's shared an identical preface outlining their common goals. "In publishing the Science Primers on Chemistry and Physics," Roscoe and Stewart declared, "the objects of the Authors has been to state the fundamental principles of their respective sciences in a manner suited to pupils of an early age."[46] In his autobiography Roscoe later referred to his primer as being written "for little boys out of the street."[47] In the preface Roscoe and Stewart declared that they weren't as concerned about giving information as they were about disciplining the mind of the student "by bringing it into immediate contact with Nature herself."[48] The emphasis was therefore put on including experiments designed to awaken and strengthen the power of observation in students.

Besides the common preface, the two primers were also designed to fit together by pointing to the connections between physics and chemistry. Roscoe's chemistry primer, listed as the second volume in the series, was divided into sections on fire, air, water, earth, nonmetallic elements, and metals. It contained a number of references to the physics primer. In a discussion of three simple experiments illustrating the nature of fire, Roscoe wrote, "And when you come to read the Physics Primer (Articles 48 and 75), you will learn still more about the *Nature of Heat*." A section on Grove's battery referred the reader to the description of this apparatus to be found in article 87 of the Physics Primer. Roscoe even included references to the introductory primer by Huxley, indicating that he had assumed it would be published by the time his primer had appeared. At one point he stated that the reader would have learned some of the common properties of earth, water, and rain in the introductory primer; later in the book he remarks that the reader already knew how to use the scales from the introductory primer.[49] Roscoe's chemistry primer could not be read on its own. It demanded that the reader apply knowledge contained in other primers.

Stewart's physics primer was listed as number three in the series. On the first page he assumed that the reader had gone through the chemistry primer. He summed up what had been learned about chemistry and used this as a springboard for defining physics. Both sciences dealt with dead matter. "You have been told in the Chemistry Primer what sort of things we have around us," Stewart affirmed. "You have seen what the chemist does; how he weighs things and finds their quantity, and also how he finds that certain things are composed, and may be split up into two or more new things; while again other things are simple or elementary and cannot be so split up." While the chemist was concerned with the weight, quantity, and composition of matter, the physicist focused on the cause of changes in the moods or qualities of dead matter. Throughout the book, Stewart referred to the other primers, as had Roscoe. He pointed to the chemistry primer (articles 33 and 17) on how to make carbonic acid gas and hydrogen. Like Roscoe, he referred to the as yet unpublished introductory primer. "You have already been told that steam is a gas like air," he wrote, "and you have learned in the Introductory Primer that you cannot see true steam." Stewart ended the book with a reminder of the undeniable connection between physics and chemistry. They converged toward the same scientific principle. Just as the science of chemistry was built "upon the principle that matter only changes form, going from one combination to another, and does not absolutely disappear, so the science of Physics is founded upon the principle that activity or energy only changes form, and never absolutely disappears."[50]

For Stewart, physics was inherently interdisciplinary. Since it was closely related to chemistry, his primer had to be read in conjunction with Roscoe's primer and in light of the yet-to-be written introductory primer by Huxley, which would spell out the connections among chemistry, physics, and the other provinces of science.

THE OTHER PRIMERS

The other primers published after Roscoe's and Stewart's contributions and before Huxley's introduction continued the strategy of referencing other books in the series to indicate the need to cross disciplinary boundaries. Some also introduced colorful metaphors. Archibald Geikie (1835–1924), geologist and historian, produced the next two primers, on geology and physical geography, published in 1873. In his autobiography Geikie recalled that their "immediate and extraordinary success was a complete surprise to me." The first edition of *Physical Geography* was exhausted in six months. Up to 1922, 513,000 copies were printed in England. It was translated into most European languages and used in Continental schools. It was also translated into several languages of India.[51] Geikie was appointed to the Scottish branch of the geological survey in 1855 and remained a part of this organization until his retirement in 1901. Although extremely devout when he was young—he admired the theologian Thomas Chalmers (1780–1847)—his zeal gave way to conventional religious observance.[52]

In his *Physical Geography* Geikie offered a dense system of references to other primers to demonstrate the unity of the sciences. There were six references to the chemistry primer and four references to the physics primer necessary to deal with his discussion of the shape of the earth, day and night, air, the circulation of water on land, the sea, and the inside of the earth. For example, the chemistry primer had already taught the reader that air was not a simple substance but a mixture of two invisible gases, that rainwater contained carbonic acid, and that rainwater and springs contained little impurity. The physics primer was relied on to explain how a thermometer worked, the freezing point, and the nature of gravity.[53] Geikie also offered an interesting metaphor to describe the connection between physical geography and the other sciences. He encouraged his youthful readers at the beginning of the book to explore the great book of nature throughout their lives. "It is this great book—Air, Earth, and Sea—which I would have you look into." Geikie made it clear that physical geography was but one "branch of science," which traced the relationship between the air we breathe and the earth we live on, just as the other sciences dealt with similar relationships across nature.[54] As Matthew Stanley has shown, the physicist James Maxwell (1831–1879) put forward a similar idea. He too believed that nature was like a book and that its individual elements should be seen as manifestations of deeper unified principles.[55] Geikie presented the book of nature analogy without the religious overtones evident in Maxwell's formulation.

Geikie's *Geology* contained a similar strategy, offering references to other primers combined with the intriguing book metaphor to illustrate the unity in nature that could only be grasped through an openness to interdisciplinarity. The chemistry primer was referred to once, the physics primer twice. His own *Physical Geography* was referenced nineteen times, indicating that Geikie intended his two primers to form a unit, similar to the pairing of the physics and chemistry primers.[56] In the opening pages Geikie developed the notion of geology as the study of the floor of stone that covered the bottom of the sea and the surface of the land. Each stone was compared to a book, and Geikie promised his reader to teach them how to "read the contents of stones." He assured his reader that learning to read stones was easier than reading books, as "in reality the chief groups of stones are very much fewer in number than the chief groups of books."[57] Once the general principles of geology were mastered, it would be possible to trace out the "marvelous History of the Earth."[58]

Two more primers appeared in 1874, a year after Geikie's: on astronomy by Norman Lockyer (1836–1920), and on physiology by Michael Foster (1836–1907). Both authors had connections to scientific naturalism. Lockyer, astronomer and journal editor, was Huxley's ally on the Devonshire Commission. Through his editorship of the journal *Nature*, Lockyer was also brought into contact with the scientific naturalists, as well as with Macmillan.[59] Foster, physiologist and politician, was trained at the medical school at University College London. A dissenter, he later abandoned evangelical religion for agnosticism. Foster was one of the demonstrators in Huxley's course of elementary biology at South Kensington. With Huxley's help, he was appointed to a praelectorship in physiology at Trinity College, Cambridge, in 1870. Here he successfully created what has been called the Cambridge school of physiology.

Lockyer's *Astronomy* did not include as many references to specific sections in the other primers as we have found in the earlier books. Instead, he began by developing an extended analogy between geography and astronomy, since both were essentially mapping enterprises. When Lockyer instructed his youthful readers to write down their school, street, parish, county, country, kingdom, continent, and planet in the opening pages of the book, he realized that he had puzzled his audience. "Some of you may think," he declared, "that I have made a mistake, and am going to write a book on Geography instead of Astronomy." But

Lockyer had not made a mistake. His aim was "to show you that where Astronomy leaves off Geography begins." Lockyer could then refer to the physical geography primer for information on the earth. When he discussed the surface of the moon, he approached it as an exercise in physical geography.[60] But in other sections of his book, he pointed to the links between astronomy and the other sciences. His examination of the physical constitution of the earth and the sun drew on the science of chemistry, while the section on the regular motion of the heavenly bodies inevitably considered the concepts of gravity and magnetism in physics.[61] By the end of the book Lockyer had depicted astronomy as being divided into branches, each with a particular connection to another science.

Foster's *Physiology* was perhaps the clearest statement of scientific naturalism next to Huxley's primer. In fact, he described *Physiology* as an introduction to Huxley's *Lessons in Elementary Physiology* (1866). He also asserted near the beginning of the book that humans could move like other living things because our bodies were like a steam engine fueled by a fire burning within us. Foster's preface pointed explicitly to a parallel between physiology and chemistry. Handling and examining the hearts of dead animals were, Foster insisted, "as necessary for the sound learning of even elementary physiology, as are actual experiments for chemistry." In the body of the book, there were numerous references to the chemistry primer when he discussed issues such as how fire burned, the need of the body for oxygen, and how the body gave out carbonic acid. There was also a single reference to a section in the physics primer on the pressure exerted by the atmosphere in order to explain how the chest was airtight.[62]

Three more primers were published in subsequent years on botany, logic, and political economy. The author of the botany primer was Joseph Dalton Hooker. He succeeded his father as director of Kew Gardens in 1865 and was a close friend of both Darwin and Huxley, as well as a charter member of the X Club. He wrote to Darwin of the "folly" of undertaking the botany primer. "The great difficulty," he wrote on February 20, 1873, "is to go to the bottom of things and yet to avoid detail—or rather to keep pointing to the bottom of things without going into it."[63] Organized around the parts, species, and uses of plants, Hooker did not offer extensive references to the other primers. Rather, he made a general statement at the beginning of the book about the relationship between botany and the other sciences. Hooker claimed in opposition to the common belief that botany was a science of observation, that it was actually an experimental science, like chemistry. Moreover, these experiments required "for the most part a previous knowledge of Chemistry and Physics; those, however, described in this Primer will need no more knowledge of these subjects than is to be found in the Primers devoted to them." Interestingly, Hooker included a section on evolution. Here he treated independent creation as a purely speculative theory, like a good scientific naturalist, while arguing that evolution was "gaining adherents rapidly, because most of the phenomena of plant life may be explained by it," as well as for other reasons.[64]

William Stanley Jevons (1835–1882) wrote the other two primers, on *Logic* (1876) and *Political Economy* (1878), the two books in the series that were not, strictly speaking, dealing with a natural science. Jevons, an economist and philosopher of science, was born into a Unitarian family. Roscoe was his first cousin. Jevons attended University College School, London, in 1850–1851, where he studied mathematics and chemistry. He spent five years in Australia in the post of assayer at a branch of the Royal Mint, returned to England, and in 1863 took up a position as junior tutor at Owens College, which led to his appointment in 1866 to a new chair of logic, mental and moral philosophy, and political economy. In 1875 he accepted the chair of political economy at University College London. He worked on the two primers shortly after accepting the new appointment.

Jevons was very happy with the sales of his *Logic*. He wrote to his sister Lucy in 1876 that it was "selling pretty well," with eight hundred copies sold in six months.[65] His discussion of deductive and inductive reasoning was intended to show that the rise of modern science was due to the rejection of Aristotelian deductivism and the acceptance of the importance of observation and experimentation. Inductive reasoning was essential for discovering the laws of nature or "what is true of many different things."[66] Logic provided the basis of sound reasoning for all of the sciences. Throughout the book Jevons gave numerous examples of good reasoning drawn from biology, astronomy, and chemistry. Published two years later in 1878, *Political Economy* made the case for the importance of the social sciences. These sciences taught how to "use our powers wisely in relieving the labors and misery of mankind." Political economy was not a dismal science, as it banished depressing things such as strikes, unemployment, bankruptcy, and expensive bread. Near the beginning of the book, Jevons situated political economy in relation to the other sciences. Like the other sciences, it helped achieve prosperity. The science of mechanics "shows how to obtain force, and how to use it in working machines"; chemistry taught how useful substances could be produced; astronomy was necessary for navigating the oceans; and geology "guides in the search for coal and metals." The social sciences were also needed to promote the welfare of humanity, such as jurisprudence, political philosophy, sanitary science, and statistics. But political economy, Jevons insisted, was also distinct from all of these other sciences. For it "treats of wealth itself; it inquires what wealth is; how we can best consume it when we have got it; and how we may take advantage of the other sciences to get it."[67]

SCIENTIFIC NATURALISM AND INTERDISCIPLINARITY

Although the textbooks that composed the Science Primers divided up scientific knowledge into separate disciplines, reinforcing the move away from natural history and natural philosophy, they were not attempting to present these disciplines as fully autonomous. Rather, a variety of strategies were used to point to the underlying unity between the sciences and the need for the scientist to move across disciplines. The most widely used strategy was the extensive cross-referencing to the other primers. In some cases, primers were designed to be read in tandem with another primer, such as the chemistry and physics primers, which shared an identical preface, or the two primers written by Geikie. Here the cross-referencing was particularly dense, indicating that these two disciplines needed to be learned almost as a unit. But the relationship between the disciplines was also conceived of in other ways. For Jevons, logic supplied the principles of reasoning for all of the sciences, while to Hooker botany relied on the experimental method like many of the other sciences. By contrast, Lockyer looked upon the disciplines as occupying adjacent territories. In the case of geography and astronomy, both of which were mapping exercises, one discipline began where the other left off. The unity of the sciences could be conceived of in terms of their shared purpose. Jevons believed that political economy had the same utilitarian purpose of all of the sciences: to promote the welfare of humanity. Finally (though Geikie was the only author to pursue this strategy), a colorful metaphor such as the book of nature could be developed to explain how the sciences all connected together. Huxley's attempt to impose unity on the primers in the introductory primer through the principle of scientific naturalism had been explored only by Foster in his physiology primer. But even if the Science Primers cannot be said to be unambiguous proponents of scientific naturalism, they were defenders of the unity of science. They point to an enduring interest in the second half of the nineteenth century in an interdisciplinary vision for science, despite the move in the direction of specialization.

Our study of the Science Primers throws new light on how we should think about the process of specialization in the second half of the century. It also forces us to rethink the role of the scientific naturalists, who are often seen as being committed to some form of professionalization and therefore to the aggressive pursuit of specialization in science. Limiting our discussion to some of the key scientists who were part of the group, it becomes clear, Huxley's championing of biology notwithstanding, that over the course of their lives many of them moved between disciplines in their research. They cannot be seen as narrow specialists.

Let us start with Charles Darwin. Darwin's early interests were in natural history. It is sometimes forgotten that Darwin's main interest while he was on the *Beagle* voyage from 1831 to 1836 was actually geology. Darwin wasn't actively looking for evidence for a theory of evolution—that was something he worked out later after he had returned. Throughout the voyage he was testing Charles Lyell's (1797–1875) geological theories, and using them to explain the geology of South America. Although Darwin later turned his attention to zoology as he developed his evolutionary theory, the *Origin of Species* still contains quite a lot on geology. Darwin never became a narrow biological specialist. After the *Origin* was published, he wrote about the origins of emotions, of morality, and of religion. In his more biological works, he discussed flowers, birds, earthworms, and, of course, humans.

John Tyndall, who was professor of natural philosophy at the Royal Institution, actually began as a mathematician. He had been a land surveyor before he decided to do a PhD in Germany at the University of Marburg under the chemist Robert Bunsen (1811–1899). His dissertation was actually on a topic in mathematics. Like earlier natural philosophers, Tyndall's varied research projects led him to move across disciplines. During his career Tyndall conducted research on diamagnetism, glaciers, radiant heat, atmospheric gases, sound, electricity, and spontaneous generation. He was the first to verify experimentally the role of specific gases in producing the earth's natural greenhouse effect. His experiments drew on an extensive knowledge of chemistry, physics, and biology. Tyndall resembles an early nineteenth-century polymath far more than a narrow specialist in physics. He was, after all, a powerful voice in favor of preserving a role for the imagination in science, especially in his well-known paper on that topic given at the British Association for the Advancement of Science in Liverpool in 1870.

However, it was Herbert Spencer who was the most ambitious polymath of the entire group. Spencer believed that evolution was at work in every stage of the development of the natural world, from the formation of the solar system by the condensation of white-hot nebular matter into planets (the nebular hypothesis), to the geological forces that determined the nature of the earth's crust, and to the development of life, from monad to man. Even the intellectual and cultural achievements of humanity, including science, were a part of the cosmic evolutionary process. Spencer laid out this vision of cosmic evolution in his multivolume *System of Synthetic Philosophy* (1862–1896), which was intended as a synthesis of all knowledge connected through the concept of evolution. There were volumes on psychology, sociology, biology, and ethics. Evolution in its broadest sense became the organizing principle for all of the sciences, including the social sciences, connecting them into a unified whole.

Even Huxley cannot be considered a narrow scientific specialist, despite his attempt to replace natural history with biology. Huxley was trained in a dissenting medical academy and was appointed as assistant surgeon aboard the HMS *Rattlesnake* in 1846. His meticulous dissections of marine animals while on that voyage helped him to establish a reputation in invertebrate morphology. But Huxley had broader research interests. As Mario di Gregorio showed in his *T. H. Huxley's Place in Nature Science* (1984), Huxley shifted from invertebrate morphology to vertebrate paleontology in the late 1850s. In the 1860s he began to work on human evolution and anthropology, becoming president in 1868 of the Ethnological Society.[68]

As the publication of the massive *Routledge Research Companion to Nineteenth-Century British Literature and Science* makes clear, every scientific discipline was shaped in some way by the literary culture of the period.[69] The chief scientific figures were familiar with current poetry, fiction, art, and theater, including the scientific naturalists. Just as we can recognize the interdisciplinarity of Victorian scientists at the general level of the relationship between science and other forms of culture, we should be willing to consider how interdisciplinarity operated within the scientific disciplines as well.

Chapter 2

THE RECALCITRANT CASE
OF ELECTROCHEMISTRY

GEOFFREY CANTOR

Writing in the *Quarterly Review* in 1834, the polymath William Whewell (1794–1866) alluded to a pressing problem that he had encountered at recent meetings of the British Association for the Advancement of Science, founded three years earlier. As the term *philosophers* was "too wide and too lofty" to describe the assembled "savants" (to use the French term), he sought an appropriate English word to "designate students of the knowledge of the material world collectively." As is well known, he coined the word *scientist* as the umbrella term to refer to those studying any of the many "departments" of science. While advocating this new general term he also noted the "inconveniences of this [current] division of the soil of science into infinitely small allotments."[1] Whewell recognized not only that the field of science had expanded considerably but that many practitioners were working in just one allotment or in a very small number of the many allotments that constituted contemporary science. Whewell thus implied that science could no longer be considered a unity but instead manifested signs of fragmentation.

By the century's end the map of the sciences had expanded and changed considerably, with clusters of Whewell's allotments coming together to form disciplines. Moreover, boundary fences had been erected between the disciplines. Individual scientists, and the specialist societies to which they belonged, were increasingly identified with these disciplines and often also included a number of subdisciplines. Science education likewise became increasingly structured along disciplinary lines, so that students were trained to be chemists, physicists, and even color chemists, rather than scientists.

The classificatory system of plants devised by Linnaeus provides an obvious analogy to the division of science into separate disciplines that emerged by the end of the nineteenth century. Linnaeus divided plant specimens into classes (of which there were twenty-four) according to the number of stamens they possessed and then subdivided each class in accordance with the number of pistils. In line with this analogy, we can inquire whether the division of scientific knowledge into disciplines is natural or artificial, Linnaeus having recognized that his system of botanical classification was artificial. The disciplinary map is likewise artificial because some natural phenomena do not fit unambiguously into a single scientific discipline but instead straddle two or more disciplines. This chapter therefore discusses the history of one such awkward phenomenon (or, more exactly, class of phenomena): the production of electricity by chemical action and vice versa (generally known as electrolysis).

As electricity falls clearly within the physicist's domain and chemical action is part of chemistry, both disciplines are concerned with the phenomena of electrochemistry, albeit—as we shall see—in significantly different ways. In order to chart how electrochemistry came to be presented on the changing disciplinary map across the nineteenth century, we will first examine early discussions of the subject, especially Michael Faraday's (1791–1867) innovative and exemplary electrochemical researches of the 1830s—which were contemporary with Whewell's coining of the word *scientist*—before turning to the presentation of electrochemical phenomena in textbooks used by school, college, and university students in Britain in the 1890s. Textbooks provide an appropriate site for appreciating how disciplines evolve since they disseminate the body of knowledge that the current leaders of a scientific discipline prescribe as necessary to subsequent generations of practitioners.

By the 1890s the disciplines of physics and chemistry had clearly emerged and the boundary between these disciplines was widely acknowledged and was thus reflected in their respective textbooks. By analyzing the presentation of electrochemistry in physics and chemistry textbooks of the 1890s, this chapter examines the significantly different presentations of electrochemistry in the contemporary disciplines of physics and chemistry. In contrast to Faraday's unified perspective on electrochemistry, dating from the 1830s, it will be shown that six decades later the subject played an interdisciplinary role, straddling physics and chemistry. However, these two disciplines highlighted substantially different facets of electrochemistry; each discipline emphasized only those electrochemical experiments, laws, theories, etc. that could be integrated within the wider contours of that discipline.

Thus we shall compare and contrast the presentation of electrochemistry in physics and chemistry textbooks. However, as we will also see, electrochemistry was discussed in a number of training manuals used by students in the industrial practices of electroplating and electrometallurgy. Three different accounts of the laws, phenomena, and practices of electrochemistry will thus be discussed—the physicists' and chemists' interpretations and the more hands-on deployment of electrochemistry by practicing electricians. The analysis of these texts will in turn raise issues about the nature of scientific disciplines.

INTERDISCIPLINARY ORIGINS OF ELECTROCHEMISTRY

It will be helpful to reflect briefly on the early history of what became known as electrochemistry in order to understand its relationship to various subject areas. The first key figure was Luigi Galvani (1737–1798), who studied medicine at the University of Bologna, where he also trained in surgery. After graduating in 1759 in medicine and philosophy, he was appointed lecturer in anatomy and honorary lecturer in surgery at Bologna. In 1791 he published *De Viribus Electricitatis in Motu Musculari* (*On the Forces of Electricity in Muscular Motion*) containing his often-cited observation that a frog's leg twitched when the circuit was completed using a metal connector.[2] To explain this phenomenon he envisaged the conductor releasing some "nerveo-electrical fluid," a subtle elastic fluid that had been stored in the frog's muscle, just as electricity could be stored in a Leyden jar. While he drew on contemporary ideas about an electric fluid, Galvani's theorizing was directed to the presence of this fluid in the body—so-called animal electricity—and especially its role in the nervous and muscular systems. Thus his research was firmly based on his training and experience in medicine and, in particular, in physiology.

As professor of experimental physics at the University of Pavia, Alessandro Volta (1745–1827) was highly perplexed when he read about Galvani's experiment, which he initially dismissed with the complaint that physicians were ignorant of electricity. However, he pursued a train of experiments with a whole frog, not just a frog's leg, which quivered when connected either to a Leyden jar or to a bimetallic connection. This motion, he reasoned, was not due to electricity stored within the animal's muscles but was rather the result of applying electricity to the animal's nerves. Realizing that the contact between two different metals could produce electricity, Volta constructed a stack of zinc and copper discs separated by a cloth soaked in brine. This produced a constant supply of electricity, originating, he argued, from the chemical action between zinc and copper. He also experimented with different metals. For Volta, chemical action produced electricity; animal electricity was irrelevant.

At its meeting on June 26, 1800, Volta's work was presented to the Royal Society of London in the form of a letter to its president, Joseph Banks (1743–1820).[3] The paper rapidly caused a great surge of interest, one contemporary describing the newly discovered phenomenon as "a fresh hare just started; the Royal Society hounds are now in full cry."[4] Among the first to respond to that cry were the chemist, natural philosopher, and publisher William Nicholson (1753–1815) and the surgeon Anthony Carlisle (1768–1840), who used the electricity generated by a voltaic pile to decompose water into hydrogen and oxygen.[5] Over the next few decades, as improved forms of battery were introduced, much research was directed to the decomposition of complex chemical substances by electricity. The domain of electrochemistry was given a secure conceptual foundation in the investigations of Michael Faraday, to whose work we now turn.

In his third series of *Experimental Researches in Electricity*, read at the Royal Society of London in January 1833, Faraday sought to determine conclusively "the identity or distinction of electricities excited by different means."[6] Here he considered five different kinds of sources and demonstrated that they displayed identical effects with respect to the production of both electrical tension (static electricity) and motion (current electricity). In addition to the forms of electricity produced by heat, by changes to the magnetic field, and by mechanical means (such as in an electrostatic generator), Faraday investigated the electricity produced by voltaic cells (batteries) and by certain animals, specifically the electric eel (at the time classified as *Gymnotus electricus*) and the *Torpedo*. These five sources all produced what Faraday identified as the same kind of force, namely electricity. Looking beyond the empirical evidence, Faraday increasingly appreciated a unity in nature and that the forces manifested in these different kinds of phenomena were quantitatively interrelated. He therefore concluded his paper by suggesting a quantitative relationship between electrical and chemical forces such that "chemical power, like the magnetic force[,] is in direct proportion to the absolute quantity of electricity which passes."[7]

Five months later he read his fifth series of *Experimental Researches in Electricity* to the Royal Society, in which he used the electricity supplied by a battery to produce electrochemical decomposition. He observed the result of passing electricity through moist conductors and also through an electrolyte in which two metal poles were immersed. He

847. TABLE OF IONS.

Anions.

Oxygen	8	Selenic acid	64	Tartaric acid	66
Chlorine	35·5	Nitric acid	54	Citric acid	58
Iodine	126	Chloric acid	75·5	Oxalic acid	36
Bromine	78·3	Phosphoric acid	35·7	Sulphur (?)	16
Fluorine	18·7	Carbonic acid	22	Selenium (?)	
Cyanogen	26	Boracic acid	24	Sulpho-cyanogen	
Sulphuric acid	40	Acetic acid	51		

Cations.

Hydrogen	1	Cadmium	55·8	Soda	31·3
Potassium	39·2	Cerium	46	Lithia	18
Sodium	23·3	Cobalt	29·5	Baryta	76·7
Lithium	10	Nickel	29·5	Strontia	51·8
Barium	68·7	Antimony	64·6?	Lime	28·
Strontium	43·8	Bismuth	71	Magnesia	20·7
Calcium	20·5	Mercury	200	Alumina	(?)
Magnesium	12·7	Silver	108	Protoxides generally.	
Manganese	27·7	Platina	98·6?	Quinia	171·6
Zinc	32·5	Gold	(?)	Cinchona	160
Tin	57·9			Morphia	290
Lead	103·5	Ammonia	17	Vegeto-alkalies gene-	
Iron	28	Potassa	47·2	rally.	
Copper	31·6				

FIGURE 2.1. Faraday's "Table of Ions" (January 1834). *Source:* Michael Faraday, *Experimental Researches in Electricity*, 3 vols. (London: John Murray, 1839–1855), 1:247.

soon concluded that "electro-chemical decomposition does not depend upon any direct attraction and repulsion [of the chemical elements] of the poles."[8] Instead he developed his theory of electrolysis in which the electricity caused the molecules comprising the electrolyte to be split into two oppositely charged components, each of which is conveyed to the appropriate pole. Investigating the numerical relationship between the electric current and its chemical action, led him to speculate that, in light of his experimental evidence, "for a constant quantity of electricity . . . the amount of electro-chemical action is also a constant quantity."[9]

In the seventh series, read in January 1834, Faraday reported a number of experiments in which he determined the mass lost or gained by the poles in electrochemical decomposition. This "irresistible mass of evidence" provided proof of his earlier hypothesis "that the chemical power of a current of electricity is in direct proportion to the absolute quantity of electricity which passes";[10] this was subsequently known as Faraday's first law of electrochemistry. He then introduced the reader to the terminology he had adopted in correspondence with William Whewell; including the terms *anode*, *cathode*, *electrochemical equivalents*, *ions*, *anions*, and *cations*.[11] He also recognized an important relationship, which was to become his second law: the quantities of different substances deposited or dissolved by a given quantity of electricity are proportional to their chemical equivalent weights. Paragraph 847 contained a summary of his results in the form of a table (figure 2.1) listing the electrochemical equivalents for the ions and cations of many substances; hydrogen (1), oxygen (8), chlorine (35.5), etc. Electricity and chemical action were now not only directly interrelated but were numerically linked in a way that was recognizable to practicing chemists.

A major and recurrent theme throughout Faraday's researches was the interrelation between different classes of what he often called "forces"—not only electricity (in its various forms, including animal electricity) and chemistry, but also magnetism, heat, light, and the gravitational force. In other words, there was a unity in nature such that in a specific interaction, if one form of force wanes, it simultaneously creates an equivalent amount of force in another form. Thus, for example, in a battery the chemical force would be lost as the battery runs down while a proportional quantity of electrical force is produced. This correlation of forces also implied that the total amount of force is conserved in any process. The conservation and correlation of forces underpinned all Faraday's scientific investigations. These concepts were also discussed by others, including William Robert Grove (1811–1896), whose *Correlation of Physical Forces* passed through six editions between 1846 and 1874.[12] Faraday later provided his most explicit discussion of conservation in an 1857 Friday evening discourse at the Royal Institution, titled "On the Conservation of Force," in which he argued that force is conserved since it can neither be created nor destroyed.

Implicit in this formulation of the principle of force conservation was the theological imperative that only God can create or annihilate force. Thus Faraday envisaged that God had introduced a fixed amount of force at the Creation.[13] That quantity of force has remained constant ever since, but it is constantly being transmuted from one form into another: say, from a chemical action into electricity, and then from electricity into heat (in passing through a wire with resistance), etc. Hence Faraday's discussion of electrochemistry was underpinned by a theological postulate that was implicit in his two laws of electrochemistry of 1833–1834. Faraday was therefore drawing on the "discipline" of religion just as the first Bridgewater Treatises, which combined authoritative accounts of modern science and orthodox Christianity, were published.[14]

By the beginning of the nineteenth century chemistry was generally treated as separate from natural philosophy and physics. For example, William Nicholson's *An Introduction to Natural Philosophy*, which passed through five editions between 1782 and 1805, was unusual among natural philosophy texts in including a section on chemistry. Yet he adopted a physicalist approach in his discussion of chemistry; he began with the influence of heat on chemical substances and worked his way up to the metals, simple earths, and alkalis.[15] Toward the end of the fifth edition (1805) Nicholson added a separate short section on electrochemistry—galvanism and voltaic pile—a subject to which he (together with Anthony Carlisle) had contributed in 1800 by using a voltaic pile to decompose water into hydrogen and oxygen.[16]

Unlike Nicholson's *Introduction*, later natural philosophy and physics texts generally excluded chemistry. For example, in his *Elements of Physics* (1827) the Scottish physician Neil Arnott (1788–1874) specified the scope of his subject by identifying its five subtopics: "Somatology and Dynamics," "Mechanics," "Hydrodynamics," "Imponderable Substances" (including electricity), and "Astronomy."[17] Moreover, he divided the totality of science into five domains: physics, chemistry, life, mind, and the "science of quantity" (i.e., mathematics). Although for Arnott chemistry was a distinct domain, separate from physics, he insisted that it "is truly . . . a modification of or superstructure on Physics, and cannot be understood or practiced by a person ignorant of Physics."[18] Likewise, in his *Dictionary* of 1842, William Thomas Brande (1788–1866) acknowledged that there are "three great, divisions of natural science; namely, *Physics, Chemistry*, and *Natural History*. Physics has for its object the properties of bodies, [whereas] chemistry studies their elementary principles."[19] Hence for both Arnott and Brande chemistry and physics (which included electricity) were distinct sciences. However, the phenomena of electrochemistry were problematic as they crossed the boundary between these two "great divisions."

Where did Faraday stand? In 1854 he explicitly eschewed the title chemist, "of which body I do not count myself one now a days."[20] A few years earlier he had expressed concern that the Royal College of Chemistry had forsaken the ideal of pursuing chemistry "upon the basis of high feeling and utility" and that chemists were in danger of becoming mired in commerce.[21] Although he distanced himself from those professional chemists who analyzed substances for financial gain, throughout the time he was researching electrochemistry he can be described as a chemist or, more precisely, a chemical philosopher. Faraday's colleague at the Royal Institution, Brande, defined the role of this type of chemist in 1842 as "to investigate the nature and properties of the elements of matter, and their mutual actions and combinations; to ascertain the proportions in which they unite, and the modes of separating them when united; and to inquire into the laws and powers which preside over and affect these agencies."[22] In offering this definition Brande may have had Faraday in mind.

As Joseph Agassi has insisted, Faraday "hated the word 'physicist' and even the word 'scientist.'" Instead, he clearly identified himself as a natural philosopher. Although the term was being contested by the early nineteenth century, natural philosophy was often conceived as including the theology of nature within its scope. For example, in his *Lectures on Natural and Experimental Philosophy* (1794) the instrument maker George Adams (1750–1795) stated that while natural philosophy had previously been considered "dangerous to natural theology and piety," it had now "become an innocent inoffensive science, a useful minister to the temple of the Lord."[23] Faraday would have agreed with Adams's characterization. He also identified with the role of an electrician, which in the eighteenth and early nineteenth centuries meant "a person who studies or is knowledgeable about the science of electricity."[24] In 1844 a contemporary used two of the relevant terms in describing Faraday as an "indefatigable and erudite chemist and electrician."[25]

DISCIPLINARY CONSTRUCTIONS OF ELECTROCHEMISTRY IN TEXTBOOKS OF THE 1890S

Before turning to the presentation of electrochemistry in British textbooks dating from the final decade of the nineteenth century, a brief mention of developments in science education principally during the period following the 1851 Great Exhibition is in order. There was much interest and concern about scientific and technical education since there was a perception that other countries, especially Germany, were far ahead of Britain in this area.[26] To help address these shortcomings, two royal commissions had been established: the Royal Commission on Scientific Instruction and the Advancement of Science, generally known as the Devonshire Commission (1871–1875), and the Royal Commission on Technical Instruction (1881–1884).[27] Both commissions took evidence from a wide range of experts and produced detailed reports recommending significant changes in both areas.

Among the important innovations of the period were the founding of new institutions where science was taught and the centrality of laboratory work in science education. While the Scottish universities already possessed an impressive record in science education, the modernization of the Mathematical Tripos examination in the 1820s enabled Cambridge University to become a beacon of science. University College and King's College, London, (founded in 1826 and 1829, respectively) offered a range of science courses. Science was also central to many of the new civic universities, starting with Owens College, Manchester (f. 1851), which in 1880 formed part of the Victoria University (and subsequently included University College, Liverpool, and the Yorkshire College, Leeds, in 1884 and 1887, respectively).

The first chemistry teaching laboratory in Britain was Thomas Thomson's (1773–1852) at the University of Glasgow. It was opened in 1818, the year after Thomson expressed to a friend his ambition "to erect a laboratory upon a proper scale to establish a real chemical school in Glasgow and to breed up a set of young practical chemists."[28] The founders of a number of later chemical laboratories in Britain were inspired by Justus von Liebig's (1803–1873) laboratory-based teaching at the University of Giessen, including the founders of the Birkbeck Laboratory at University College, London, which opened in 1846. There is little evidence of chemical laboratories in schools prior to the 1860s, one of the few exceptions being Queenwood College near Stockbridge (f. 1847), where the chemist Edward Frankland (1825–1899) taught. In 1851 Heinrich Debus (1825–1915) joined the school, having previously obtained his doctorate in chemistry at the University of Marburg. Among the other early teachers at this Robert Owen–inspired school were the physicist John Tyndall (1820–1893) and the mathematician Thomas Archer Hirst (1830–1892). Frankland, Tyndall, and Hirst all subsequently studied in Germany.

William Thomson (later Lord Kelvin, 1824–1907) was responsible for creating in 1855 the first physics teaching laboratory, which was also at Glasgow University. Between 1866 and 1872 similar physics laboratories were founded at Oxford, Edinburgh, Owens, and the two principal London colleges, followed by the Cavendish Laboratory in Cambridge in 1874. These innovations led William Thomson to assert in 1885, "The physical laboratory system has now become quite universal. No University can now live unless it has a well-equipped laboratory."[29] We now turn to the portrayal of electrochemistry in textbooks and manuals published in the 1890s, starting with those directed towards physics students.

Physics Textbooks

One of the foremost physics textbooks of the period was *Lessons in Elementary Physics* by Balfour Stewart (1828–1887), who was professor of physics at Owens College from 1870 to 1887. First published in 1870, *Lessons* had been revised and reprinted several times and was widely used in colleges and universities. In his preface Stewart set out his aim, which was "to bring before the student, in an elementary manner, the most important of those laws which regulate the phenomena of nature."[30]

In 1885 Stewart added "a short sketch of the more prominent practical applications of electricity which have recently been made."[31] Following Stewart's death in 1887, later editions were revised by William Winson Haldane Gee (1857–1928), who was then a demonstrator at the college and had worked as Stewart's assistant. Gee's revisions to the chapters on electricity and magnetism were thus included in the fourth edition (1895) discussed here.

The main chapters of *Lessons* covered such standard topics as the laws of motion, force, and energy, followed by sound, heat, radiant energy (principally optics), static electricity, and magnetism. Batteries and electrolysis were among the topics discussed in the penultimate chapter, "Electricity in Motion." The first lesson in that chapter dealt with batteries: their history, the structure and chemical operations of the simple cell, and the specific forms of battery developed by Grove, John Frederic Daniell (1790–1845), Robert Bunsen (1811–1899), and Georges Leclanché (1839–1892).[32] In a later section three questions are posed about measurable quantities associated with a voltaic battery: how to determine its electromotive force, its resistance, and the strength of the current it produces.[33] The chemical effects of currents also featured prominently in a short section dealing with the various effects produced by it, including electrolysis; Faraday's laws and Theodor von Grotthuss's (1785–1822) theory of the conductivity and decomposition of water were mentioned. The book also included a very brief discussion of electrochemical equivalents and the determination of the value for copper (followed by a calculation of the amount of copper deposited in three hours by a current of 10 amperes).[34]

The penultimate lesson contained several diverse applications of electricity, including examples drawn from electrochemistry and electrometallurgy. Electroplating and storage batteries—principally Grove's gas cell—were briefly mentioned.[35] The final lesson focused on the overarching theme of energy, recurrent throughout the book, especially the conversions of energy involving chemistry, such as the quantity of energy released in the form of heat when certain substances were burned in oxygen. This example of conversion led the authors to speculate that "the electro-motive forces are really those which cause heat when chemical combination takes place."[36] Stewart, who also wrote a textbook on the conservation of energy, thus presented the student with an energeticist interpretation of physics and brought both chemistry and electricity (and their interrelationship) into the wider circle of energetics.[37]

The more advanced, laboratory-based students of physics could instead turn to *Practical Physics* by Richard Tetley Glazebrook (1854–1935) and William Napier Shaw (1854–1945), both of whom had been demonstrators at the Cavendish in the early 1880s. In arranging their textbook, Glazebrook and Shaw drew heavily on their experience of teaching practical physics to students registered for the first part of the Natural Sciences Tripos at Cambridge University. *Practical Physics*, which was first published in 1885, reached its fourth edition in 1894, the one that will be discussed here. Addressing teachers and students of practical physics, the authors sought "to place before the reader a description of a course of experiments which shall not only enable him to obtain a practical acquaintance with methods of measurement, but also as far as possible illustrate the more important principles of the various subjects."[38] The dominant theme throughout was therefore how to make accurate measurements of physical quantities, such as the length, area, mass, density, temperature, and electrical resistance of physical bodies and the focal lengths of lenses. As Graeme Gooday has noted, "the commitment of the emergent community of professorial experimentalists to precision measurement [w]as the definitive vehicle of disciplinary progress in physics."[39]

Apart from a passing reference to "voltaic or galvanic batteries" as a means of maintaining an electrical potential,[40] the only substantive discussion of electrolysis in *Practical Physics* began with a short section addressing the question of how to determine the reduction factor (k) for a tangent galvanometer, k being the ratio between the strength of the earth's magnetic field and the field strength produced by a unit of current.[41] By introducing a battery and a cell consisting of two copper plates in a copper sulfate solution into the circuit, the authors showed how the quantity of electricity flowing through the galvanometer can be calculated. This calculation drew on the electrochemical equivalent of copper and the mean deflection of the galvanometer, together with the mass of copper deposited on the cathode after fifteen minutes. From this the value of k was determined. There followed a second section in which Faraday's laws were used to determine the electrochemical equivalent of the substances deposited or released during an electrolysis experiment.[42] In contrast to Stewart's energeticist overview of physics, Glazebrook and Shaw focused on the uses of the theory of electrochemistry to achieve accurate physical measurements.

Chemistry Textbooks

By contrast with the rather limited role of electrochemistry in physics textbooks, contemporary chemistry texts directed at students generally used electrolysis as a fundamental tool for analyzing the chemical composition of substances. Among the most popular introductory chemistry texts was Henry Roscoe's (1833–1915) *Lessons in Elementary Chemistry: Inorganic and Organic* (1866), which was "widely adopted and translated into nine languages";[43] the sixth edition (1892) is cited here. In his text-book Roscoe, who was professor of chemistry at Owens College from 1857 to 1885, sought "to arrange the most important facts and principles of Modern Chemistry in a plain but precise and scientific form, suited to the present requirements of Elementary Instruction."[44] The individual chapters were principally concerned with identifying elements and compounds and describing their chemical properties. Electrochemistry was not accorded a separate section; instead, it was deployed to address a number of specific chemical problems, including how to determine the composition of water. Roscoe thus asserted that "the most striking method of demonstrating the composition of water analytically is by splitting it up into its constituent gases by means of a current of volta-ic electricity."[45] He then described this experiment using three or four Grove cells. Another example was the isolation of fluorine, which had recently been achieved by Henri Moissan "by the electrolysis of perfectly anhydrous hydrofluoric acid containing a little acid potassium fluoride to enable it to conduct electricity."[46]

In contrast to Roscoe's *Elementary Lessons*, William Ramsay's (1852–1916) *A System of Inorganic Chemistry* (1891) was a far weightier tome directed at university students. Ramsay, who had been appointed professor of chemistry at University College, London, in 1887, based his textbook on his teaching. He considered that chemistry students should become familiar with the vastly expanded knowledge of inorganic chemistry over the preceding few decades and appreciate the way the "periodical arrangement of the elements" had brought order and rationality to this rapidly developing science. He also complained of what he saw as the tendency of teachers to introduce students to "the physical aspect of chemistry . . . only when necessary [in order] to explain modern theories." Instead, "I hold that a student should have a fair knowledge of a wide range of facts before he proceeds to the study of physical chemistry, which, indeed, is a science in itself."[47] Thus in his *System* Ramsay sought to convey the principal facts of inorganic chemistry in an orderly manner, starting with the elements before moving to the compounds—the halides, oxides, sulfides, selenides, tellurides, borides, nitrides, etc. Toward the end of the book Ramsay briefly described alloys and also mentioned the rare minerals and how they could be identified by spectrum analysis. In the final section, however, he discussed industrial processes that required knowledge of chemical principles. "The student, having acquired the requisite acquaintance with facts, is now better able to appreciate these principles."[48]

As with Roscoe's text, Ramsay did not devote a specific chapter or section to electrochemistry. Moreover, he paid little attention to electrochemical theory; neither Faraday's laws nor the concept of electro-chemical equivalents were mentioned.[49] Instead, electrolysis entered into his account of specific chemical procedures, especially those involved in the acquisition of pure samples of chemical elements. For example, in discussing fluorine, he asserted that electrolysis "is the only method of preparing fluorine" and described the procedure using "a solution of potassium fluoride in dry hydrofluoric acid cooled to a low temperature." Ramsay's discussion was accompanied by a diagram of Moissan's appa-ratus.[50] Likewise, he cited the electrolysis of water (or "fused oxides or hydroxides") as one of the principal methods for obtaining oxygen.[51] The section on alloys similarly included examples of the use of electrolysis, for example, the production of an alloy of manganese and mercury.[52] Thus Ramsay paid little attention to the general theory of electrolysis, which he treated as just one of many techniques for decomposing compounds.

Other Textbooks and Manuals

As Robert Bud and Gerrylynn Roberts have argued, during the middle decades of the nineteenth century, chemistry underwent a significant change resulting in its separation (despite some overlap in subject mat-ter) into two branches that reflected different conceptions of the subject. While the academic and the practical had been closely fused earlier in the century, largely owing to the insistence of a group of Scottish chem-ists that chemistry was basic to manufacturing, later academic chemists tended to portray their subject as a science that they firmly differentiated from its practical applications in industrial contexts. The founding of the Chemical Society (1841), the Royal College of Chemistry (1845), and the Institute of Chemistry (1877, three years after the Institute of Physics) helped to cement the preeminence of the academic chemistry community.

The teaching of chemistry at universities and other academically oriented institutions therefore concentrated on the principles of chemis-try. "The academics," write Bud and Roberts, "while recognising the es-sential role of practical experience, nevertheless asserted the primacy of academic knowledge."[53] Many chemistry textbooks, including Roscoe's *Lessons* (but not Ramsay's *System*), reflect this prioritizing of knowledge of the general principles governing chemical phenomena, with relatively little attention paid to their practical applications. In teaching chem-istry, the main aim was to instill an understanding of the principles of chemistry. Technological topics were introduced only "at the end of their courses, while most pupils stayed only for the early elementary and fun-damental lessons."[54]

While academic chemistry had largely separated itself from technical education by the closing decades of the nineteenth century, a number of colleges were founded to provide artisans with education in technical subjects. Central to this movement were the City and Guilds Institute of London and its educational institutions, including Finsbury Technical College (f. 1883). Similar colleges serving local artisans were soon established throughout the country, with City and Guilds offering examinations that led to recognized qualifications. By the end of the century, 2,460 classes in "technological and manual training" were offered under the aegis of the City and Guilds Institute, and 15,557 candidates sat examinations in 1900.[55]

The teaching of electrochemistry was therefore not confined to physics and chemistry departments at institutions like the University of Cambridge and Owens College. Manufacturing chemists and others involved in such activities as electroplating also studied chemistry and electrochemistry in a variety of colleges and schools. For example, Silvanus Thompson (1851–1916), who served as professor of physics at Finsbury Technical College from 1885 to 1916, included electroplating among the topics he taught to his evening class. The subject was also taught at Finsbury over a period of thirty-five years by E. Rousseau, the instructor in the physical workshop.[56]

By the 1890s there were a number of textbooks with titles like *Chemistry for Engineers and Manufacturers* and *The Art of Electro-metallurgy*, both of which included discussions of electrolysis. The first of these—with the subtitle *A Practical Text-book*—was directed to "Practicing Engineers, Managers of Works, and Technical Students desirous of obtaining some knowledge of Chemical Technology."[57] Published in two volumes in 1896, it was written by Bertram Blount (1867–1921), a consulting chemist to the Crown Agents for the Colonies, and Arthur George Bloxam (1866–1940), a consulting chemist and head of the Chemistry Department at the Goldsmiths' Company's Technical and Recreative Institute, which had opened in 1891 and sought to promote "technical skill, knowledge, health, and general well being among young men and women of the industrial working, and artisan classes."[58] The first volume, subtitled *Chemistry of Engineering, Building and Metallurgy*, included a discussion of the conversion of chemical into electrical energy. Here the authors noted that "certain chemical reactions, capable of causing evolution of heat, can be made to take place under such conditions that a portion of the heat appears as electrical energy." They proceeded to calculate the energy of a Daniell cell (of 1.07 volts), 50.1 calories being required to dissolve 65 grams of zinc; the electrical energy corresponding with the chemical energy was nearly 207,900 electrical units of energy. This was followed by an extensive discussion of the energy configurations of other kinds of cell.[59] The volume also mentioned the use of electrolysis in various chemical processes, such as the winning and refining of copper, the reduction of aluminum, and the electroplating of silver.[60] In the second volume, which was devoted to the *Chemistry of Manufacturing Processes*, Blount and Bloxam mentioned several processes that employed electrolysis; for example, its use in producing such substances as sodium hydroxide (caustic soda) and potassium chlorate.[61]

Directed to a far more specialized readership, George Gore's (1826–1908) *The Art of Electro-metallurgy; Including All Known Processes of Electro-deposition* (1877).[62] Gore's varied career included working as a practitioner of medical galvanism, a chemist in a phosphorus factory, and a teacher of chemistry and physics at King Edward's School, Birmingham. Like a number of contemporary commentators, he considered that Britain was falling far behind other countries, especially Germany, because of a lack of scientific and especially technical education.[63] A close friend of Josiah Mason, the Birmingham industrialist, he became a committed advocate of both technical education and government support of scientific research.[64] He wrote several other books and contributed a large number of papers on electrochemistry to the *Electrician*, the principal weekly journal devoted to electrical engineering, and to other specialist periodicals. In the introduction to his book, Gore clearly set out its aim and intended readership: "I have endeavored to produce such a book as would be useful to scientific students, to practical workers in the art of electro-metallurgy, gilders, platers, &c., and to all persons who wish to obtain in a compact form, an explanation of the principles and facts upon which the art of electro-metallurgy is based, the circumstances under which nearly every known metal is deposited, and the special details of technical workshop manipulation in the galvanoplastic art."[65] Following a historical sketch of his subject, Gore devoted a substantial chapter to its theoretical underpinnings, its principles, and laws, including Faraday's laws of electrochemistry. He then turned to the "Practical Division," which comprised the majority of the book. After some general considerations, he focused in turn on the deposition of specific substances, devoting a chapter each to gaseous metals, electronegative metals, noble metals (gold, silver, etc.), base metals (copper, nickel, etc.), earth and alkali earth metals, alkali metals and metalloids (including carbon, oxygen, and nitrogen). Thus Gore organized his book by defining the chemical characteristics of substances. The final two sections were devoted to issues that an artisan would encounter in undertaking electrometallurgical manipulations, such as the design of the room where such apparatuses should be located, the methods of cleaning objects used by the artisan, and the relative merits of different types of battery.

THE ENDURING INTERDISCIPLINARITY OF ELECTROCHEMISTRY

The phenomena and theories of electrochemistry have a complex disciplinary history. With his discovery of the behavior of a frog's leg, Galvani evoked the notion of animal electricity and thus located the topic within the scope of physiology and, more broadly, medicine. Yet the reactions of animal tissue to electricity and the production of electricity by certain species of fish, such as the *Electrophorus*, received only passing mention in physics and chemistry textbooks of the 1890s as it fell outside these disciplines. Instead it can be found in some contemporary physiology textbooks, such Augustus Waller's (1856–1922) *An Introduction to Human Physiology* (1891). Waller, who was a lecturer in physiology at St. Mary's Hospital London, devoted a whole chapter to the subject, which included not only discussion of electric fish but also of the electrical properties of muscles, nerves, hearts, and eyes.[66]

A related but highly controversial intervention that challenged the contemporary disciplinary map of the late 1830s was the purported production of mites of the genus *Acarus* in an electrochemical experiment conducted by Andrew Crosse (1784–1855), a reclusive gentleman experimenter. Crosse claimed to have observed these acari emerge when he tried to create artificial crystals of silica by dripping a chemical solution (of silicate of potassa and hydrochloric acid) over a porous stone from Mount Vesuvius containing red oxide of iron, while subjecting the stone to an electric current from a voltaic battery. As Jim Secord has shown, with considerable difficulty some contemporaries tried to assimilate Crosse's creation of these living creatures to existing disciplines, such as entomology or geology, while some others considered that it heralded an exciting new discipline—electrobiology. However, natural philosophers in particular insisted that living beings had no place within their physicalist worldview, and they (including Faraday) dismissed Crosse's experiment as a groundless fantasy. Moreover, many Christians dismissed Crosse's claim to have created life out of inert matter as utterly untenable, as it was tantamount to atheism.[67]

Following Volta's work, the phenomena and theories of electrochemistry were firmly captured by the subject areas of chemistry and natural philosophy, which were generally considered separate domains, with physics slowly emerging from the latter. Working principally in the 1830s, the key figure of Faraday is interesting because he adopted a natural philosophical perspective that included chemistry, while eschewing its more commercial aspects. By about mid-century, however, chemistry and the newly branded discipline of physics were widely viewed as separate disciplines, both of which included electrochemical phenomena and theory, albeit in different ways. This divide is clearly visible in the physics and chemistry textbooks of the 1890s, by which time there also arose a market for texts directed at artisans and others working with electroplating and other electrochemical processes.

While there was some diversity within each of the three groups of textbooks, physics texts tended to focus on electrochemistry as a means of providing a viable and reasonably constant source of electricity that could be used for other purposes. We see this in Balfour Stewart's *Lessons* and also in *Electricity for Public Schools and Colleges* (1877) by Walter Larden (ca. 1855–1919) of Cheltenham College, who devoted a whole chapter to cells and batteries.[68] Physics textbooks to accompany laboratory work, exemplified by Glazebrook and Shaw's *Practical Physics*, utilized electrochemical phenomena to measure certain electrical quantities accurately. Chemistry textbooks, by contrast, were not interested in electrochemistry as a source of electricity. Instead, such texts as Roscoe's *Lessons* and Ramsay's *System* were principally concerned with electrolysis, which they used as a means of decomposing compound substances. The industrial uses of electrochemistry were sometimes mentioned—albeit briefly—in physics and chemistry textbooks. For example, Stewart's textbook devoted just one page to the topic, while Ramsay's contained only a cursory mention.[69]

The role of electrochemical theory in the 1890s textbooks varied considerably. Chemistry textbooks usually made only passing reference to Faraday's laws and ignored the theory of electrolytic action. (An exception is George Newth's [1851–1936] *A Text-Book of Inorganic Chemistry*, which included a discussion of Arrhenius's theory of electrolytic decomposition.)[70] By contrast physics textbooks often paid attention to the theory of electrolysis. For example, Stewart offered a full statement of Faraday's three laws in his *System* and also provided students with an account of the mechanism that Theodor Grotthuss had proposed in 1805 to explain the decomposition of water, whereby the oxygen anions are progressively shuttled to the anode and the hydrogen cations to the cathode, each anion and cation combining with an oppositely charged particle at every step of the way.[71] In his *Elementary Lessons in Electricity & Magnetism* (1881) Silvanus Thompson likewise discussed not only the laws of electrolytic action but also the theories of Grotthuss and Rudolf Clausius (1822–1888), who integrated Grotthuss's hypothesis with the kinetic theory of liquids.[72] Turning to texts directed to the practical uses of electrochemistry, Blount and Bloxam's *Chemistry* included only a passing reference to Faraday's laws, whereas Gore in his *Art of Elec-*

tro-metallurgy provided a full account of Faraday's explanation of electrolysis (which was based on Grotthuss's hypothesis). Gore was enthusiastic about Faraday's electrochemical researches and emphasized the importance of his laws, describing them as a "great truth."[73]

Between the 1830s, when Whewell contemplated the division of the sciences and Faraday undertook his innovative electrochemical experiments, and the 1890s, the disciplines of physics and chemistry grew significantly and also became more clearly differentiated from each other. Both encompassed electrochemistry; yet, as we have seen, each laid claim to different aspects of electrochemistry, with some overlap. This dual appropriation of electrochemistry is instructive because it shows how parts of a small domain of scientific knowledge were shared between two major disciplines during the Victorian period. In one sense electrochemistry had by the 1890s become interdisciplinary—crossing the boundary between physics and chemistry. However, electrochemical phenomena did not map straightforwardly onto either discipline, but straddled their boundary. (Moreover, some phenomena were no longer deemed relevant to either discipline; for example, *Electrophorus* was demoted to an interesting historical footnote.) This would indicate that disciplines are not natural groupings of phenomena but are historically contingent and of human, and thus social, construction.

In focusing on disciplines, we should not ignore the third constituency, consisting of those who used electrochemistry principally for practical, often industrial, purposes, such as electroplating. These practitioners did not adhere to any one scientific discipline but selectively used the knowledge gleaned from both physics and chemistry, together with the specific technical knowledge of the practical processes they performed, such as undertaking electrolysis on an industrial scale or electroplating a large batch of cutlery. Therefore Gore's *Art of Electro-metallurgy* included a long and detailed "special technical section" that included such topics as the layout and contents of the depositing room, the construction of vats for silver solutions, and how the surfaces on which metals are to be deposited should be scrupulously cleaned.[74] Gore also included a section on how various chemical substances should be safely handled and what should be done in case of accidents.[75] The existence of this third constituency shows that the map of knowledge of the physical world was not confined to the contemporary scientific disciplines. It also included the practical knowledge of artisans, such as those working in the electroplating industry. Although such artisans drew selectively on the disciplines of physics and chemistry, they possessed their own specialist knowledge and skills, which were not greatly shared by physicists or chemists. Instead they drew on such manuals as Blount and Bloxam's *Chemistry* and Gore's *Art of Electro-metallurgy*. While not denying the historical significance of scientific disciplines, the importance of the practical knowledge and expertise possessed by artisans should also be fully acknowledged.[76]

II

SYNTHESIZERS

Chapter 3

EXPERTISE ACROSS DISCIPLINES

Charles Darwin's Engagement with Natural History

JANET BROWNE

In 1953 the philosopher Isaiah Berlin popularized the view that intellectual history could be divided between thinkers who adopt a single overarching vision and those who pursue multiple aims.[1] He playfully drew on Archilochus's characterization of hedgehogs (specialists) and foxes (pluralists) in order to discuss the ability of some individuals to bear down on a single problem and others to think laterally and draw connections. Reading Berlin nowadays one gets the impression that he was flippantly referring to styles of conversation at the college high tables of Oxford University. He afterward claimed he did not put forward the proposal as seriously as people took it.[2] However, this same notion reformulated as insider and outsider thinking, specialist and generalist, is today enshrined in Western cultural practice, especially business management. Something of the same can be seen in science.

A scientific insider knows very many details about one specific topic whereas an outsider, it is thought, can bring fresh and unexpected vision to bear, perhaps by applying principles derived from far-removed areas of inquiry or overturning established rules of procedures.[3] One well-known example in history of science is that of Watson and Crick bringing their nonspecialist, foxy, outsider perspective to molecular biology in 1953 to elucidate the structure of DNA. This achievement was possibly in Berlin's mind when he wrote. Other examples can be found, not least in the contemporary high-tech world. The point is important for today's university administrators, managers, and funding bodies who seek to identify and support innovation; and also for philosophers and historians of science who explore the origination of new ideas. What sort of practices encourage innovation? Is interdisciplinarity the answer? Is this the same process as what we used to call cross-fertilization? Indeed, in the sciences it would seem that interest in transdisciplinarity is the new normal.[4]

Categories such as these break down in the case of Charles Darwin (1809–1882). He was both a pluralist and a specialist, capable of deeply detailed and expert research in a variety of natural historical domains and yet also wedded to one big idea as an explanation for multiple phenomena. As Darwin conceived it, the idea of natural selection unified and explained the entirety of the living world and could account for its myriad details. In Berlin's typology, he was hedgehog *and* fox.

How might this odd beast fit into a nineteenth-century world that was vigorously dividing into professional intellectual disciplines, even though these disciplines were less firmly established than the present day and typified by relatively fluid boundaries?[5] Evidence for this subdivision in Britain is not hard to spot. In the years 1800–1850 the generalist Royal Society of London gave rise to the more specialized Geological Society and Astronomical Society, and the natural history–based Linnean Society of London gave birth to the Zoological Society. The Horticultural Society began in 1804, and the Geographical Society in 1830. The Statistical Society was established in 1834, the Botanical Society of Edinburgh in 1836, and the Microscopical Society in 1839. The Ethnological Society of London was founded in 1843. Societies dealing with science as a specialty in itself also sprang up: the British Association for the Advancement of Science was established in 1831, in which the annual meetings and reports were divided into nascent disciplines.[6]

A Society of Engineers was established in 1854. In other domains, the legal profession began to restructure itself into specialisms, as did medicine, the military, and education. No doubt there are many historical reasons for this demarcation of professional expertise in Britain during the industrializing nineteenth century. Victorianists can discern a cultural leitmotif of differentiation running through the period: for example, nineteenth-century society began to arrange itself into distinctive social strata according to occupation and prosperity, the manufactories of the day adopted the economic principle of the division of labor to maximize efficiency, and the bourgeois domestic home was increasingly subdivided into specialized areas of activity. Natural philosophy also began to compartmentalize into a series of separate intellectual disciplines that were made manifest in new institutional structures: some in physical form through premises, staff, publications, and elections to membership, and others in intellectual form with increasingly self-defined standards and values. As Lightman and Zon have shown in their recent study on the origins of modern disciplines in Britain, the impulse to differentiate and specialize has its origin in the Victorian era.[7]

Even though he was a member of several of these learned societies, read their journals conscientiously, and occasionally went to their meetings, Darwin did not seemingly perceive this ease of movement across institutional boundaries as a deliberately interdisciplinary pursuit, either within science or beyond it. Instead, using the terminology of his time, Darwin mostly thought of himself as a naturalist: as a thinker who took the whole of nature as a field of inquiry. Sometimes, as a young man, he called himself a geologist, and this was a category that had felt right to him at the time.[8] In an annual directory of British householders published in 1845 he described himself as a "farmer" (amusing to modern historians, but indicating his status as a person who owned land and employed agricultural workers and domestic staff).[9] He never called himself a botanist, despite singular contributions to that field and earning the esteem of several notable botanical professionals such as Asa Gray (1810–1888), professor of natural history at Harvard University, and Joseph Dalton Hooker (1817–1911), assistant director and then director of Kew Gardens. As far as is known, Darwin did not call himself a biologist either. The term was by then already in use, introduced around 1800 principally by European reformists such as Jean-Baptiste Lamarck (1744–1829) and Gottfried Reinhold Treviranus (1776–1837). In 1856 George Henry Lewes (1817–1878) referred to the category of "biologist" in a review article on heredity in the *Westminster Review* ("for a biologist of the nineteenth century").[10] Darwin lived at a time when that label usually referred to men interested in physiology, as was Lewes.[11] The word shifted in meaning during the second half of the century, coming to denote the study of the life sciences, as shown, for example, in 1864 when Ernst Heinrich Haeckel (1834–1919) labeled colleagues in Germany as "we biologists" (uns Biologen).[12] Thomas Henry Huxley (1825–1895) defined the field for British attention in 1876.[13] Darwin used the word *biologist* once about someone else in a short paper in 1877.[14]

As a naturalist, Darwin ranged over a number of different fields: geology in the widest sense, to include tectonics, paleontology, erosion and sedimentation, and the geology and zoology of coral reefs; zoology, including taxonomic work on barnacles and problems of naming and classification; botany, including orchid fertilization, plant physiology, and so forth. Throughout, his intense interest in the interactions of living beings with their habitats has led many subsequent scholars to call him one of the first ecological thinkers.

This wide scope provides the basis for thinking of him as pursuing an interdisciplinary program. Famously, the notion of natural selection was derived by him in 1838 from a comprehensive reading program that brought him face to face with Thomas Robert Malthus's (1766–1834) *An Essay on the Principle of Population* (first published in 1798, though Darwin likely read the sixth edition, 1826). This fact is one of the strongest indications of a wide-ranging outlook that embraced economics and political theory as well as natural history, as noticed decades ago by Robert Young.[15] Another is that *On the Origin of Species* (1859) comprised a body of tightly connected observations drawn from many different fields. Right from the start of his research program, Darwin searched for a single comprehensive law to interlink the phenomena of nature. So there is much to suggest that he engaged very widely with nineteenth-century thought in general and that the *Origin* reflected these researches. Furthermore, the readership for the *Origin* came from many diverse corners of the Victorian world, such as economics, politics, literature, and linguistics. The book was eventually read by members of many social classes and in translation. It was highly influential in stimulating cross-fertilization and connections among diverse fields.

The wide-ranging scope of Darwin's research and its published product does not, however, mean that he was a broad-brush generalist or a remnant of an earlier era when natural philosophers dabbled in a bit of everything. His proficiency in the area of geology was recognized by the award of the Wollaston Medal in 1859, and his monographs on barnacle morphology were awarded the Royal Medal of the Royal Society in 1853. In 1878 he was elected corresponding member of the French Académie des Sciences for his botanical work. A full list of his awards and memberships in societies is given in Francis Darwin's (1848–1925) *Life and Letters of Charles Darwin* (1887).[16] He was evidently perceived as an expert in a number of different fields, although in no way regarded as a polymath. He in turn recognized many of his contemporaries as experts in their own fields of inquiry, from gamekeepers to Cambridge University professors. As a cultivated man, he also read the quarterly review journals, political commentaries, biographies, and general literary works. He was not alone in this. Perhaps Alexander von Humboldt (1769–1859) is the most obvious parallel, possessing exceptional breadth of cultural vision and expertise. Another might be Robert Chambers (1802–1871) who juxtaposed many intellectual domains in the service of a theory of transmutation, but with less evidence of expertise.[17] The term *interdisciplinarity* can as easily suggest a form of superficial, broad-stroked knowledge.

What might it have meant to be an "expert" in any area of nineteenth-century science? While this is perhaps not immediately relevant to the history of interdisciplinarity, it is necessarily significant in the emergence of disciplines. The Victorian turn toward specialization suggests that the various societies and clubs that were established for particular subject areas acted as gatekeepers to govern who might be thought of as an "expert." Huxley and the polymath William Whewell (1794–1866) both ranged widely over nature and culture. Whewell, initially self-identified as a mathematician and mineralogist, was talented enough to serve as an officer in the Geological Society of London, and then took to mechanics, physics, astronomy, and philosophy, while also finding time to compose poetry and translate Goethe.[18] This intellectual range might well be called interdisciplinary but was probably in its day regarded as exceptional expertise in many domains. Another wide-ranging contemporary thinker was Baden Powell (1796–1860), the mathematician and evolutionist.[19] Whewell and Powell were much the same generation as Darwin and Alfred Russel Wallace (1823–1913), and were theological professionals, as Darwin and Wallace were not. Religious commitment to explaining God's works seems a likely factor in the variety of topics taken up by the former, but not a determinant in the case of Huxley, Wallace, or Darwin. What links these thinkers together is that they held to the belief that nature could be understood as a unified whole—that the natural world was one world, even though its many parts appeared to be dissimilar. In this sense, the natural philosophers of the early years of the nineteenth century were all-embracing in their outlook and aimed in their work to understand several areas of thought (that we retrospectively label as disciplines) long before the concept of interdisciplinarity emerged.

So can we call Darwin's approach an "interdisciplinary" activity? Maybe. It might be more historically accurate to regard his activities as interconnected thinking. He applied one crucial insight to many fields. Wallace, the cofounder of the concept of natural selection, also pursued research that was similarly detailed in many different natural history fields in order to justify the same all-embracing theory. Both men seemingly felt that to propose such a major conceptual shift as evolution by natural selection required evidence of the applicability of the theory over a wide array of natural history topics. For them, the aim was to bring many apparently disparate concerns together as support for what Darwin called "one long argument" and Wallace jokingly referred to as "a mountain of facts."[20] Philosophically speaking, they both understood themselves more or less to be following Whewell's prescription for ascertaining a true theory of nature through the consilience of inductions, in that an induction obtained from one class of facts should agree with an induction obtained from another, different class. The more consiliences there were, then the more likely the theory was to be true.[21] In Darwin's case, a model—minus the religion—was derived from William Paley's (1743–1805) *Natural Theology; or, Evidences of the Existence and Attributes of the Deity* (1802), in which Paley presented material from several natural historical domains as evidence for God's beneficent design permeating the given world.[22]

It seems useful therefore to explore the ways that a single individual such as Darwin could operate as an expert in several fields *and* as a synthesizer in a social and intellectual context that exhibited increasingly compartmentalized zones of knowledge. Did freedom from the need to earn money release him from grinding through a specialized career trajectory, as was mainly the story for contemporaries such as John Tyndall (1820–1893)?[23] This chapter covers the manner in which Darwin was able to create authoritative specialized information that was acceptable to metropolitan experts through somewhat quaint home-based observations and networks of knowledge; how he published meticulous scholarly work in the journals of newly distinct disciplines, and at the same time participated in the more general natural history community of popular periodicals and magazines; while his primary aim was to support one giant synthesis that reached across many intellectual boundaries and was published in the *Origin of Species* and later works. The justification for focusing on these three areas is to indicate, in a compact manner, the actual practice of a person customarily designated as uniting many domains of thought. To revert to the earlier remark made by Isaiah Berlin, the sections explore the interplay between hedgehog to fox. The puzzle is whether such practices could be called interdisciplinary.

OBERVATIONS AS EXPERTISE

Darwin is often characterized as one of the last "gentlemen naturalists"—as a man whose private wealth allowed him to work at home on any topic that caught his attention.[24] This was true enough, yet does not fully explain the intellectual reach and drive that he brought to his work. Darwin was a superb observer and practical experimentalist who designed many effective techniques to probe the natural realm. Carried out with simple tools and procedures, his observations provided insights into phenomena as diverse as pigeon ancestry, plant fertilization, the role of earthworms in recycling soil, and the fly-catching and digestive habits of insectivorous plants.

He had been a hands-on amateur naturalist from very early days.[25] The *Beagle* voyage turned this knowledge into a structured worldview, in part by providing opportunities to explore the ideas expressed in Charles Lyell's (1797–1875) *Principles of Geology* (1831–1833), the philosophical manifesto in geology that Darwin took with him on the voyage, and in part by enabling him to expand his practical knowledge of nature on a global scale.[26] On the *Beagle* Darwin learned to think big while at the same time engaging with numerous particularized instances of natural phenomena. It seems likely that this twofold approach to the natural world—think big and observe small—became second nature to him and can be seen as an intellectual theme through the rest of his life. Darwin especially learned from Lyell that many small geological events, if occurring repeatedly over eons of time, could add up to substantial effects. Darwin applied this dictum not only to geology but to all spheres of living natural history; ever after, he sought the small changes that might accumulate into differentiated organic phenomena. It is not too strong to claim that this was the single most important methodological commitment that he made, one that accompanied him from youthful speculations to mature reflection.

Settled in his country home, Down House in Kent, from 1842 Darwin took the opportunity to establish himself as an investigative naturalist with plenty of time to run long-term projects if he wished. Other naturalists of his generation experimented at home, to be sure. Yet Darwin had the financial stability to make his house and grounds into a private laboratory or observatory, and his study into a center for the production of scientific papers, books, and correspondence: all of his daily scientific life was carried out with natural selection in mind. His son Francis stated that "it was his [Darwin's] habit to work more or less simultaneously at several subjects."[27] He was always alert to the tiny fact, the unobserved point that might contribute to his theory. The cross-breeding of plants, domestic dog behavior, honey bee flight paths, the equine mind, human child development—these were each grist to his investigative mill. For example, in 1863 and 1864 when he was unwell, Darwin chose to observe twining plants in flowerpots from his sickbed. He saw the tendrils circle about in search of something to fasten onto. Soon he was placing sticks in different places for the shoots to catch onto. "It is all I am good for; I can just do an hour or two's work, when I can do nothing else," he told his village acquaintance Rev. John Brodie Innes (1815–1894).[28] In this instance, simple experimental interventions were fueled by the larger question of the adaptive strategies of plants. Bit by bit he built up a file of unique observations on climbing plants that he published in 1865 in the Linnaean Society journal. Domestic incidents had been transformed by him into high science.

A much longer observational project was that on barnacles. This was carried out on living and fossil barnacles and occupied him from 1846 to 1854, some eight years in all.[29] It began as the small task of identifying an unusual cirripede that he had collected in Chile during the *Beagle* voyage. After the voyage ended, Darwin retained in his possession most of the marine invertebrates; he hoped to work on them, reimmersing himself in the enthusiasms that had been fostered under Robert Edmond Grant's (1793–1874) influence long ago at Edinburgh University.[30] But the voyage had ended ten years ago. His friend Joseph Hooker inadvertently supplied the direct stimulus for taking up this new project by remarking casually in a letter in 1846 that he was not inclined to trust individuals who had not worked closely on classifying species.[31]

Hooker explained afterward that he had not meant Darwin, but Darwin took the remark to heart, feeling that he did not have the acknowledged expertise that Hooker valued.[32] Darwin's *Beagle* specimens, after all, had been identified by others and his reputation at that point in time was mostly as a geologist. Darwin's barnacle work can therefore be regarded partly as the manifestation of an urge to establish himself as a card-carrying taxonomist, and partly also to establish a firm reputation as an expert that would help stave off attacks that he was too speculative when he published his future book on evolution. He was well aware of the accusations of wild speculation directed at the anonymous author of *Vestiges of the Natural History of Creation* (1844), and recognized the spread of contemporary fears about transmutationary arguments.[33] Thoughts such as these no doubt underpinned the determination with which he tackled what became a huge taxonomic project.

In order to classify one aberrant barnacle, Darwin found he needed to understand much more about the comparative morphology of the various subgroups. "I work out mouths and cirri carefully, muscular structure & tunics of the sac, & some of the structure of the viscera," he told the comparative anatomist Richard Owen (1804–1892) in 1846.[34] Another colleague, John Edward Gray (1809–1875) at the British Museum, advised him to make a complete survey of the living and fossil barnacles and facilitated the loan of specimens from the museum's national collections. Most unusually at that time, Darwin was therefore able to call on his preexisting social credibility as a geologist and author to examine these specimens at Down House instead of visiting the museum to inspect them.

He carried out the barnacle study very thoroughly. Notably, he sent letters far and wide to naturalists soliciting information or access to desirable specimens, such as his former assistant on the *Beagle* voyage, Syms Covington (1816–1861), who now lived in Australia. Covington collected and sent Darwin some pedunculated barnacles that could not be found in any museum collection closer to home. Darwin's work continued daily, to the extent that his children thought that "doing" barnacles was something all fathers did. Darwin never regretted dedicating time to this pursuit but poked fun at himself for the length of the investigation.[35]

What lifted this project out of the ordinary was Darwin's interest in functional anatomy. Early on in the study he realized that some species of barnacle were not hermaphroditic—as was usual—but seemed to possess individuals that were either male or female. This observation was based on astonishingly detailed microscopical observations. Perplexingly, one or two genera, *Ibla* and *Scalpellum*, possessed minute rudimentary males that lived almost as parasites on the females. Some of the hermaphroditic species also possessed minute extra males. He believed he was seeing the evolutionary divergence of separate sexes from an unknown hermaphrodite ancestor. Yet in his eventual publications he did not allude to this dramatic explanatory theory. His purpose in these disciplinary volumes was instead to display high-level taxonomic expertise. As already mentioned, his barnacle monographs won him the Royal Medal of the Royal Society in 1853.

As life moved on, Darwin's experimental investigations mostly came to focus on plants, with occasional forays into other organisms such as insects and worms. Natural selection theory provided him with an absorbing reason to explore plant adaptations. A special concern of his was to identify instances of coadaptation between insects and flowers to ensure cross-pollination. His monograph on orchids (1862) pursued this point in exceptionally detailed fashion, based on concentrated observational research.[36] Where most people might assume that the fantastical flower parts of orchids were the handiwork of a creator to beguile human beings with their beauty, Darwin saw them as an assemblage of ad hoc evolutionary adaptations to entice insects inside and facilitate the deposit of pollen onto their backs and sides for transport to another flower. "I cannot fancy anything more perfect than the many curious contrivances," he told his botanical friend Hooker. The book generated warm praise from fellow botanists. To Darwin, it was an answer to Paley's heavenly designer. By then, well into the controversy over the *Origin of Species*, he was frank about natural selection guiding his interpretations. He called his orchid book "a flank movement on the enemy."[37]

Less discussed in the literature are his observations on the dimorphism of wild primroses, carried out during the summer of 1861 on specimens grown in his garden and greenhouse. Two kinds of primrose flowers were already well known, one called "pin-headed" that possesses a long female style in the center and the other the "thrum" form with a short style. Each kind is self-fertile, but Darwin thought to cross-pollinate them in a carefully designed exercise to show that pin-to-thrum matings were more successful in numbers of offspring than pin-to-pin or thrum-to-thrum. A discrepancy in the size of the pollen grains as seen under a microscope had suggested to him that this might be the case. He made more than a hundred crosses, hand-fertilizing the plants with a paintbrush and then collecting and counting the resultant seed-pods. He opened the pods and weighed the seeds from different batches, having decided that weight would indicate number of seeds, which in turn would indicate relative fertility. He established that if pins were allowed to fertilize thrums, the plants were more productive of seeds than any other mating. From this well-organized study, Darwin concluded that the two forms of flowers were an adaptation to facilitate outbreeding, almost as if the basic self-fertile condition was differentiating into two functional sexes. He sent a very original paper on the topic to the Linnean Society of London in November 1861, which was eventually published as a short book, *The Different Forms of Flowers on Plants of the Same Species* (1877). As with twining plants, he had created something meaningful out of apparently insignificant observations.

In practical terms Darwin's experiments in his home or garden were neither more nor less than any other naturalist with similar social advantages might have pursued. Yet he brought to them an overriding preoccupation with the consequences for evolutionary theory. In this sense he was a remarkably synthetic thinker This feature of his practical work remained to the fore throughout his long life. Every experiment was devised in order to discern its relevance for evolution by natural selection. He often tried things out to see what might happen. "The love of experiment was very strong in him," said his son Francis Darwin after his death, "and I can remember the way he would say 'I shan't be easy till I have tried it,' as if an outside force were driving him."[38] That outside force was the notion of natural selection. He was clearly working across domains to identify the same functions or explanations at work in nature. This might well have been a conscious cross-disciplinary effort.

CORRESPONDENCE AS EXPERTISE

Might Darwin's correspondence networks reveal something extra about his reach across established intellectual boundaries? It hardly needs saying that Darwin was a prolific correspondent. Some fifteen thousand letters to and from Darwin have been preserved to the present day, and presumably others must have been lost or destroyed. It is useful here to recall that for him letters did much more than merely communicate news and views. It is reasonable to think of Darwin's correspondence as a scientific networking device, much like today's internet sites, that served him as one of the basic means by which he located information and adopted and transmitted common value systems. This correspondence took him far and wide into various natural history communities in Britain and abroad: into elite scientific society; into the world of merchants, gamekeepers, kennel hands, and natural history suppliers; the hidden knowledge systems of female naturalists; the web of colonial administrators, diplomats, schoolteachers, and officials; and an ever-increasing circle of individuals from many levels of society who, he felt, could contribute to his research. Mostly Darwin initiated the connection. As he became famous, these circles expanded to include people who wrote to him spontaneously about his published views or matters of the day such as religion or ape ancestry. Darwin courteously replied to all. It is fair to say that the material covered in his letters after the *Origin* was published was highly eclectic, fully representative of the world of Victorian culture at large. There are other nineteenth-century scientific archives that similarly reflect a wide range of contemporary areas of interest and divergent points of view, such as John Tyndall's networks of communication, but the Darwin correspondence stands out for the scale of its reach and range (and preservation).

As is increasingly recognized by historians, letters comprised one of the leading communication technologies available to natural philosophers in the nineteenth century. For some Victorians, a large-scale colonial correspondence comprised a research methodology, as in the systematic hunt for data on the tides undertaken by William Whewell.[39] Communication via letter was also an established way to bring potential scientific theories to the attention of others and to claim ownership of those theories before print publication. Letters helped to standardize terminology. They speak to historians about patronage networks, social differentiation, and gender issues, and reveal the ways in which communities of scholars disseminated a common culture. It has been said of early modern communication networks that the globe became smaller when people in different countries began to exchange information. For Europeans and settler communities, the notion of far-flung regions being joined together under one systematic political, social, and economic umbrella was made real.[40] Letters moving back and forth generated trust and authority between correspondents, the essential elements of social capital.[41] In the context of this volume on interdisciplinarity, they also were a primary means of delineating specialized domains in science; uniting members of new disciplines with a common set of principles, terminology, and values; and bringing together those individuals who wished to push their particular view of science ahead, for example, as in the self-selecting members of the X Club described by Ruth Barton.[42]

Most of all, and sometimes easiest to forget, correspondence used to be one of the main ways that scientists collected, processed, and disseminated data. In the observational sciences, it was a crucial facilitating factor in the mobilization and standardization of information, especially in the middle decades of the nineteenth century. For Darwin, letters exchanged between contemporaries were not just a daily fact of life but comprised a great deal of what it meant to be a man of science.

Correspondence networks—at least, scientific ones—therefore created an epistemic space in which disciplines could be built and the boundaries of class, nationality, education, and gender might occasionally be overcome. It is notable how much Darwin corresponded with individuals outside the elite circles of London science. His network of contacts in the poultry world, for example, was extensive, although he only knew one or two of them in person.[43] The majority were respectable middle-class householders who bred and exhibited show-quality poultry. The point that drew Darwin to them was that they had expertise in areas that he did not. Women were sometimes included in these networks. He corresponded with some 115 women—not so many as one might hope among his three thousand correspondents, and none of them formally engaged in science, but still an important group.[44] Artisans also feature in his correspondence as bearers of unique local knowledge, a parallel to the case described in Anne Secord's studies of botanical correspondence between William Jackson Hooker (1785–1865), director of the Botanic Gardens at Kew, and workmen in the cotton mills of Lancashire.[45]

Making contact with William B. Tegetmeier (1816–1912) was an important step forward. For a decade or more Tegetmeier kept Darwin informed by letter about specialist points on poultry, provided live birds from his own collection, arranged for Darwin to receive carcasses from people he knew, and answered every conceivable question as best he could. With Tegetmeier's assistance, Darwin kept ornamental pigeons and a few specialist chicken breeds at Down House in the 1850s for research purposes. These were domesticated organisms that helped him articulate analogies between artificial selection (that is, produced by human intervention) and selection in nature, and additionally provided information on the results of cross-breeding. Most of Darwin's knowledge of the art of pigeon breeding was thus derived from Tegetmeier, and this was acknowledged both in *Origin* and *The Variation of Animals and Plants under Domestication* (1868). Tegetmeier understood the value of his social connection to Darwin and occasionally made use of it in print by referring to their correspondence. In turn Darwin was generous in his praise and evidently respected the depth of Tegetmeier's knowledge.

Darwin latched onto experts in every field. Often these individuals were his social equals, as was the Cambridge mathematics professor William H. Miller (1801–1880), who knew far more about geometry than he did and could answer complicated questions about the construction of honeycombs, composed of many hundreds of hexagonal wax cells.[46] Some were grander than he, as in his correspondence with Lady Dorothy Nevill (1826–1913), the London political hostess. Darwin approached Nevill to ask if she might be able to spare him some rare tropical plants from her greenhouses, and a correspondence ensued. Others were hidden from history but nonetheless knowledgeable. There are many letters in Darwin's extant correspondence from strangers who had read his publications or had noticed something that they thought might interest him. Once or twice, women wrote to challenge him on his views about women's place in society, including Antoinette Brown Blackwell (1825–1921), a critic of the competitive, cut-throat element in natural selection.[47] Casting his net abroad, Darwin acquired a dizzying array of introductions to gardeners in the British colonies,[48] especially the botanical brothers Roland (1840–1916) and Henry Trimen (1843–1896), working in South Africa and Sri Lanka, respectively, each of them a mine of information.

Networks like these were the crankshaft of his intellectual machinery. Yet even though the range and variety of inquiries encouraged Darwin to move far beyond his own knowledge base, it is clear that he regarded the process as a personal supply system, providing him with the information he needed to synthesize results and continue his own projects. He did not give much back except a friendly word or two in return and, if needed, some gentlemanly patronage. This was not a circulation of knowledge: it was mostly a one-way movement of information toward Darwin. It was easy for him to use social position and his reputation as a man of science to ask questions and expect a reply.

This correspondence network was made possible by the era and social spaces in which he lived. They reveal the many tendrils of the British administrative system coiling around the globe, enhanced by the newly efficient procedures of the postal system. When the *Origin* was published, Darwin's aims shifted from gathering to influencing. While declining to appear in public to defend natural selection after 1859, a barrage of his letters nevertheless made their way into the outside world, letters that encouraged, supported, nudged, explained, disagreed, thanked, consulted, advised. He used letters to persuade and reassure. He used them to tilt reviewers in more favorable directions, arrange translations, make new contacts, gather support. During the 1860s he sent some five hundred letters a year to weather the storm of controversy.

These overlapping networks of letters therefore indicate something of the vibrant world of specialized knowledge in the nineteenth century and the various ways in which it could be accessed. Darwin had a gentleman's power to ask for information, the power to soothe and disarm critics, to persuade those who did not normally contribute to science to help him, and to blur national, geographical, gender, and social boundaries as he did so. He was able to touch on many topics and gather specific information from a large number of domains, be a charming and confiding correspondent, and to appreciate the information that he acquired, without feeling the need to articulate any expressly interdisciplinary aims.

Publishing as Expertise

Was the world of print much different? Every year of his life after returning from the *Beagle* voyage Darwin published material on natural history, some lengthy, some less so. The total number of items (roughly calculated) is impressive: sixteen books and some 266 shorter papers comprising scholarly articles, letters to editors, printed questions, pamphlets, and many brief writings.[49] Together these publications reveal the variety of his scientific interests and the range of publishing vehicles that he felt was appropriate for the kind of knowledge he was presenting or accessing.

This productive life in print bears some examination. When he wrote, Darwin usually had a specific audience in mind. He did not write textbooks or any popular versions of evolution by natural selection, as did Wallace, Huxley, Asa Gray, John Lubbock (1834–1913), or George Romanes (1848–1894). He did not write children's books or natural history primers, as did Arabella Buckley (1840–1929) or Phillip Henry Gosse (1810–1888). Nor did he condense his *Origin of Species* into a more readable form, as his friend Lyell had done with *Elements of Geology* (1838). Darwin's books were mostly accessible in the sense that he wrote engagingly in the first person, without the use of technical language, and that the structure of each book was narrative, following a trajectory almost as in a novel, rather than taking the format of a scientific disquisition. Yet it is clear that he anticipated that his books and articles would be read mostly by educated gentlemen of science. The most accessible of his publications was his *Journal of Researches* (1839), part travelogue and part descriptive natural history of the countries visited during the *Beagle* voyage, modeled on Humboldt's *Personal Narrative* (1814–1829), which had captured his imagination at Cambridge University. Later titles, especially *The Expression of the Emotions in Man and Animals* (1872) and *The Formation of Vegetable Mould through the Action of Worms, with Observations on Their Habits* (1881) dealt with attractive natural history material and were described by contemporary reviewers as appealing.

It is sometimes claimed that *On the Origin of Species* was accessible to the general reader.[50] Darwin's personalized writing style and the anecdotal air of much of the cited evidence certainly made the book seem companionable and manageable. Moreover, the general thesis—the need to remove the deity as an explanatory force in nature—was easily understood by any reader. But the actual mechanism of descent with modification as Darwin described it was complex and not fully intelligible except by experts. It was not written as a generalist book, intended for wide circulation, as was Robert Chambers's *Vestiges of the Natural History of Creation* (1844). Darwin explained this to Huxley at the time of publication: "Exactly fifteen months ago, when I put pen to paper for this volume, I had awful misgivings, & thought perhaps I had deluded myself like so many have done; & I then fixed in my mind three judges, on whose decision I determined mentally to abide. The judges were Lyell, Hooker & yourself."[51] Even so, the book enjoyed an extremely wide readership, stretching from poets like Alfred Lord Tennyson (1801–1892) to cultural critics like Karl Marx (1818–1883).[52]

Except for *Journal of Researches*, which was intended for a general educated audience, and perhaps *The Formation of Vegetable Mould through the Action of Worms* (1881), most of Darwin's scientific books were directed to newly professionalized and intellectually demarcated scientific audiences. It was a bonus for him if nonscientific readers engaged with those professional texts. By contrast, when he wrote short pieces for natural history periodicals (as distinct from the journals of learned societies), he was deliberately aiming to engage the wider public. It sounds obvious, but he selected the medium that would reach his desired audience. The burgeoning print marketplace of Victorian Britain facilitated his ability to choose.[53]

Books first. In the standard bibliographical handlist to Darwin's publications, Richard Freeman states that a complete enumeration of editions of Darwin's books published in England during his lifetime amounts to 116 items. To this may be added the American printings and translations into eleven languages before his death in 1882.[54] Over his lifetime he was contracted to five commercial publishers in the UK. Three of these handled one title apiece. For the rest, Darwin mostly published his books with the firm of John Murray, based in Albemarle Street in London. He was introduced to Murray (1808–1892, the third John Murray of that name) by Charles Lyell (1797–1875) in the hope that the firm would produce a second edition of Darwin's *Journal of Researches*. The relationship proved successful, the revised *Journal* was produced in 1845, and Murray was Darwin's choice thereafter for many of his publications, including the *Origin of Species*. Theirs became a key partnership in the making of Victorian knowledge.[55] In the United States the New York company of William Appleton provided the same service, becoming the primary firm in bringing Darwin's writings and Darwinism in general to North American shores.

Murray's firm did a great deal in the nineteenth century to bring scientific books to the attention of the public, as well as being one of the most active and commercially self-aware publishers of travels, biographies, poems, seasonal albums, and novels. The firm published the celebrated travel guides, Murray's Handbooks for Travelers, from 1836. It also owned and published the major literary review journal *Quarterly Review*, in which Murray's titles were reviewed alongside other books of the period, thereby maintaining the sense that science and natural philosophy were expected companions in the cultivated domain. This coverage might perhaps be called interdisciplinary but more accurately reflects what Robert Young persuasively called the "common context."[56] John Murray promoted science personally by maintaining a number of prominent figures on his lists, such as Lyell, and publishing the annual report of the British Association for the Advancement of Science.

Of the rest, Henry Colburn published the first edition and an independent reprint of Darwin's *Journal of Researches* in 1839. Smith, Elder & Co. published his three geological books and the magnificent edited set *Zoology of the Voyage of the Beagle*, in five parts, with hand-colored lithographs, over the span 1838–1843. For the latter, he applied for and received a British government grant to produce the illustrations. In that venture, Darwin explicitly aimed to bring his achievements as a collector to the expert natural history community comprising university professors, museum professionals, and independent gentlemen scholars who were members of scientific societies in Europe and North America. The *Zoology* was, so to speak, his passport to the zoological community and presented new or remarkable species in the conventional cataloging manner. His geological books served the same function in geology, and his narrative *Journal of Researches* gave him a public profile. He was an ambitious and talented young man consciously using the different (often overlapping) markets of the newly established specialized learned societies and the expanding Victorian publishing world to make an impact.[57]

A few of Darwin's books were outliers in this publishing pattern and reveal other elements of the scientific publishing world. His specialized volumes about the morphology and taxonomy of living barnacles were privately published in 1851 and 1854 by the Ray Society. This society was founded by George Johnston (1797–1855) in 1844 to honor the name of John Ray (1627–1705) and existed solely for the printing and distribution of specialist natural history classifications to members who were entitled to a copy of each work as published. The society had no premises or meetings.[58] Volumes were published by subscription and produced on a tight budget. Darwin corresponded with Edwin Lankester (1847–1929), the society's secretary, in the early 1850s about the illustrative plates that he considered essential. On March 4, 1851, he requested nine plates instead of the allocated eight, and a tenth in color he would pay for himself.[59] *Living Cirripedia* has ten plates, none colored. The jobbing natural history artist George Brettingham Sowerby (1788–1854) engraved them under Darwin's direction. Sowerby's brother James de Carle Sowerby (1787–1871) engraved the plates for Darwin's *Fossil Cirripedia* (1851, 1854).[60] Darwin paid for these, as documented in his account books. The two volumes about fossil barnacles were published by the Palaeontographical Society, established in London in 1847 as a subscription venture similar to the Ray Society for specialized fossil taxonomic monographs. Darwin did not expect any of these monographs to be of interest to anyone expect a specialist; they were a record of his competence as an expert.

In later decades he paid to print a couple of pamphlets. The most well known of these was a collection of Asa Gray's review essays about the *Origin*, originally published in North America in the *Atlantic Monthly* during 1861. Keen to spread Gray's theologically conciliatory view of evolutionary theory, Darwin arranged to have the reviews reprinted and distributed in Britain at his own expense, under the title *Natural Selection Not Inconsistent with Natural Theology* (1861). He was mildly irritated that Murray refused to accept this commission, saying it was not commercial. Consequently, Darwin paid to have it printed by the London firm of Trübner's who specialized in the American book trade. He distributed it personally from his home, each copy accompanied by a letter. This episode shows Darwin eager to ensure that any advantageous, socially reputable views about his theory should be made available and the lengths to which he would go to achieve this.

His scientific articles reflect much the same pattern as his books, although more explicitly directed to disciplinary societies. Among the first of these were extracts from Darwin's letters written during the *Beagle* voyage, privately printed by his mentor John Stevens Henslow (1796–1861) as a pamphlet, read aloud in the Cambridge University

Philosophical Society in 1835, and mentioned in the *Proceedings of the Geological Society of London*.[61] After his return from the *Beagle* voyage, he announced his theory of the formation of coral reefs and made public his ideas about crustal uplift and subsidence in seriously scientific articles submitted to the Geological Society.[62] A major theoretical paper was published in 1839 in the Royal Society's *Philosophical Transactions*, offering a new interpretation of the "Parallel Roads" of Glen Roy in the Scottish Highlands, a series of terraces along the sides of mountains whose origins perplexed geologists. Although Darwin later regarded this paper as "a great failure," it reveals not only his method of theorizing,[63] but also his intention to establish himself as a major player in science by publishing in the most prestigious scientific society and periodical of the day. He was elected to the Royal Society that same year.

These scientific papers were, for the most part, published in the privately funded journals produced by the learned societies of Britain, societies that aimed to establish clear distinctions between one discipline and another. Members of such societies were mostly gentlemen of science, the predominantly Oxbridge-educated Anglicans who controlled British science during the first half of the century, as described by Thackray and Morell, and included a scattering of museum professionals, commercial and industrial magnates, politicians, landed gentry, and aristocrats.[64] Some members had paid employment in science, like Richard Owen, curator of the Royal College of Surgeons, or George Biddell Airy (1801–1892), the astronomer royal from 1835. Others possessed private income like Darwin or were servants of the state in a scientific capacity, such as Edward Sabine (1788–1883), an army officer and member of British exploring expeditions in early life and then president of the Board of Longitude. They were united by a common educational background, a well-defined set of moral values, and respect for knowledge.[65]

Darwin continued to publish specialized essays and articles, often in later years as precursors to books. The most famous of these is the paper on the theory of natural selection jointly published with Alfred Russel Wallace in the zoological portion of the *Journal of the Proceedings of the Linnean Society* in 1858. For several weeks after that paper was delivered, Darwin corresponded with his friend the botanist Joseph Hooker about whether he should publish the long version of his theory in the *Journal*.[66] He decided not to and launched into writing what would become *On the Origin of Species*.

A different perspective is offered in considering his brief contributions to mass-market periodicals. These show that he was eager to publish also in the more public realm of generalist, commercially produced magazines. Collectively, these brief printed contributions can best be described as requests for information, and in this sense can be envisaged as a harvesting operation. As revealed by John van Wyhe's *Shorter Publications*,[67] Darwin invited information on microscopes, donkey's stripes, dog's feet, lizard's eggs, cherry blossoms, an edible fungus found in Tierra del Fuego, and many other topics.

This speaks to the way that natural history—as distinct from high-level theoretical science—was practiced in Victorian Britain. The readers and contributors to these periodicals came from a variety of backgrounds, including established naturalists, landowners, animal breeders, horticulturists, and knowledgeable amateurs, all of whom engaged with topics of mutual concern in an increasing abundance of periodicals.[68] What attracted Darwin was the structured access to a wide variety of special knowledge. Here, in the printed letter columns of a magazine, Darwin was more or less an intellectual equal with other readers, although of course his name was well known from *Journal of Researches* and the controversies surrounding the *Origin*. In periodicals such as these he could broadcast small natural history inquiries to a community of knowledgeable individuals and engage in discussion.

Casting Darwin's work into the framework of interdisciplinarity is probably too modern a maneuver to explain his personal intellectual reach, even though we can see that he contributed as an expert to several scientific disciplines, such as geology and botany, and even though his magum opus, *On the Origin of Species*, synthesized a huge range of natural history topics and appealed to many different readers. As documented here, there were plenty of times when Darwin moved across clearly demarcated fields, plenty of times when he integrated materials drawn from different areas. Yet there were also other times when he sought to display his deep expertise in a restricted specialty domain. His personal track seems to suggest that he did not recognize himself as engaged in interdisciplinary research. The term *naturalist* gives Darwin, Wallace, and likely others too a suitably generous label to encompass their varied research activities. This seems more appropriately Victorian than the word *interdisciplinarity*.

In retrospect, then, it is likely that the exceedingly wide readership and response to the *Origin of Species*, and the modern practice of incorporating evolutionary maxims in almost all humanities and social science disciplines has colored our perception of how questions about interdisciplinarity in Darwin's work might be answered historically. There can be no doubt that the *Origin* was significant to a huge range of markedly different communities stretching across the globe. Indeed, Darwin's book has come to be regarded as a supreme example of interdisciplinary impact. In the closing pages he evoked this comprehensive ambition by suggesting that the study of embryology and psychology would be transformed, and that "light will be thrown on the origin of man and his history."[69]

In regard to this volume on interdisciplinarity, it is perhaps helpful to keep these aspects of Darwin's trajectory in mind when we consider the structure of science in the nineteenth century. The emergence of disciplines was clearly a crucial step in the professionalization of science. Individuals grouped together expressly to create a number of learned societies that would function as governing bodies for new or redefined intellectual domains. They created disciplinary boundaries by fostering professional codes of engagement through meetings, publications, awards, and in later decades regulatory devices such as examinations. Freshly crafted boundaries could not create employment, but they indicated where expertise might be located, so that figures such as Huxley or Hooker could rise through a succession of poorly paid but well-regarded positions. Such aspects of nineteenth-century science are well noted in the historiography.[70] In Darwin's case it appears that he simultaneously did and did not participate in these processes. His private income made it possible to separate himself whenever he wished. He could pursue some of his professional concerns just as easily by paying to publish what he wanted, by soliciting information via letter, by running a household that gave him plenty of time to observe and experiment and ponder on his results. Unlike Tyndall he did not have to grade exams or lecture for a living.[71] Nevertheless the science in which Darwin engaged was changing. He needed to participate in the new disciplines in order to produce and maintain his intellectual authority—and he enjoyed doing so. As science increasingly diversified into tighter and more structured fields, he took advantage of his ability to move fluidly across a great range of natural inquiry while also actively contributing to the consolidation of specialized domains such as zoology or geology. Perhaps others were also able to do this while the disciplines were solidifying. Those who called themselves naturalists seem the most likely candidates to be both fluid and specialized. Could this be a characteristic feature of natural history that survived even as the boundaries of disciplines became more rigid during the following century?

Chapter 4

WITHOUT A DARWINIAN CLUE?

Henry Thomas Buckle and the Naturalization of History

IAN HESKETH

On February 8, 1858, the physicist John Tyndall (1820–1893) wrote to his good friend, the mathematician Thomas Hirst (1830–1892), about an entertaining dinner party that he had attended the previous week. As well as meeting the celebrated traveler David Livingstone (1813–1873), Tyndall was fortunate to sit next to someone named Henry Thomas Buckle (1821–1862). He had "recently published a book that has caused a very great sensation in London," Tyndall wrote. It was a "history of civilization in England." He was, according to Tyndall, "the lion of the literary season." Tyndall went on to describe in great detail a "thoroughly good humored" debate between the two men that was "energetic in the extreme" and eventually "broke out into open war." Tyndall himself was well known by this time for his engaging dinner conversation and his general debating abilities. In conversation with Buckle, however, Tyndall was glad just to hold his own. "He swept at me sometimes like a billow," Tyndall explained; "he lifted me but did not overwhelm me. I never lost my balance for an instance." This was the "second great conflict" Tyndall had with Buckle, he informed Hirst, "and from each I had the good fortune to escape with a whole skin."[1]

There are numerous accounts of similar encounters with "the myriad-minded Buckle" at this time.[2] This is because, as Tyndall suggested, Buckle was rather suddenly thrust into London's intellectual spotlight after the 1857 publication of his popular *History of Civilization in England*. In hindsight, Leslie Stephen (1832–1904) argued that the book was responsible for one of the two great intellectual shocks of the 1850s, the other being Charles Darwin's (1809–1882) *On the Origin of Species* (1859). And at the time, according to Stephen, "Buckle's performance perhaps seemed the most important."[3] This was because Buckle's *History of Civilization* sought to be, in the words of an anonymous reviewer in the *Quarterly Review*, "the Novum Organum of historical and social science."[4] Indeed, Buckle's work claimed to be able to explain the development of human civilization itself, not as the result of an unfolding of Providence or as the result of political contingencies, but rather as the result of natural laws that undermined any notion of chance or free will. History, Buckle argued, was subject to the same sort of laws that determined all other natural processes, and he utilized the burgeoning science of statistics to uncover those laws in the human past.[5]

The typical narrative about Buckle, however, is that, despite the rather sudden interest in his book, he was just as quickly forgotten. Historians, on one hand, rejected Buckle as a mere "philosopher of history," and, as I have argued elsewhere, went about articulating a very different science of history that involved not discovering laws of development but rather uncovering an array of factual details for increasingly specialized studies.[6] Men of science, on the other hand, rejected Buckle's environmentalism, which suggested that the right circumstances could level the distinctions between race and gender, a position that became difficult to sustain after Darwin's *Origin of Species* gave such a strong scientific foundation to the importance of heredity in shaping historical development. From this perspective, Buckle was so soon forgotten because he was without, as Stephen put it, a "Darwinian Clue."[7]

However, even while Buckle may have been largely forgotten by the time Stephen was writing near the end of the nineteenth century, when we look beyond the views of his loudest critics, his work was more central to shaping subsequent debate about the intersection between history and science than has been typically recognized. Assuming that Buckle was antiquated by the arrival of the *Origin of Species* fails to recognize the importance of Buckle's assertion that the human past should be studied like any other physical process. He, moreover, argued that the focus of such a scientific study should be on the progress of civilization itself, which for him entailed explaining the rise of the European intellect in connection with physical and social circumstances. His own narrative of this development highlighted the struggle between humans and their physical environments on one hand, and between scientific and theological forces on the other, and thereby inspired other large-scale natural histories as well as the imagined histories of science that came to justify the authority of their authors.[8] Buckle is therefore a wonderful example of a prominent Victorian who was working at the intersection of history and the historical sciences, and whose work pushed others to think about the connection between them in new ways.

Indeed, Buckle's grand historical narrative was particularly appealing for those men of science often identified as scientific naturalists, men of science such as Tyndall who were looking to wrestle cultural authority from the Anglican establishment by producing their own histories of civilization that explained the rise of science as part of a larger battle between the forces of science and theology.[9] Buckle's *History of Civilization* provided just such a narrative outline that placed men of science at the forefront of further progress. It turns out, however, that over time Buckle's approach began to look too materialistic and offered a message that became too radical for scientific naturalists to embrace fully as they sought to achieve a certain amount of respectability throughout the 1860s and early 1870s. In particular, Buckle's suggestion that morality was not a progressive force in history was one that did not sit right with just about anyone, but it particularly disgruntled Tyndall, who believed that his own scientific upbringing was shaped by his moral education. But even in this regard, it was Buckle's work that engendered a wider discussion about how one naturalized the history of morality without undermining the respectability of the man of science. As we'll see, Buckle's work became a common point of reference for Victorians looking to make naturalistic claims about the historical and contingent, even if those claims were ultimately in opposition to Buckle's own.

THE SCIENCE OF HISTORY AND THE HISTORY OF SCIENCE

Part of the reason why the reception of Buckle's *History of Civilization in England* has been misunderstood is because the book has typically been analyzed in the context of the development of history as a discipline. This is somewhat understandable given that Buckle explicitly formulated his project as an intervention in the field of history. In the opening pages of the book, for instance, he argued that contemporary historians had become too specialized and antiquarian and had failed to keep up with "the most successful cultivators of physical science." There was, according to Buckle, an "unfortunate peculiarity" when it came to the "history of man." "Although its separate parts have been examined with considerable ability, hardly any one has attempted to combine them into a whole, and ascertain the way in which they are connected with each other."[10] It was therefore Buckle's plan to engage in this overdue historical synthesis and bring history up to the level of the physical sciences by uncovering the laws that govern the history of civilization. But the book also made important claims about the relationship between natural and human history, and about the relationship between the development of science and civilization. Moreover, the nature of Buckle's grand historical synthesis meant that it had implications not just for historians but for anyone interested in the historical process, and therefore largely transcended the emerging boundaries between history on one hand and the historical sciences on the other.

After the first chapter Buckle's next task was to explain those forces that determined human history. He argued that the early history of humanity was determined by "physical agents," meaning "climate, food, soil," and what he called "the general aspects of nature." Humans living in climates that made the cultivation and distribution of food and wealth difficult—climates that Buckle identified as being largely in the Southern Hemisphere—developed at a slow rate such that their imaginations ran wild. It was from within this state that the "earliest civilizations belong," argued Buckle, but those societies were unable to advance beyond a rudimentary level. The peoples living in northern climates, however—and here Buckle largely referred to Europe—who had more favorable physical environments with which to cultivate food and distribute it more widely, were able to transcend daily concerns of subsistence, shelter, and the weather. Their intellect progressed at a quicker rate and reached a stage whereby its further progress was no longer determined by the physical environment.[11]

Once Buckle established that the progress of European civilization had transcended physical processes, he argued that further progress was determined by mental laws alone. It was possible, he said, that these laws could be moral or intellectual in nature. Buckle believed, however, that because morality itself was stationary, while "intellectual truths" were constantly changing, it was those changing intellectual truths that shaped the progress of European civilization. Those truths were then determined, argued Buckle, by the "intellectual condition of the age." This "condition," for Buckle, was to be explained by understanding the general dissemination of what he called the "skeptical" or scientific spirit, in contrast to those forces that sought to arrest progress, which were represented by the "protective" or metaphysical/theological spirit. The reason Buckle titled the book *The History of Civilization in England* is because he believed that the English nation represented the ideal case, where the skeptical spirit had slowly but gradually overcome the protective. But in the case of England progress was established in order to compare its development to other European nations whose progress occurred much less gradually, much less naturally. The rest of the first volume, then, was largely a comparison of the intellectual progress of England and France, and much of the analysis centered on the struggle, in both nations, between science and theology, or to use Buckle's language, between the skeptical and protective spirits.[12]

Buckle argued that early on the intellectual development of France and England was roughly the same. Like the rest of Europe both nations suffered during the "dark ages," not because of Catholicism per se, but rather because "the conditions which regulate the progress of modern civilization" were such that a credulous religion could exert such a powerful hold on the population at large.[13] It was in the sixteenth century when things began to change. Human reason began to rebel against the protective spirit, and it was at this moment when the great struggle began that Buckle sought to trace in order to uncover the "rise of that secular and skeptical spirit to which European civilization owes its origin."[14] "From that moment," he argued, "there has been an unceasing struggle between these two great parties,—the advocates of inquiry, and the advocates of belief; a struggle which, however it may be disguised, and under whatever forms it may appear, is at bottom always the same, and represents the opposite interests of reason and faith, of skepticism and credulity, of progress and reaction, of those who hope for the future, and of those who cling to the past."[15]

The key moment in this great struggle was the Protestant Reformation, which was the direct result of a "general spirit of inquiry, of doubt, and even of insubordination, [that] began to occupy the minds of men."[16] Protestantism therefore "was a normal movement" and represented the "legitimate expression of the wants of the European intellect."[17] The movement, moreover, helped advance the cause of the sciences, particularly in England, where they were "founded not on notions of old, but on individual observations and individual experiments."[18] In England there was an intimate connection between the new spirit of the Protestant theologians and "the rise of the Baconian philosophy" that stressed the need to establish a new form of knowledge based on the "right of private judgment."[19] Buckle then described the importance of Francis Bacon (1561–1626), Robert Boyle (1627–1691), and Isaac Newton (1642–1727) for taking advantage of this new freedom while bringing such seemingly supernatural forces as the physics of the universe under natural explanation.[20]

In France, by contrast, the power of the clergy was much stronger than in England, meaning that the progress of knowledge there was "abnormal," occurring in fits and starts rather than gradually. Because the powerful clergy "were able to withstand the Reformation," everything in France "assumed a more theological aspect than in England."[21] Buckle was particularly contemptuous of Louis XIV's (1638–1715) reign (1643–1715), which he argued led to an incredible age of decay. Not only did the French not produce a single scientific advancement of note for several generations, they also completely ignored what was happening across the English Channel. However, so many years of oppression ultimately led to the French Revolution, which brought about sudden advances in French science, such as Antoine Lavoisier's (1743–1794) work in the realm of chemistry and the geological and anatomical discoveries of Georges Cuvier (1769–1832) in the realm of natural history. Buckle also highlighted the anatomist Xavier Bichat (1771–1802), whose works Buckle believed marked a new epoch in the history of science as they were associated with such "vast and magnificent schemes" as uniformitarianism, the transmutation of species, and the nebular hypothesis.[22]

If Buckle's analysis of France ended on a particularly high note, he was much more reserved in his final consideration of England, whose "normal" development began to stall in the nineteenth century. This was because, Buckle argued, the English were becoming too obsessed with minor, specialized advancements and were no longer seeking to imagine grand, speculative laws of nature. He was particularly scathing about the most prominent of English naturalists who were opposing the magnificent schemes of nature produced by the French. They had, moreover, "degraded their own noble science by making it a handmaid to serve the purposes of natural theology,"[23] a not-so-subtle critique of the current direction of English science. Therefore Buckle's *History of Civilization* was very much an attempt to put English science back on its normal trajectory by imagining a new vision for further progress, one that would make history itself the next and last terrain for science to colonize.

Scientific History and Scientific Naturalism

When the *History of Civilization in England* was published in 1857, it did indeed become an instant literary sensation. As the *Dublin University Magazine* put it: "Mr. Buckle has taken the world by surprise."[24] Readers were quite simply stunned at the scale of the project and the amount of learning that was on display. "It is long since any single volume has come before us characterized by so much learning," noted a reviewer in the *British Quarterly Review*, "and, on the whole, by so much power."[25] Meanwhile, the *Gentleman's Magazine* maintained that Buckle was clearly "one of the most deep-read scholars, one of the keenest enquirers, and one of the most original thinkers of the day."[26] For the liberal Anglican theologian Mark Pattison (1813–1884), writing in the *Westminster Review*, Buckle's *History* was "the most important work of the season" and "perhaps the most comprehensive contribution to philosophical history that has ever been attempted in the English language."[27] Tyndall was therefore not exaggerating when he told Hirst that Buckle was the "the lion of the literary season," a phrase that was itself often repeated with reference to Buckle.[28]

Despite the initial praise that Buckle received, it was soon recognized that his work failed to convince its many readers about its proposed laws of history.[29] Historians almost immediately rejected Buckle's particular attempt to make history a science.[30] The book was also generally criticized from conservative and religious quarters of the periodical press, not necessarily for attempting to make history a science, but rather for Buckle's dismissal of morality as a progressive force in history and for his suggestion that the relationship between science and religion was necessarily conflictual.[31]

There were many other readers of Buckle's *History*, however, who found that its promise of bringing human history under the control of a scientific method and framework was particularly attractive. One such group was the burgeoning scientific naturalists, who appreciated Buckle's criticisms of natural theology's overbearing role in shaping contemporary English science. For instance, while he was trying to finish writing his manuscript on "Natural Selection," Darwin found the time to read Buckle's work and thoroughly enjoyed it. "Have you read Buckle?" Darwin wrote to his good friend, the botanist Joseph Hooker (1817–1911), in March 1858. "I think you would be interested in it."[32] Darwin found the book, he told Hooker, "*wonderfully* clever & original & with astounding knowledge."[33] When the second volume was published in 1861, just a few months before Buckle died, Darwin again wrote to Hooker to express the pleasure he felt on reading the work: "There is a noble love of advancement & truth throughout," he wrote, "& to my taste he is the very best writer of the English Language that ever lived."[34] The geologist Charles Lyell (1797–1875), meanwhile, also found the book "full of talent." He told his friend, the American George Ticknor (1791–1871), to purchase a copy of the book for himself as well as for the Boston Public Library, for which Ticknor was a trustee. "For so bold a book," Lyell wrote, "it is having a considerable sale." More than anything, according to Lyell, its "great merit" was that it was "setting people thinking."[35]

Indeed, during the period immediately following the publication of *The History of Civilization*, it is apparent that Darwin and his close group of supporters saw Buckle as a useful ally. This was particularly clear after Thomas Henry Huxley (1825–1895) was elected to the Athenaeum Club. With Huxley now a voting member, Hooker and Lyell conspired to put Buckle up for membership as well. Hooker wrote to Huxley to get his "help to swamp the Parsons & get Buckle in." It was important to mobilize Huxley's support because the conservative members of the club threatened to blackball Buckle for his seemingly heretical views. In the event, the conspirators were able to get enough support and Buckle was elected. The victory was, according to Lyell, "a good demonstration in favor of freedom of opinion."[36] It was soon after his successful election to the Athenaeum that Buckle was invited to deliver a prestigious Friday Evening Discourse at the Royal Institution.

If Buckle's invitation to lecture at the Royal Institution was symbolic of how he was being welcomed by the scientific community, the response to the lecture suggests that the welcome was particularly short-lived. Despite the fact that the lecture theater was sold out and Buckle's well-known rhetorical skills were fully on display, the subject matter along with his particular argument proved far too radical for those in attendance. During the lecture he expanded on a theme that was implicit in the *History of Civilization*, namely that the progress of civilization in England had stalled because men of science had narrowed their methodology to such an extent that facts were being obsessively collected without the requisite speculation, that the deductive was being ignored in favor of the inductive. He made this claim, moreover, by relying on a fairly radical analogy between the feminine and the deductive and the masculine and the inductive. Most significantly, he argued that it was women who kept the deductive spirit alive, and that it was time for men of science therefore to give more scope to the imagination and embrace the necessary feminine aspects of scientific discovery as well. Buckle's comments seemed to bring together just the right combination of radicalism and ignorance to upset just about everyone, including those scientific naturalists who were seemingly ready to welcome him into their ranks.[37]

Tyndall was particularly disappointed with the lecture. While he admitted that it was "a most wonderful discourse, wonderful from the vigor and abandonment with which [Buckle] threw himself into it," he ultimately felt that the lecture itself "did no good—false it is most assuredly, but then it lifted one to a level of chivalrous antagonism."[38] He soon after had a conversation with his mentor Thomas Carlyle (1795–1881) about Buckle and reported in his journal Carlyle's view that Buckle was "a weak, watery, unfruitful creature out of whom no good can come," an observation he passed along to Huxley.[39] Meanwhile, Huxley at this time began referring to Buckle derisively as "Buckle the Great," and declined a dinner invitation from him, stating that he had "too much of the Arab about me to eat a man's salt and then pitch into him."[40] Even Darwin, who was so impressed with the *History of Civilization*, was disappointed on meeting Buckle in person, as was Hooker.[41] Darwin continued to appreciate Buckle's writing, but he admitted that he grew more skeptical of the particular facts of history that Buckle purported to illuminate. Darwin would, moreover, give a very different account of the mental development of women in his *Descent of Man* (1871).[42]

THE NATURAL HISTORY OF CIVILIZATION

When Buckle died in 1862, soon after his second volume was published the previous year, it is safe to say that he was no longer "the lion of the literary season." But while Leslie Stephen would want us to believe that he slid into obscurity, the fact of the matter is that Buckle became an important point of reference as scientific figures of various stripes sought to grapple with a naturalized account of human history. It is, of course, well known that Darwin referred only cryptically to humans in his 1859 *Origin of Species*. However, by the mid-1860s focus very much turned to the deep history of humanity and how it might relate to an evolutionary account of life.[43] For someone who was apparently antiquated by Darwin, it is surprising how often Buckle's work was referred to in these discussions.

The American chemist John William Draper (1811–1882), for instance, finally published the book-length version of a lecture he had originally presented at the British Association meeting in Oxford in 1860, which promised to discuss the "intellectual development of Europe with reference to the views of Mr. Darwin and others."[44] Draper's presentation actually provided the pretext for the now famous exchange between Samuel Wilberforce (1805–1873) and Huxley concerning the latter's simian ancestry and the former's bad manners. Draper's paper itself, however, actually had little to say with reference to Darwin and, according to Hooker, was "all a pie of Herb[ert] Spencer [1820–1903] and Buckle without the seasoning of either."[45] The book version, *History of the Intellectual Development of Europe* (1864), tends to support that view, as Draper argued that his purpose was to "consider in what manner the advancement of Europe in civilization has taken place, to ascertain how far its progress has been fortuitous, and how far determined by primordial law."[46] Like Buckle, Draper argued that intellectual development had indeed been determined by law, and he sought to show the various stages through which it had occurred, stages that were analogous to birth, growth, maturation, old age, and death.

The similarities between Draper's and Buckle's respective histories were undeniable. The *Saturday Review*, for instance, found that Draper's attempt "to reconstruct the fabric of history as a . . . scientific system of life in their widest and most comprehensive unity, was the end for which

Mr. Buckle strove to connect together the facts from every province of inquiry, and to demonstrate the harmony of their results." It could now be said, according to the anonymous reviewer, that Draper and Buckle were part of a movement that could take "its distinctive place in the intellectual cycle of English thought."[47] For Sheldon Amos (1835–1886), writing in the *Westminster Review*, by extending Buckle's analysis of intellectual laws to all of Europe, Draper had effectively taken "the lamp of knowledge" from Buckle and continued on in the race to push the boundaries of science itself.[48]

The other book that became closely associated with Buckle in this period was W. E. H. Lecky's (1838–1903) *History of the Rise and Influence of the Spirit of Rationalism in Europe* (1865). Unlike Draper, Lecky readily acknowledged his debt to Buckle as he explicitly sought to contribute to an analysis of the progress of civilization by focusing on what he called the growth of the "spirit of rationalism." This was Lecky's attempt to bring further clarity to Buckle's "skeptical spirit" by more explicitly narrowing its meaning to scientific forms of thought. Like Buckle, he too often found himself describing a struggle between science and theology. But unlike Buckle, he was careful to separate dogmatic theology from Christianity and was therefore able to present a view of scientific progress in concert with a liberal Anglicanism, a view that was much less contentious. It followed that while the periodical press stressed the "family resemblance" between Buckle's and Lecky's books, Lecky was praised for his "fair and moderate judgement" in comparison to Buckle's "passionate and often one-sided declamations."[49]

Buckle's book also came into conversation with a set of works that were concerned with the early development of civilization and which focused on the practices of so-called savages. As we have seen, Buckle was not terribly concerned with humanity's early history, as he largely bracketed prehistory out of his main narrative by arguing that civilization only really emerged in societies that overcame the power of nature. John Lubbock's (1834–1913) *Pre-historic Times* (1865), however, sought to focus specifically on what he called "uncivilized races" in order to recognize the immense progress humans have made since emerging from the "savage" state that those races represent.[50] Intellectual progress for Lubbock was therefore a much lengthier process than for Buckle and could not be understood without making a comparative analysis between the civilized and savage states central to it. This comparison for Lubbock also made clear the important connection between the rise of science and a more virtuous humanity, in contrast to the debased practices of the savages of prehistory.[51]

E. B. Tylor's (1832–1917) *Researches into the Early History of Man-*

kind and the Development of Civilization (1865), published in the same year as Lubbock's *Pre-historic Times*, started from a similar premise that the practices and beliefs of contemporary European society could be truly understood only by tracing their origins back to "the condition of rude and early tribes."[52] Thus it was important for Tylor to recognize that civilization was "a process of long and complex growth." In uncovering the cultural practices of early humans, Tylor explicitly admitted that he would, where possible, identify general laws of development. In this regard he would follow "the late Mr. Buckle [who] did good service in urging students to look through the details of history to the great laws of Human Development which lie behind." However, Tylor also stressed that such general laws could often obscure more than they illuminate. He therefore argued that it was also necessary to examine "the complication of events" that gave rise to certain phenomena. In this regard Tylor was also critical of Buckle's "attempt to explain, by a few rash generalizations, the complex phases of European history," which he found to be "a warning of the danger of too hasty an appeal to first principles."[53]

Given Tylor's explicit invocation of "the late Mr. Buckle" as an important influence on his thinking about the nature of historical processes, readers of his work, as well as of Lubbock's and Lecky's works, took notice of their relevance to Buckle's earlier history of civilization. Hooker, for instance, in a letter to Darwin on July 13, 1865, wrote that he was "reading Lubbock very carefully," finding it "most excellent," as well as Tylor, which was the most interesting to him "since your orchid work that I have read." He was also in the process of reading Lecky, which he found "extremely interesting & instructed." He was therefore convinced that as soon as "the next holiday" he "must read Buckle, of which I have only read part of vol. 2."[54] Hooker did not explain why he should go back and read Buckle, but it is clear from the context of the letter that Lecky, Tylor, and Lubbock had made Buckle's book relevant again. Meanwhile, Darwin responded later in the month that he found it amusing that both he and Hooker were currently reading, or had recently read, the same books. "It is curious how we are reading the same books," Darwin wrote. He had already read Tylor on Hooker's earlier recommendation and was "deeply interested by it," and was "heartily glad" that Hooker "liked Lubbock's book so much." Moreover, "we intend to read Leckie [*sic*] & certainly to reread Buckle, which latter I admired greatly before."[55]

At the same time as Darwin and Hooker were rereading Buckle in the context of the new histories of civilization that appeared in the mid-1860s, so was Alfred Russel Wallace (1823–1913), who had himself only just recently presented his views on the evolution of early humans.[56] Darwin had earlier written to Wallace to inquire whether the latter had read Tylor, Lecky, or Lubbock, three works that he highly recommend-ed.[57] Wallace responded that he had read Tylor and Lubbock, and was currently reading Lecky. He was disappointed with Tylor because of the "absence of any definite result or any decided opinion on most of the matters treated of." He liked Lubbock's book very much, but did not like the concluding chapter where Lubbock sought to appeal to religious concerns with his naturalist method by suggesting that the progress of science itself might reduce sin. "Why are men of science so afraid to say what they think & believe?" he wrote. About Lecky, Wallace said that he approved of it, "though he is rather tedious & obscure at times."[58]

Perhaps most important, Wallace wrote that he believed that "most of what [Lecky] says has been said so much more forcibly by *Buckle* whose work I have read for the second time with increased admiration although with a clear view of some of his errors." Those errors, however, did not al-ter Wallace's opinion of Buckle's book. It was, he argued, "unapproach-ably the grandest work of the present century, & the one most likely to liberalise opinion."[59] Entirely independent of Darwin and Hooker then, Wallace was also rereading Buckle. Tellingly, all three recognized that Buckle's work was far from reliable or accurate. But they also all believed that Buckle was worth rereading nonetheless, even in a Darwinian con-text that supposedly invalidated many of Buckle's key claims. This was because Buckle's *History of Civilization* clearly influenced the grand his-torical literature of the mid-1860s, which was obvious to readers of those works, whether the focus was a history of rationalism or a prehistory of "savage" culture. Indeed, for scientific naturalists, Buckle's work repre-sented an important example for how to go about bringing together a historical and scientific analysis of the natural world.

The Natural History of Morals

In 1867 Longmans, Green, and Company took advantage of this re-newed interest in Buckle by reissuing his *History of Civilization* in three volumes, ten years after the work first appeared. This led to the publica-tion of an important review article, written anonymously by the Scottish journalist and assistant editor of the *Daily Mail*, James Macdonell (1841–1879), that sought to consider an aspect of Buckle's book that had not received nearly the attention it deserved in light of recent developments: Buckle's argument concerning the role of morality in the development of the intellect. While Macdonell would go on to criticize Buckle's under-standing of morality, he stressed that it was Buckle's work that made this conversation about morality possible. In this regard, Macdonell argued that Buckle was right to think about the way in which the intellect de-veloped over time, but that he was wrong to separate morality from that process. Buckle was mistaken, Macdonell argued, in assuming that moral principles have been static whereas scientific principles have changed over time. The fact of the matter was that the main principles of both science and morality were already firmly established in the ancient period, and that their progressive development had to do with how they were put in practice. He also argued that it made little sense to separate the moral from the intellectual given that it is "the same mind that performs both acts."[60] Intellectual development was determined by moral development, according to Macdonell, and vice versa.

Having made the case that both the development of morality and the development of the intellect were dependent on the given conditions of time and place, Macdonell argued that there was a law that could explain their codevelopment: the law of natural selection. Macdonell then went on to describe the evolution of the intellect in concert with morality, from their origin out of the self-preservation instinct found in the primeval tribe up to their highest expression in the established virtues of European civilization. His point was that it was possible to discern a trace of modern scientific and ethical principles in primeval forms found deep in the past, and that an evolutionary account could make sense of the most difficult of subject matters, namely the origin and development of morality itself. It may be true that Buckle's account of morality and the intellect were derived from a pre-Darwinian framework. At the same time Macdonell was inspired by Buckle's attempt to think about the historical relationship between morality and the intellect in light of more recent scientific developments. "Next to a book at once full of ability and of truth," Macdonell wrote, "the most valuable is a book at once full of ability and of error; since its power gives a stimulus to thought of which a feeble work is incapable, and opens up prospects of which the writer did not dream. Such is emphatically the case with much of Buckle's *History*, one of the faultiest, and yet one of the greatest books of his generation."[61] In this regard, Buckle inspired what is quite likely one of the first attempts to provide an evolutionary account of morality, preceding more well-known works that were published subsequently by Herbert Spencer, Walter Bagehot (1826–1877), Lubbock, and Darwin.[62] Moreover, the essay itself attracted considerable attention. At Oxford, it was, according to Macdonell's biographer, "almost elevated for a time to the rank of a text-book."[63] Meanwhile, the editor and historian John Morley (1838–1932) responded to Macdonell in the *Fortnightly Review* in praise of the connection Macdonell raised, via Buckle, between morality and social conditions.[64]

There was, moreover, another way in which Buckle's discussion of morality came to inform men of science, and this goes back to the debate Tyndall had with Buckle in 1858. As Tyndall explained to Hirst, the subject matter that caused the two men to engage in such a heated exchange concerned "the influence of moral agents in the advancement of the world."[65] While Tyndall was initially excited by Buckle's work, he was immediately opposed to Buckle's claim that morality played no role in the development of the intellect, and by implication in the development of science. This was because this claim entirely contradicted Tyndall's own development, which he articulated as a moral journey that led him from his poor background to his current scientific vocation and ultimately shaped the kind of science that he produced.[66] Therefore, on a deeply personal level, Tyndall believed that Buckle's argument about morality was simply false. As he explained to an American correspondent in 1870, "I think my own single example would demonstrate the futility of all attempts to sever intellectual progress from moral influences, as Buckle tried to do some years ago."[67] Buckle therefore helped Tyndall articulate a moral vision of science and scientific identity that was in opposition to the amoral one of *The History of Civilization in England*.

However, Tyndall's construction of scientific progress was otherwise similar to Buckle's. In his public address at the British Association in 1870 Tyndall stressed that it was important for the man of science to utilize his imagination and therefore go beyond the typical inductive reasoning associated with science in order to advance knowledge of the natural world. This was not unlike Buckle's argument in his Royal Institution lecture about the importance of deductive reasoning in scientific discoveries. But whereas Buckle made what may have been a reasonable observation within the context of a contentious debate about women, femininity, and the scientific method, Tyndall stressed the ultimate moral foundation that was necessary for the man of science to engage in such speculative thinking. He referred in particular to the patient, careful, and self-effacing scientific labor of Darwin that enabled the discovery of the theory of evolution by natural selection.[68]

Tyndall expanded on this view in his famous Presidential Address at the British Association meeting in Belfast in 1874 when he situated evolution, the atomic theory of matter, and the conservation of energy in a sweeping history of science.[69] The address itself caused a fair amount of controversy and is now typically remembered as contributing to the so-called conflict thesis, which posited that the relationship between science and religion was one best defined as conflictual. Indeed, Tyndall often referred to Draper, one of the supposed chief architects of the conflict thesis, particularly to his work on the *History of the Intellectual Development of Europe*, in order to describe moments of conflict in the history of science and religion. It must be said, however, that he could have just as easily cited Buckle in these instances, as his *History of Civilization* told much the same story. While there are certainly subtle distinctions among the conflict histories of the three men—Tyndall's concerned the conflict between science and theology, Draper's science and Catholicism, Buckle's skepticism and metaphysics—they all argued that in the future theology would need to cede further terrain to science.[70] But whereas Buckle's narrative concerning the progressive advancement of science was based on an explicit distinction between moral and intellectual labor, Tyndall's stressed the important moral framework that guided modern scientific work. "Science itself derives its motive power from an ultra-scientific source," Tyndall argued, one that was decidedly moral in nature. In order to further illustrate his meaning, Tyndall reminded his audience about how "Mr Buckle sought to detach intellectual achievement from moral force." For this, Tyndall explained, Buckle "gravely erred; for without moral force to whip it into action, the achievement of the intellect would be poor indeed."[71]

It was Tyndall's point that men of science should be trusted to deal appropriately with subjects previously regarded as belonging to the realm of the theologian precisely because their inquiries were not amoral, but were in fact guided by a "moral force." While Tyndall may have ultimately reproduced the basic thesis of Buckle's *History of Civilization*, Buckle also taught Tyndall how *not* to speak about morality. Severing morality from the scientific endeavor would not advance the larger goal of the scientific naturalists, who were in an outright struggle for cultural authority. Establishing the inherent respectability of their science and identities as men of science was an important aspect of that larger struggle and could be signaled by explicitly rejecting Buckle's separation of moral and intellectual development.[72] Tyndall therefore may have followed Buckle by connecting the current practices of scientific naturalism to a progressive and grand history of scientific development, but he did so by rejecting the distinction between morality and science that Buckle believed was so central to it.

WITHOUT A DARWINIAN CLUE?

Buckle's death in 1862 meant that he was unable to finish his *History of Civilization* project, which he anticipated reaching ten volumes and comparing all of the major civilizations. Even while writing the second volume, however, Buckle recognized the project's inherent problems and knew that he would never be able to finish it, even before his own premature death made that a certainty.[73] Our ability to grasp the work's significance is hindered by the fact that Buckle himself was unable to continue advocating on its behalf but also by the fact that it was written at a time when there were no clearly established boundaries between history and the historical sciences on one hand, and between the humanities and the sciences more generally on the other. That said, Buckle's intervention was driven by a perception that history was becoming far too specialized and was therefore becoming a discipline that would be out of sync with future scientific developments. At the same time, he saw English science also failing to make the grand speculative gestures that were being made in France largely because the dictates of natural theology reduced science to a series of Baconian inductions. Buckle is therefore an intriguing figure to consider in the context of a volume devoted to the theme of Victorian interdisciplinarity. He wanted to save both history and science from becoming too specialized and narrow, and he believed that making history itself a science would solve both of these problems while bringing unity to forms of knowledge that he worried were only going to become more fragmented in the future.

If it is true that historians largely rejected Buckle's attempt to align their discipline more closely with the historical sciences, it does not follow that all men of science did the same. Indeed, the argument here is that Darwin and his close group of supporters never really saw Buckle's work in opposition to theirs—quite the opposite. Buckle produced a book that they initially read as contributing to their struggle on behalf of the freedom of scientific thought against the power of the Anglican clergy. They then reread that book in the mid-1860s as new scientific histories had extended and refined Buckle's project and as studies of the prehistory of humanity sought to push the origins of civilization further back in time. Buckle then gave impetus to an early attempt to apply Darwin's theory of natural selection to an understanding of the origin and development of morality. He also became a useful foil for Tyndall, whose "man of science" persona was motivated by deeply moral concerns to uncover the truth of nature. Perhaps more than anything, Buckle laid the foundation for a long-running discourse in the nineteenth century about the inherent conflict between science and theology, a conflict that was historicized in his *History of Civilization in England* as a struggle between the skeptical and protective spirits. Buckle may have been without a "Darwinian Clue," as Stephen would have it, but that does not seem to have been a factor in his reception among scientific naturalists, who found other aspects of his thought to embrace, refine, and, indeed, reject.

III

PRACTICES AND DISPLAYS

Chapter 5

GOOD HEALTH ON DISPLAY

Sanitary Science at the International Health Exhibition

ELSA RICHARDSON

The International Health Exhibition (IHE) was formally inaugurated by the Duke of Cambridge (1847–1904) on May 8, 1884, who praised it as an "honest and worthy endeavor made to promote good health, which is one of the first conditions necessary for the happiness and prosperity of a nation."[1] Before delivering his speech the duke had taken a guided tour of the vast exhibition that boasted, among other things, a working dairy, an anthropometric laboratory, model kitchens and schoolrooms, a reconstruction of a city street as it would have been before the Great Fire of London replete with five-story houses and fountains, appliances for manufacturing clothes, machinery of all kinds, ventilators and other hygienic devices, bakeries, restaurants, exhibits of historical and modern dress, an aquarium, ornamental gardens, and a bathhouse. Spread across a number of locations in South Kensington, it also featured national pavilions from France, Belgium, Japan, China, and India showcasing food, clothing, and housing from countries around the world.

The "Healtheries," as it came to be known, was among the most successful exhibitions of the nineteenth century. By the time the exhibition closed at the end of October, having had its run extended by two months, it had won wide acclaim, attracted around four million visitors, and mobilized thousands of participants.[2] Directed by Sir Philip Cunliffe-Owen (1828–1894), who had overseen the surprisingly popular International Fisheries Exhibition of 1883, it represented a new approach to exhibition making in Britain. After the failure of the second London International Exhibition in 1862, which was generally held to have been a rather poor imitation of the Great Exhibition, a move was made toward smaller, themed events that emphasized visitor satisfaction over costly spectacle.[3] According to Ernest Hart (1835–1898), editor of the *British Medical Journal* and head of the exhibition's executive council, public health was an ideal topic on which to test this innovative new method because of its "altogether modern origin" and "chiefly English growth." Envisioned as a celebration of the "science and art of sanitation," the IHE hailed late nineteenth-century Britain as a pioneer of clean water, nutritious food, and safe housing.[4]

The exhibition blended edification and entertainment. The public could attend nightly lectures on a wide range of topics, peruse stands displaying the latest sanitary innovations, or spend time in a temporary library stocked with over five hundred books. They could also, if they pleased, spend an evening strolling through gardens illuminated by brightly colored lamps sampling the delights of the many food and drink stalls set up in the grounds, visit shops selling furs and elaborately feathered hats, or let loose at one of the many dances and concerts staged every week. Observing the presence of such diversions, some commentators grumbled at the elevation of worldly pleasures over practical instruction, while others rankled at the hypocrisy of a money-making enterprise masquerading as an honorable public health initiative. An article in the *Western Mail*, for instance, dismissed the claim that the IHE was "scientific in character" as "humbug" and urged its organizers to face the truth that "people do not go to these places to be instructed, but to be amused." Visitors were, the article continued, quite right to adopt this attitude, as the exhibition was far more "likely to benefit the health of hard-worked Englishmen and Englishwomen by enabling them to enjoy a pleasant holiday than by stuffing tracts on sanitary appliances down their throats."[5] Even favorable reviewers expressed confusion at how some of what was put on display related to the subject of health.

"What," asked an article in *The Graphic*, "has a smock-frocked artisan making clay pipes, a genuine pork-butcher driving a noiseless sausage machine," and "a couple of Highland lassies knitting woolen stockings" have to do with health?[6] One answer lay in how the IHE was funded. Sponsored by the City and Guilds and Livery Companies, with exhibits provided by over two thousand businesses, organizations, and individuals, its content was, at least partly, shaped by commercial interests. In his account of the rise of consumer culture in Victorian Britain,

Thomas Richards identifies the Great Exhibition as the first "coherent representational universe for commodities" where "spectacle excelled at making symbolic virtue out of economic necessity." The transformative power of spectacle, which Richards describes as having "exalted the ordinary by means of the extraordinary," was clearly at work in the vast arrangements of drainpipes, dried goods, and toilet brushes on display at the IHE.[7] Its organizational principles were, however, quite distinct from those that had governed the 1851 exhibition. While its predecessor had celebrated feats of manufacturing prowess and industrial progress, the IHE explored the role played by the environment—broadly defined by its organizational committee as including food, dress, dwelling, and workplace—in the maintenance of good health. Answering their own question, *The Graphic* explained that "the scheme of the Exhibition is not merely to illustrate methods for the preservation of health from a medical and scientific point of view, but to exhibit in the widest and most popular sense everything in connection with our daily and even hourly lives, and which in any way relates to our physical condition."[8] Visitors were invited to view the familiar world in a new light, to examine it through the lens of sanitary science.

Millions took the opportunity. In addition to attracting Londoners of all classes, the delights of IHE drew people from around the country to the capital, and railway companies offered discounts to all districts within sixty miles of the city.[9] Widely reported in the press, satirized by magazines like *Punch* and *Funny Folks*, and at one point even discussed in Parliament, the Healtheries was a national sensation.[10] Thus far the story of this remarkable exhibition has attracted very little interest from historians of health and medicine, but it has featured in a number of studies dedicated to the cultures of display in the nineteenth century. The most fulsome account to date appears in Anthony David Edward's *The Role of International Exhibitions in Britain, 1850–1910* (2008), but mention of it is also made in Peter H. Hoffenberg's *An Empire on Display: English, Indian and Australian Exhibitions from the Crystal Palace to the Great War* (2001) and Julie K. Brown's *Health and Medicine on Display: International Exhibitions in the United States* (2009). Particular attention has been paid to the Old London Street, which proved so popular that it was kept up for the International Inventions Exhibition of 1885, with essays by Annmarie Adams and Wilson Smith attending to different aspects of its design, construction, and reception.[11] Making the case for it as a valuable object of historical research, Adams has argued that the plan of the IHE (figure 5.1) "functioned as a lucid sketch of health issues as they were understood by the British middle class in 1884; like other large public exhibitions, the health fair simplified and clarified complex urban questions,

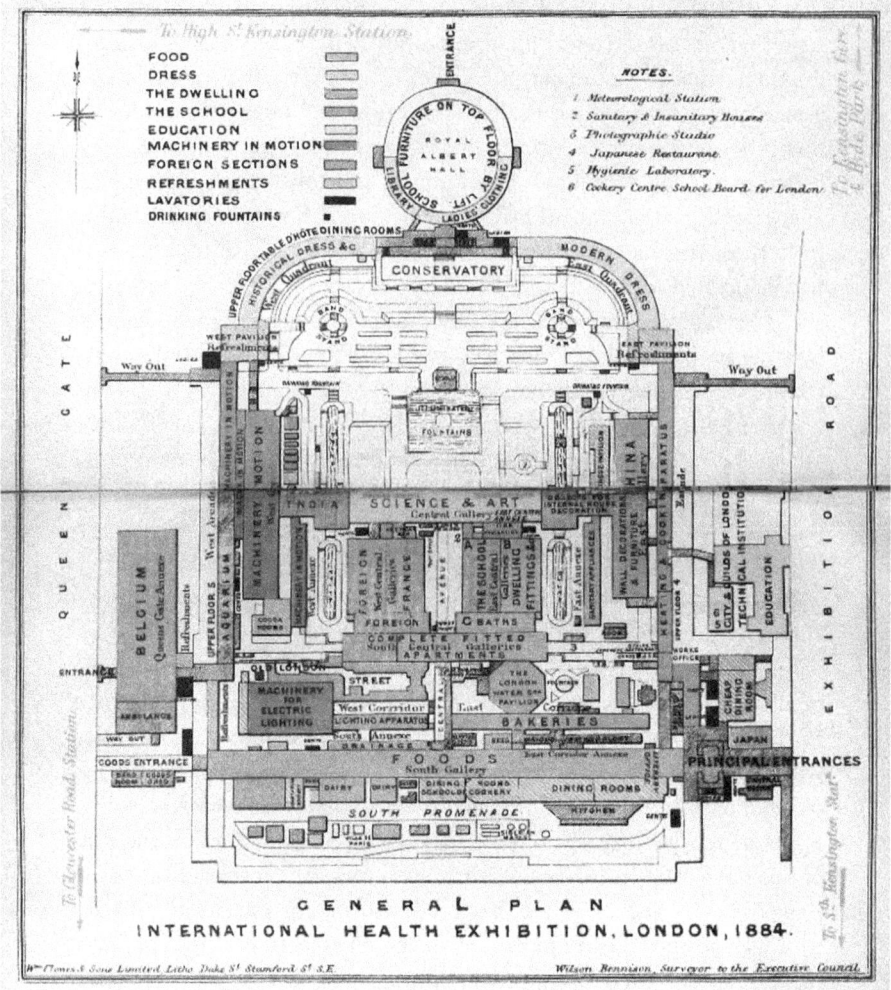

FIGURE. 5.1. "General Plan of International Health Exhibition." *Source: International Health Exhibition Official Catalogue* (London: William Clowes and Sons, 1884), 1–3.

codifying relationships of power, disease and recreation in physical form at a single moment in time."[12] Though this goes some way to describing its significance, the Healtheries offers more than simply a snapshot of attitudes to health in 1884. Just as important are the cross-disciplinary relations, conflicts, and entanglements that it makes visible.

The IHE was, as this chapter reveals, a riotously interdisciplinary space where different registers of knowledge and diverse forms of expertise circulated. It was a site in which laboratories sat alongside working dairies; where drainpipes and cooking appliances jostled against ice cream stalls and Japanese restaurants; where live experiments in bacteriology, instructive guides to good nutrition, and lectures on the dangers of corsets all vied for attention. The design of the exhibition flattened out distinctions between disciplinary fields, so that medicine and engineering, sanitary science and elementary education, agriculture and manufacturing all had potentially equal parts to play in the pursuit of health. Reflecting on his decision to forgo the "new scientific word 'hygiene,'" Hart explained that the term *health* was preferable because it was "at once English and popular," and therefore less likely to alienate potential visitors.[13]

Yet as a concept health proved difficult for the exhibition's organizers to define: seemingly simple and yet open to multiple, complementary, and contradictory interpretations. Historians examining the formation of fields like psychology and sociology have begun to challenge any straightforward narrative of professionalization and specialization by pointing instead to the importance of cross-disciplinary connections, collaboration, and popularization in shaping academic subject areas. Nineteenth-century culture was, as Bernard Lightman and Bennett Zon have argued, interdisciplinary to the core. Not only were the sciences and the humanities widely held to be "complementary rather than oppositional," Victorians also "conceived of almost everything in terms of something else which represented it" so that knowledge emerged as a constellation of differently, and sometimes disparately, connected ideas.[14] Attention has been paid to manifestations of this enthusiastic cross-contamination in literature and art, in the poetic ambitions of scientists and the scientific longings of poets, in the popular press and specialist periodicals. The sources that underpin such an analysis are, for the most part, text-based, and this perhaps explains why exhibitions have been largely left out of this story. Which is not to underplay the significance of text to events like the IHE; as Verity Hunt has demonstrated in relation to the Great Exhibition, guides and pamphlets were integral to how visitors forged narratives and made meaning from the vast spectacle.[15] Along similar lines, this chapter recognizes the wealth of printed material produced by the Healtheries as a key site of interdisciplinary exchange, indivisible from the visual culture of the exhibition.

Approaching the IHE as an interdisciplinary space provides a number of insights into the organization of knowledge in Victorian Britain. Most obviously, the exhibition's thematic drive—an exploration of health in all its myriad complexities—drew attention to the relations between its different elements and demanded that such interconnections be taken seriously. This democratic schema was underpinned by the principles of public health, with late nineteenth-century London positioned as an exemplar of what could be achieved through sanitary engineering, medical interventions, improved nutrition, and educational advancement. At the heart of this triumphalist narrative lay sanitary science, a discipline described in 1857 as that "which deals with the preservation of health and the prevention of disease in reference to the whole community, as contradistinguished from medical science . . . which has for its aim the restoration of health when lost."[16] More so than any other field, sanitary science relied on the consent and active participation of ordinary people: to get vaccinated, to ventilate their homes, to eat well, and so on.

The IHE was dedicated to increasing public awareness of the scientific rules of health in the hope, as an article in *The Lancet* put it, of arousing "the public to the important matter it treats" so that they might "materially assist the sanitary authorities of the country in the discharge of duties which can only be adequately performed with the cheerful co-operation of the masses."[17] With this in mind, it is possible to view the exhibition as embodying and exercising a form of coercive power, as seeking to manage the social body through discourses of hygiene and health. This reading chimes in with foundational works on visual culture like Guy Debord's *Society of the Spectacle* (1967) and Jonathan Crary's *Suspensions of Perception: Attention, Spectacle, and Modern Culture* (1999), which exposed the ideological work of spectacle, namely its capacity to discipline, alienate, and subdue its audience. Along similar lines, Tony Bennett has argued that the emergence of museums and art galleries through the nineteenth century was driven by an "exhibitionary complex" that taxonomized both "objects for public inspection" and "the public that inspected."[18] Useful as these theoretical models are, they do not account for the interpretative freedom that the design of the IHE made possible: moving through the space, visitors produced individual visions of health synthesized from the cornucopia of visual, material, textural, and experiential information on offer. Demanding more than passive consumption, the success of this particular exhibition relied on the public's ability to make connections among very different forms of knowledge based on their shared contribution to public health. Paying particular attention to the tensions that emerged between registers of expertise and disciplinary authority, this chapter explores the version of sanitary science produced at two locations in the IHE: the Health Laboratories and the Vegetarian Dining Room.

THE HEALTH LABORATORIES

Visitors to the IHE entered the Old London Street through Bishopsgate—a scaled-down model of one of eight ancient gates to city that

THE BUILDER, MAY 17, 1884

View looking West

View looking toward the BISHOPSGATE

A. BERESFORD PITE, *delt.*

THE INTERNATIONAL HEALTH EXHIBITION, SOUTH KENSINGTON

MR. G. H. BIRCH, A.R.I.B.A., *ARCHITECT.*

THE OLD LONDON STREET.

FIGURE 5.2. Views of the exhibition from different angles. "The International Health Exhibition, South Kensington: The Old London Street." (*Left*), "View Looking West," (*right*), "View Looking towards the Bishopsgate." *Source:* Reproduction of a woodcut by A. Beresford Pite, May 17, 1884, Wellcome Collection, Public Domain Mark.

had been demolished in 1760—past a re-creation of the famous Cock Tavern of Leadenhall Street, on toward a block of Fleet Street houses once known as the "Three Squirrels," through the narrow passage of Elbow Lane, and out to face a house erected on Goswell Street during the reign of Elizabeth I (figure 5.2). Composed of twenty-five buildings, painstakingly reproduced using historic drawings, Old London was peopled by costumed performers and street vendors selling refreshments and souvenirs. It was an immersive experience that encouraged its audience to, according to Wilson Smith, "suspend disbelief, to imagine themselves carried back in time to pre-Fire London." In Smith's account this "illusionary absorption in the picturesque" offered a "pleasurable contrast to the displays of modern production and consumption" that characterized the rest of the IHE. However, while visitors may have experienced the transition between the two spaces as a "jarring, anachronistic intrusion of the real," this was certainly not the aim of the exhibition's organizers.[19]

Instead, the model historic city was intended to illustrate the sanitary follies of the past. The street's designer George H. Birch (1842–1904), an architect who specialized in church restoration, justified the street's inclusion in an exhibition dedicated to health on the grounds that it usefully represented "the manner in which our forefathers were housed and the sanitary conditions under which they lived."[20] Stepping from the tightly packed, overcrowded, and badly ventilated environs of Old London into the airy spaces of the main exhibition, visitors were encouraged to reflect on the remarkable progress that had lately been made in fields as diverse as plumbing, domestic architecture, dress, and food hygiene. The show's design communicated, as Annemarie Adams puts it, a "clear message to the fairgoers: living conditions in 1884 were much better than they had been in the past."[21] Here, as with the Great Exhibition, evidence for the upward advance of the nation could be found in the vast assortment of commodities on display, but where its antecedent celebrated the wonders of modern manufacture, the IHE had loftier ambitions. Bristling at the suggestion that the "Health Exhibition [was] but a form of World Fair," physician and committee member Henry W. Acland (1815–1900) countered that if "it be so, it is a World Fair resting on the solid foundation of Scientific Knowledge practically applied, and good for instructed and uninstructed alike."[22]

Committed to providing rational entertainment for the "middle and hard-working classes," the IHE charted the dazzling ascent of sanitary science.[23] Many key reforms, those that addressed water supply and sewerage, as well as those dedicated to establishing boards of health and appointing medical officers, originated in the 1850s, but it took until the end of the century for sanitary science to emerge as a coherent discipline. Marked by a shift from the streets into the laboratory, over this period public health increasingly fell under the jurisdiction of biologists and chemists. Driven by the pioneering work of figures like the German physician Robert Koch (1843–1910) and the British surgeon Joseph Lister (1827–1912), germ theory became the dominant paradigm for understanding not only how disease spread within the population, but also how to sustain individual and national well-being. Popularizing new conceptions of the human body that were grounded in the language of organism, bacillus, and pathogen, emergent fields like bacteriology threatened the primacy of established medical disciplines in defining health.

It is of some significance that an exhibition dedicated to bodily well-being "construed in the widest possible sense," defined its purpose as "in no sense medical."[24] Among the many elements identified as "effecting the conditions of a healthful life"—from good food and clean stoves, to elementary education, waterproof clothing, and even the insights of meteorology—medical care remained conspicuously absent.[25] Where traditional medical disciplines were largely excluded from the exhibition, in contrast the new sanitary sciences were showcased in two dedicated laboratories: the Biological Laboratory operated by William Watson Cheyne (1852–1932), a leading bacteriologist working in applied public health research, and the Hygienic Laboratory headed by William Henry Corfield (1843–1903), professor of hygiene and public health at University College London.

Described in the official exhibition literature as "model laboratories for public health," they were staffed by assistants and provided with "all the instruments, materials and methods" necessary to their normal operation.[26] As chairman of the exhibition, Hart insisted on their inclusion and later led efforts to establish them on a more permanent basis using revenue generated by ticket sales.[27] The Biological Laboratory was composed of a "large room—the Laboratory proper—and three small ones, two for incubators etc., and one for cleansing purposes," where weekly public demonstrations of the latest research into "modes of investigating bacteria in air, water, and soil" were to be held.[28] Visitors to the exhibition were among the first, outside of the scientific world, to witness the pioneering research taking place at Koch's laboratory in Berlin. Under the leadership of Cheyne, whom Michael Worboys credits with spreading "Kochian methods and models" in Britain, the makeshift laboratory schooled audiences in the rudiments of microbiology.[29] Located nearby was the Hygienic Laboratory well stocked with "three working-tables with bottle racks above them, and drawers and cupboards for apparatus underneath . . . a furnace with a sand bath on the top for evaporating purposes," sinks, and a supply of gas.[30] Furnished with the equipment used in the "Physical, Chemical and Microscopical examination of Water, Air, Soils, Foods, Disinfectants [and] Sanitary Appliances," it demonstrated laboratory techniques from a "Public Health point of view."[31] Assisted by Charles E. Cassal (1858–1921), a trained chemist and the public analyst for Kensington, Corfield experimented with a range of methods designed to assess the composition of air, detect adulteration in food, and measure the toxicity of household goods.

Prominently positioned in one of the busiest galleries, the laboratories were among the few explicitly scientific displays at the IHE, where visitors could witness live demonstrations of key processes like the cultivation of bacteria, the examination of soil samples, and the analysis of drinking water, or they could attend weekly lectures delivered by guest speakers on a range of sanitary subjects. Offering a glimpse into their day-to-day operations, the laboratories worked to distinguish themselves from more sensational iterations of popular science—performed by itinerant showmen, staged in music halls, mechanics institutes, and dedicated venues like the Panopticon of Science and Art on Leicester Square—by promising to eschew theatrics in favor of simulacra. Though their purpose aligned with what Iwan Morus has described as the function of popular scientific performance, namely to make "science and its products real to their audiences," the laboratories imagined that audience as one composed of not only the general public, but also the academic establishment.[32]

Cheyne, it was reported, held "demonstrations especially intended for the medical profession on Thursday afternoons at 4 pm," and on Fridays Corfield welcomed "public analysts, medical officers of health and chemists" to attend his experiments.[33] More than entertainments, the Health Laboratories were imagined as sites of interdisciplinary encounter, where the first principles of the new sanitary science might be thrashed out. These efforts were supported by a display on loan from the French government of instruments and materials from the laboratory of Louis Pasteur (1822–1895). As a notice in *The Lancet* enthused, as well as giving a fulsome "history of his researches and discoveries" the display also featured such talismanic objects as the "crystals of tartaric acid" that first sparked his interest in microbes and a "large vessel charged in 1861, which still contains unaltered the production of a fermentation of 200 grams of sugar acted upon by a minute particle of yeast."[34] The legitimacy of the working laboratories was sustained, in part, by their proximity to this venerated institution, to the celebrity enjoyed by Pasteur and to the decades-long history of germ theory that he represented.

Though distinct from one another, there were significant crossovers in the kind of work performed by the Health Laboratories and the way they were framed by the exhibition literature as contributors to a shared scientific endeavor. This is evident from the text published to accompany the IHE, *Public Health Laboratory Work* (1884) that was coauthored by Cheyne, Corfield, and Cassal. In addition to serving as a record of the laboratory work undertaken over the duration of the exhibition, the book was intended as a primer for the general reader. Divided into two parts, with the first dedicated to explicating the basics of bacteriology—those "minute living bodies of various shapes and characteristics, which have in some instances been shown to be the cause of disease"—and the second to the rudiments of chemical analysis, it extended the reach of live demonstrations and lectures beyond the temporary structures of South Kensington.[35] Sanitary science emerged from the pages of *Public Health Laboratory Work* as a distinctly interdisciplinary endeavor, one sustained by different kinds of knowledge—biological, chemical, bacteriological, epidemiological—committed to shared methodologies, and reliant on a broad range of techniques for its investigations. The rise of the laboratory through the closing decades of the nineteenth century has typically been read as a catalyst for specialization, whereby increased instrumentation and advancing technologies encouraged ever finer differentiation between the disciplines. But the Health Laboratories complicate this picture.[36] Visitors to the IHE, where "health was accepted as a synonym of hygiene," were presented with a vision of sanitary science as a collaborative, pan-European project dedicated to the creation of a clean, disease-free modern world.[37]

In a glowing report on the work of Cheyne and Corfield, *The Lancet* predicted that "any a disbeliever in 'germs' will have his skepticism either lessened or removed" after visiting their laboratories.[38] Germ theory's triumph was tempered, however, by lingering disquiet over where responsibility for public health lay and what discipline was best equipped to improve it. To start with, the Biological and Hygienic Laboratories were far from united on a number of key issues. Where Cheyne placed emphasis on the close study of bacteriological pathogens, his colleagues down the hall played down the significance of microbes in favor of environmental factors. Bacteriology and sanitation could not always be easily disentangled from one another—in the case of tubercular meat, for instance, or in the choleric contamination of drinking water—but the connection was not sustained at the IHE, and this disunity precipitated a broader disciplinary rift.

As Charles DePaolo has noted in relation to *Public Health Laboratory Work*, a text that would come to serve as a model for the manuals used by medical officers of health, "microbiology" eventually outgrew the "textual parameters of sanitary science."[39] Reporting on the laboratories also challenged the image of Britain as a world leader in the field. In a communiqué from the Fifth International Congress of Hygiene held that year at The Hague, *The Lancet* complained that "England is more feebly represented than any other nation" and questioned why the gathering was not staged "during the Health Exhibition." At the close of the IHE, *The Times* urged that the laboratories be established on a more permanent basis to provide "sanitary research and teaching . . . of a kind similar to that which is already afforded by various foreign institutions."[40] Away from the bombast of the official exhibition literature, the ambiguities and uncertainties within sanitary science came more sharply into view.

Visitors would likely not have been cognizant of these disciplinary tensions, but they may have detected some equivocation in the exhibition's design. Though it celebrated the advance of sanitary science, its organizers dedicated themselves more fully to the exploration of public health outside of the laboratory. Far greater attention was paid to the topology of the average home—drainpipes, fireplaces, stoves, and windows—than to cutting-edge scientific spaces. According to Adams, this insistence on the culpability of the architecture spoke of both "faith in visible, even spatial, explanations for internal states" and of a lingering uncertainty "regarding the way diseases spread."[41] The germ theories of Koch and Pasteur seemed to have offered a definitive answer, but at the IHE the question of how to improve and sustain public health remained open to broader interpretation. At play in the cornucopia of objects displayed, in the crowded visual and rich textual language of the IHE, was what Katherine Pandora has usefully termed "an ethic of miscellaneity" that "facilitated open-ended forms of knowledge circulation" and allowed "populist versions of science to flourish."[42] In other words, while the operators of the Health Laboratories imagined themselves as experts tasked with educating the uninstructed masses in the importance of their pioneering research, the rambunctious interdisciplinarity of the exhibition undermined such efforts by continually opening up alternative avenues of exploration and different solutions to the problem of health. One of these could be found in the Vegetarian Dining Rooms.

The Vegetarian Dining Room

In testament to its extraordinary popularity, over the course of 1884 the IHE was regularly mocked by *Punch*. In his "Insane-itary Guide to the Health Exhibition," Arthur a'Beckett provided readers with an "instructive" itinerary for their trip (figure 5.3). They should begin by regarding "exhibits in South Gallery from a scientific point of view, especially the cases devoted to biscuits," then attend a lecture on "Cheap Cookery" and the "deep mystery of Toad-in-the-Hole," before observing the "fascinating" habits of herring in the aquarium, and finally "running up to the Library and reading all the books you discover there on health." Having laid out an exhausting eight-hour slog of edifying pleasure, the article concludes with the admission that "if you really want to enjoy yourself, hang Science, and stick to the Band and Oil-lamps."[43] The satirical guide reflected the two criticisms most frequently leveled at the exhibition: that many of the displays appeared to have no direct relation to health and that most visitors avoided its educational exhibits in favor of the many entertainments on offer. Some accused the executive committee of having pandered to, rather than elevated, its audience. As one bad-tempered commentator queried, "What have all these brass bands; the nocturnal fetes; these flower shows and flower sales; these Chinese Courts . . . these American juleps; ice creams; and sherry cobblers that I hear about got to do with Health and an Exhibition in connexion therewith?"[44]

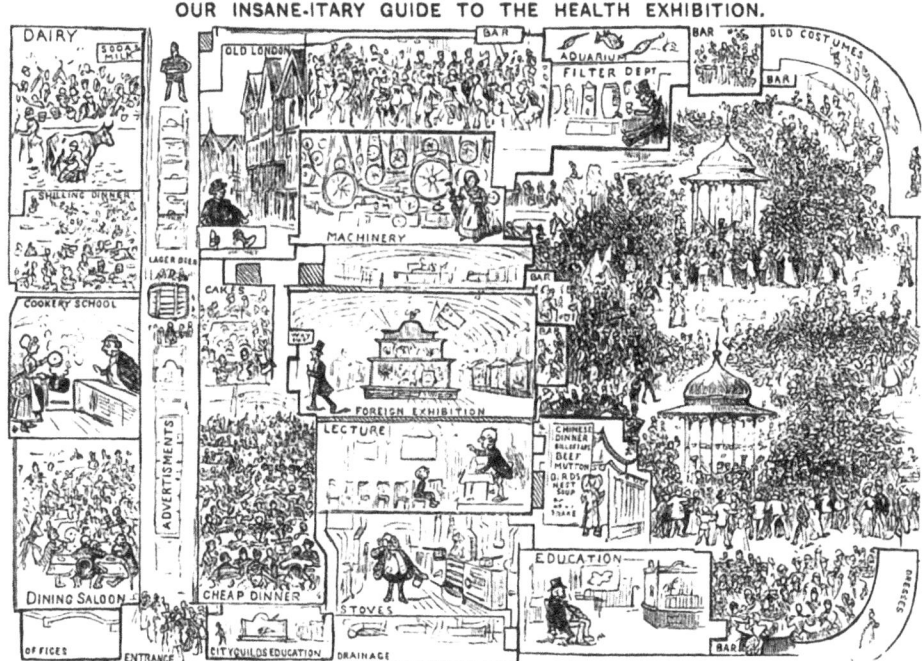

FIGURE 5.3. Satirical guide to the exhibition. *Source:* Arthur a'Beckett, "Our Insane-itary Guide to the Health Exhibition," *Punch, or the London Charivari*, August 30, 1884, 98.

The cartoon map that accompanied *Punch*'s itinerary played to suspicions that the IHE offered its visitors amusement above all else: a hoard of tiny figures cluster in the illuminated gardens, mill around the bandstands, and squeeze into bars, while in the lecture theater a speaker delivers his talk to an audience of one and galleries displaying the latest innovations in drainage stand empty. Most crowded on the map are the dining establishments, and an alternative schedule of simply "eating and drinking" is provided at the end of the article.[45] The range, quality, and affordability of the food on offer at the exhibition was much commented on in the press. It was possible to secure, an article in the *London Journal* enthused, "a capital dinner for sixpence or a shilling, and a tea for sixpence which could hardly be provided for the same money by the thrifty housewife at home," while *The Graphic* remarked on the "numerous refreshment departments designed to suit all palates and all purses."[46] Though these efforts were widely praised, the gastronomic abundance on offer at the exhibition suggested to its critics that visitors were more engaged by the pursuit of alimentary pleasures than by a quest for greater knowledge.

Hungry exhibition goers could make their way to a large dining saloon operated by Messrs. Bertrum & Roberts, where it was possible to observe the work of the kitchen; grab a cheap coffee from one of the many stalls dotted around the galleries; or enjoy mutton chops imported frozen from New Zealand in the grillroom.[47] Those who were feeling adventurous could sample tofu and sake from the Japanese delegation or dine in the Chinese restaurant where, as *Ward and Lock's Ready Guide to the Health Exhibition* (1884) reassured its readers, "instructions in the use of chopsticks" were provided.[48] Dining at the exhibition involved more than choosing among different restaurants; instead food was one of the IHE's central themes. Early in the planning stage a dedicated subcommittee was formed that divided the subject into three key areas—"animal and vegetable products," the "chemistry and physiology of food," and "cookery"—to be overseen by a cross-disciplinary group of experts.

Notable advisers to the committee included the prominent journalist George Augustus Sala (1828–1895); William Thiselton-Dyer (1848–1928), director of the Royal Botanic Gardens; the physiologist Frederick William Pavy (1829–1911); and Henry Thompson (1820–1904), professor of pathology at the Royal College of Surgeons.[49] Food emerged from these deliberations as complex, multidimensional, and essential to public health. There were displays of prepared foodstuffs like "preserved fruits" and daily demonstrations of "economic cooking" in model kitchens, held alongside lectures on the chemistry of adulteration supported by illustrations of "food constituents and equivalents, tables, diagrams" and laboratory work on the role of common "animal parasites" in the production of diseases of the digestive system.[50] While *Punch*'s "Insane-itary Guide" depicted restaurants as sequestered spaces where fun might be had away from the dry edification on offer elsewhere at the exhibition, for its organizers consumption and education could not so easily be extricated from one another.

The question of what to eat, according to the official guide, was one best answered by science: until the "last quarter of a century" food had been conceptualized only in relation to cost, taste, and availability, but recent advances in fields like chemistry, biology, and bacteriology had revealed the "various properties of different kinds of food," the "effects of different processes of cooking," and how the right food might contribute to "preservation of the health of human beings."[51] The IHE approached food as a new frontier in scientific exploration and interdisciplinary collaboration. Alongside lectures on subjects like "The Science of Cookery" and "Parasites and Meat," there was a conference on food adulteration sponsored by the Institute of Chemistry, the Hygienic Laboratory experimented with new ways of detecting contamination, and a display by the Parkes Museum of Hygiene analyzed the chemical composition of popular products like Bovril and Cadbury's Cocoa.[52] Scientific eating was also the focus of the exhibition's model workman's kitchens, operational dairies, and bakeries, where lectures on subjects like "Pure Milk" and "Salt and Other Condiments" introduced listeners to the language of fats, minerals, and acids.[53]

According to Alexander Wynter Blyth (1845–1921), then medical officer of health for Marylebone, it was necessary to begin instructing the public in the rudiments of nutritional science because the "school of dietetics" would soon supplant all other branches of medicine. Writing in a pamphlet titled *Diet in Relation to Health and Work* (1884), he predicted that when the "science of diet" was better understood, physicians would forgo drugs and begin treating "maladies by cutting off certain foods, by surfeiting with others."[54] For Blyth, this revolution depended on the kind of interdisciplinary work made possible by the IHE, and his pamphlet emphasized the role of diverse scientific expertise—apparent in his commentary on subjects like "molecular life," the storage of sugar in the body, and "nitrogenous substances"—over practical dietary instruction.[55] Other handbooks published by the IHE adopted a similar tactic, with the biochemist Arthur Gamgee (1841–1909) holding forth on the "Physiology of Digestion," the public analyst James Bell (1825–1908) offering his thoughts on the science of adulteration, and the pharmacist John Attfield (1835–1911) illustrating the chemical composition of nonalcoholic drinks.

These publications approached food as an object of empirical inquiry, but the exhibition offered up a more complex account of the relation of diet to health. Instead of enshrining food as a purely scientific matter, its abundance of restaurants and refreshment stalls, cooking demonstrations, and displays of products revealed it as subject to conflicting knowledge claims. Dietary knowledge is often formed, as Corinna Treitel suggests in her study of the rise of the natural diet in Germany, in "messy and non-hierarchal ways," and the arrangement of the IHE reflected this horizontality.[56] Formulated by an integrative public discourse in which specialist research and household wisdom played an equal role, the Victorian diet exposed tensions between differently valued forms of expertise. On the one hand, the late nineteenth century witnessed the emergence of medical technologies, like the stomach basket and the gastrograph, which exposed the workings of the stomach to scientific analysis for the first time, but on the other, the question of what to eat and what not to eat remained open to broad contestation.[57]

Visitors to the exhibition might chance on a lecture on the chemical composition of animal fat at the Hygienic Laboratory, but they might also browse the popular dietetic texts stocked by the library, sample new health-food products, and learn about different cultures of consumption from around the world. Eating at the exhibition was, moreover, an embodied experience that provoked pleasure and disgust, as well as a site of intellectual inquiry. Authority over dietary health was diffused across multiple locations at the IHE, including a surprisingly popular meat-free dining room. Operated by the Manchester Vegetarian Society, it served affordable sixpenny dinners through the day and ran a program of events in the evening. Described by the guide as an "educational restaurant," it adopted a highly didactic approach to dining.[58] A stock of improving literature was available to buy from the front desk, there were daily demonstrations of the rudiments of meat-free cooking, and patrons were encouraged to attend weekly lectures on food reform. Reporting on the success of the dining room, the Vegetarian Society's journal enthused that many customers, having "experimented" with a fleshless lunch for the first time at the exhibition, inquired after "pamphlets" detailing its health benefits and "cookery books containing similar recipes."[59] The IHE presented vegetarians with an opportunity to evangelize, but importantly it also allowed the movement to promote the diet on the basis of health and to align itself with the emerging science of nutrition.

Since its inauguration in 1847, the Vegetarian Society had attempted to convince people of the enormous influence exercised by food choices on health, and the invitation to take part in the exhibition appeared to confirm the role of that the fleshless diet might have to play in improving the well-being of the nation.[60] In truth, the restaurant's status within the Healtheries was far more ambiguous: it may have been operated with the education of the masses in mind, but it was not placed on equal footing with other exhibitors. From the beginning its organizers complained that the exhibition committee had attempted to marginalize and minimize the efforts of the society: not only had they been less than generous in their allocation of space for the dining room—made obvious by the potential customers turned away every day—but they had also excluded them from participating in the lecture series.[61]

As a report in the *Dietetic Reformer and Vegetarian Messenger* complained: "Able and willing friends were ready to place their services at our disposal have been refused . . . all offers on our part have been declined, however advantageous such lectures might have been for public health."[62] Dismissed by *The Lancet* as offering little more than "full scope for the airing [of the Vegetarian Society's] whims," the meat-free dining room remained apart from the vision of scientific eating constructed by the exhibition as whole.[63] Where the vegetarian diet was discussed in the literature of the IHE, it was condemned as nutritionally inadequate, likely to cause gastric illness, or simply unsuitable for the cold British climate.[64] This reflected a broader skepticism among orthodox physicians regarding the health claims made by the movement, which were generally in line with the *British Medical Journal*'s assessment: "Vegetarians, as a rule, are not healthy folks. They suffer from dyspepsia, flatulence, bad breath, and anemia. They burden their stomach with masses of crude stuff, and practically deprive themselves of fat and oil; and while they daily grow thin and nervous, they think they are improving in health."[65] The decision to prevent the Vegetarian Society from contributing to the exhibition's pedagogical schema offers an example of what Thomas Gieryn has termed "boundary-work," wherein an attempt is made to demarcate scientific from nonscientific knowledge in the public realm in order to "enlarge the material and symbolic resources of scientists or to defend professional autonomy."[66] By casting the meat-free restaurant as a novelty, a concerted effort was made on the part of the IHE to avoid having to include vegetarianism in its interdisciplinary vision of public health.

Despite being left out of the official program, the Vegetarian Dining Room still managed to establish itself as an educational space. It held its own lecture series, which featured medically trained advocates like Anna Kingsford (1846–1888), and arranged for Thomas Allinson (1858–1918)—a well-known vegetarian physician and advocate of whole-meal bread—to provide "public free medical consultations daily" on how to "cure disease by diet."[67] Reflecting on the restaurant's success in *Fifty Years of Food Reform: A History of the Vegetarian Movement in England* (1898), Charles Walter Forward (1863–1934) estimated that over the course of the IHE around "161,000" meals were served and postulated that all that "good cooking" had transformed public perceptions of fleshless dining. Indeed, so many had been convinced to "make a trial of the diet" that the exhibition precipitated a boom in the number of vegetarian restaurants operating in the capital.[68] In 1878 London had only one meat-free eatery, but by 1890 there were over thirty to choose from, and similar establishments could also be found in Birmingham, Glasgow, and Manchester. A dedicated magazine, the *Weekly Star and Vegetarian Restaurant Gazette*, reported on the industry, and in 1889 the London Vegetarian Restaurants Association Ltd. was established to allow investors to buy shares in meat-free businesses.[69] Evident elsewhere in the founding of the London Vegetarian Society in 1888 by Arnold Frank Hills (1857–1927)—a millionaire sports enthusiast who championed the diet as an essential aid to hiking, athletics, and cycling—the rise of the vegetarian restaurant marked a shift in how the decision to forgo animal products was conceptualized.

Where adherence to the vegetarian diet had once been understood as an alimentary expression of radical affiliations or unconventional beliefs, by the end of the nineteenth century it might also simply signal commitment to good health. Contributing to a wider popular reform movement, evident elsewhere in the rise of dietetics and a growth of physical culture, the meat-free meal was increasingly sold as a necessary aid to bodily efficiency and mental vitality. The IHE played a key role in this transformation by providing, as Forward noted, both the business "impetus" and "hygienic" authority on which to establish vegetarianism as a rational dietary regime.[70] Though the question of whether to eat meat or not was not treated seriously by the exhibition, this did not prevent visitors from reading the vegetarian dining room directly through the lens of sanitary science.

The IHE produced a vision of good health that was complex, contradictory, and, at its core, interdisciplinary. As an emerging field, sanitary science was sustained by the cross-disciplinary work of bacteriologists, chemists, and engineers, collaborations that were showcased in model laboratories, public lectures, and affordable publications. Beyond the insights of microbiology and chemical analysis, the concept of health invited a far broader interpretation and a more expansive understanding of interdisciplinarity. In his pamphlet for the exhibition, Blyth took up the modern city as a fitting metaphor for the elaborate structures and complex operations of the human body. The healthy body is, he insisted, similar to a well-ordered town, where blood moves unimpeded through the streets and where every citizen has a strictly "appointed place and vocation," from the workers in the liver busy manufacturing "bile and glycogen" to the "officer" who sits below a "thatch of hair" responsible for carefully directing the intelligence. The bustling metropolis necessitates a clear division of labor because the "internal work" of the body relies on the speedy "distribution" of food among its many inhabitants and different jobs demand specific forms of nourishment.[71] Given his role in public health, it is unsurprising that Blyth looked to the urban environment to comprehend physiological systems, but his emphasis on cooperation and shared responsibility also reflects the understanding of public health that guided the IHE. Speaking at the formal opening of the exhibition, James Paget (1814–1899)—surgeon to Queen Victoria and president of the Royal College of Surgeons—distinguished between the two very different approaches to health: individual and national. While it was perfectly possible for a man to "live idly, comfortably and long," this "unproductive" health contributed nothing to the collective well-being and a "nation composed of such people" would soon "decline and fall." The aim of the IHE was ultimately, Paget continued, to encourage the kind of "healthy man" who lives "vigorously," "does the largest amount of the best work that he can" and "leaves healthy offspring."[72]

By recasting health as an expansive category—encompassing architecture, nutrition, engineering, elementary education, dress, and countless other elements—he insisted on the culpability of the general public for its maintenance. The model of health set out by Paget—productive, reproductive, and responsible—was a stark expression of late nineteenth-century biopolitics, but the experience of attending the exhibition may have facilitated other readings. Designed to forge connections between everyday actions and "their bearing on health," its democratic schema also flattened out distinctions among different kinds of knowledge and expertise, so that vegetarian cooking might be placed on equal footing with water purification or germ theory read alongside the latest innovations in corset design.[73] Moving through the space, visitors to the exhibition constructed their own understanding of health from an abundance of highly varied visual, textural, and experiential materials. The concern voiced in some reviews of the exhibition, that the average attendee was likely forgo its more edifying displays in favor of the entertainment offered by lantern shows and cheap bars, speaks to the anxiety that this interpretative license provoked in some quarters.

Chapter 6

VICTORIAN PHYSICS AND ITS INTERDISCIPLINARITIES

IWAN RHYS MORUS

What kind of discipline was physics in the nineteenth century? The question is more difficult to answer than might be expected. On the one hand, clearly practitioners regarded themselves as exemplars of disciplined science, and in many ways their historians have agreed. The history of modern physics' emergence from natural philosophy during the course of the nineteenth century has typically been cast as a history of increasing discipline by and through a kind of disembodiment. As physics solidified into a discipline during the course of the century, bodies and sensations were banished to the margins of its culture. Undisciplined bodies were held to be unreliable sources of knowledge precisely because sensations were too erratic and individual to be good indicators of the real nature of things. The trajectory for Victorian physics identified by the historiography of the past quarter century has been one that leads to an increasing emphasis on laboratory discipline and metrology.[1] In this process bodies were displaced and replaced by instrumentation that could be calibrated and standardized, and bodies were disciplined into the routines of reliable experimentation. As early as 1840, William Whewell (1794–1866) in his *History of the Inductive Sciences* coined a new term for this new breed of disciplined practitioners: they were physicists.[2]

But Whewell's coinage is an indication of the extent to which physics fails to conform to the conventional history of nineteenth-century disciplinary formation. For most of the century physicists were deeply resistant to calling themselves physicists. Michael Faraday's (1791–1867) response to the neologism was relatively restrained: "Physicist is both to my mouth and my ears so awkward that I think I shall never be able to use it."[3] William Robert Grove (1811–1896) was more bitter about a word where "four sibilant consonants fizz like a squib."[4] Far later in the century, William Thomson (1824–1907) was just as dismissive—practitioners should "refuse to accept so un-English, unpleasing and meaningless a variation from old usage as 'physicist.'"[5] Practitioners typically preferred to describe themselves as experimental philosophers, or more commonly as natural philosophers. Victorian physics was also notably lacking the usual institutions of disciplines. The British Association for the Advancement of Science (BAAS) had a Physical and Mathematical Sciences Section, but there was no Royal Physical Society to mirror the Geological, Astronomical, and Chemical Societies, for example.[6] Until the *Philosophical Transactions of the Royal Society* divided into A and B in 1887, there was nothing resembling a specialist disciplinary journal for physics either. In many respects the most notable feature of Victorian physics was a tendency to disciplinary transgression. Throughout the nineteenth century men of physics were more interested in overstepping the boundaries of their discipline in different ways than they were in policing them.

Given the apparent resistance to the constraints of discipline, how then might the disciplinary orientation of Victorian physics best be described? Was physics an interdisciplinary discipline? Or might it be better described as over-disciplinary, for example, or cross-disciplinary? Did its practitioners want to be found between or above the disciplinary field—what did they think physics owed to, or was owed by, other disciplines? In this chapter I want to take seriously the relationship between discipline and disciplines by looking at how, in different contexts, attitudes toward disciplinarity were embodied in the bodily practices and self-fashioning performances of physics. In other words, what did different orientations toward discipline look like in the doing? In this way I hope to be able to probe at some of the ways in which the different practices and identities of physics throughout the century continued to draw on the tradition of natural philosophy that physics was supposed to have displaced. There is a growing historiography that pays attention to the place of sensation and bodily performance in Victorian physics, building on David Gooding's work on Michael Faraday, and Otto Sibum's work on James Prescott Joule.[7] More recently, historians such as Jimena

Canales and Chitra Ramalingam have interrogated the problematics of sensationalism and spectacle in the production and dissemination of experimental physics.[8] Here, I want to investigate how thinking about embodiment can offer a way of understanding the relationship of different iterations of physics to different visions of disciplinarity.

To do this I will focus on four examples of the different disciplinarities of physics and their embodiments spread across the nineteenth century. In the first of these, I will look at Grove's articulation of the theory of the correlation of physical forces as an explicit attempt to redefine the place of physics on the cultural map of Victorian science in the face of disciplinary fragmentation. His intervention was both an attempt to open up a new space for physics as a correlator of other scientific disciplines and a redrawing of the hierarchy of the sciences that could be mapped onto new institutional formations. Correlation offered a new way of performing physics. John Tyndall's (1820–1893) performances as Faraday's successor at the Royal Institution from the 1860s on offered a different model of what physics was meant to look like—and hence how its disciplinary relationships might be understood. Tyndall's lectures on sensitive flames, for example, understood as performances that deliberately played with sensation, were attempts to practically embody Tyndall's own views of physics as a particular kind of discipline that appealed to the imagination. This was physics as a model of how disciplined performance could communicate the order of nature and the embodied authority of its practitioners.

Like Tyndall, James Clerk Maxwell (1831–1879) and his successors at the Cavendish Laboratory were faced with the problem of making their view of physics' discipline conform to existing institutional spaces. Maxwell and John William Strutt, Baron of Rayleigh (1842–1919), worked hard to establish rigorous disciplinary regimes at the Cavendish designed to make their students' bodies fit for physics as they saw it. At the same time, the Cavendish's regimes had to be made fit for the purposes of a university committed to a particular vision of liberal education and mental training. Different models of disciplinarity and its functions had to be made conformable. The ether physics that Maxwell developed as part of his solution to the problems of disciplinary conformity was also central to Oliver Joseph Lodge's (1851–1940) account of physics' disciplinary place. The ether and the novel technologies that were associated with it provided Lodge with a vehicle for extending the reach of physics' practices. In all these cases, the disciplinary status of physics was at stake. They can be understood as attempts to reconform the disciplinary map of Victorian science and the place that physics occupied. By looking at them as instances of how different accounts of disciplinarity were actively performed by their authors, we can get a better sense, I hope, of what disciplinary formation meant for practitioners and how they viewed the discipline of physics' orientation to other disciplines.

A SORT OF HIGHER COUNCIL

When William Robert Grove published his essay *On the Correlation of Physical Forces* in 1846, it was a statement of intent at a critical juncture in his own career in science. Since early 1841 Grove had been professor of experimental philosophy at the London Institution, established in 1806 as a competitor to the Royal Institution. In that capacity Grove had been in a comparatively safe institutional position, holding one of relatively few established professorships in the metropolis. He had a secure place on London's scientific map. In 1845, however, he resigned his professorship in order to resume professional practice as a barrister.[9] To mark the occasion, the lectures that he had been regularly delivering at the Institution were published. The essay on correlation was far more than simply a farewell note. It was an explicit intervention in debates about disciplinarity and the nature of physics in which Grove had already made some forceful interventions. It was an attempt to carve out a space for himself in the field of physics that set out to redefine the grounds of the field and its relationship with other disciplines. As well as marking his departure from the London Institution, the publication marked Grove's election in 1846 as a council member of the Royal Society.

Grove had started his tenure at the London Institution with a lecture on the progress of physical science since the founding of the institution in 1806. He sketched out some of the elements of what would become his theory of correlation a few years later and concluded with an emphatic declaration of the place discipline occupied in putting science at the service of the state.[10] But as an essay in *Blackwood's Magazine* the following year made clear, the problem with disciplines, in Grove's view, was that they lacked discipline. If a discipline had the potential to promote "an *esprit de corps*, which forms a bond of union to each section," and impose "moral discipline in its ranks," in practice disciplinary societies promoted self-promotion. Specialist societies "do harm by the cliquery which they generate, collecting little knots of little men, no individual of whom can stand his own ground, but a group of whom, by leaning hard together, can, and do, exercise a most pernicious influence."[11] The problem with disciplines was that they led to fragmentation. There needed to be a discipline of disciplines that made sure that the rest were kept in proper order. It was a powerful statement of disciplinary discontent.

On one level it is possible to understand correlation by remembering its origins as a course of lectures. Correlation offered a way of organizing a scientific performance by providing a series of experiments with a unifying theme. It is clear from contemporary accounts that the final published essay mirrored the course of six lectures on which it was based quite closely. Spectacle was built into the concept. It offered a strategy for instantiating the relationship between a variety of different experimental practices and technologies seamlessly in front of an audience. In one experiment, for example, he placed a daguerreotype plate and a grid of silver wire in a glass-fronted box and connected them through a galvanometer and a Breguet helix to form a circuit. When the shutter was opened and light fell on the plate, it generated electricity, causing the galvanometer needle and the helix to move: "Thus, Light being the initiating force, we get chemical action on the plate, electricity circulating through the wires, magnetism in the coil, heat in the helix, and motion in the needles."[12] This was correlation made visible on a tabletop. It was a way of performing continuity by offering a grand narrative that made the stuff of other disciplines into components of the performance.

Both in its performance and as a text, correlation was a powerful statement of physics' disciplinary hegemony. Physics in Grove's showing was not just a discipline, or even simply interdisciplinary, but an over-discipline. Its place was not with or between other disciplines but above them. If his perspective on things were adopted, then correlation meant that all other scientific disciplines came under the purview of the physical sciences. The doctrine of correlation demonstrated that "no physical phenomenon can stand alone; each is inevitably connected with anterior changes and is inevitably productive of consequential changes, each with the other, and all with time and space; and either in tracing back these antecedents, or following up their consequences, many new phenomena will be discovered, and many existing phenomena hitherto believed distinct, will be connected and explained."[13] The correlation of physical forces was a kind of obligatory passage point through which all scientific disciplines needed to be channeled if their conclusions were to be properly appreciated.[14] This was a philosophical declaration that physical science was an essential overseer of the intellectual output of all other scientific disciplines—and Grove made it clear that he thought correlation applied in the organic as much as the inorganic world. In that respect physics was simply different in kind from other disciplines—or at least there had to be space in physics for someone with the philosophical acumen to look from above at what others were doing.[15]

Grove's philosophy had its politics as well. Just as Grove was contemplating his departure from the London Institution and the implications that shift might bear for his place in metropolitan science, he was appointed to the Royal Society's council in 1846 and soon after made a member of the committee charged with overseeing the reform of the society's charter and statutes. One way of approaching Grove's reform project is to understand it as a campaign to transform the Royal Society into an institution for correlating the sciences. In Grove's view reform was a prerequisite if the society was going to resume its proper role as a body capable of overseeing the activities of specialist, disciplinary societies. It would become the kind of English academy he had advocated in his essay in *Blackwood's Magazine*.[16] In the aftermath of their successful reform campaign that curtailed the president's power of patronage by limiting the number of fellows to be elected each year, Grove and his fellow reformers established the Philosophical Club in 1847. Its explicit purpose was to oversee the direction of the society. As Grove's ally and fellow member of the Philosophical Club, Edward Forbes (1815–1854), expressed it, both the Royal Society and the disciplinary societies would be taught "to look at the Philosophical as a sort of higher Council or guardian angel of them all."[17]

A key aim of the Philosophical Club was juxtaposition. This was a campaign to bring about the reunification of the metropolitan specialist societies under the aegis of a reformed Royal Society. As Grove expressed it, juxtaposition was to be a solution to the problem that "the cultivators of Science suffer from the isolated character of the principal Scientific Societies that there was a want of 'espirit de corps' and central authority among Scientific men and a consequent temptation to schism and charlatanery."[18] There were concerted efforts throughout the 1850s to lobby for juxtaposition. Philosophical Club members who held positions of influence in the specialist societies were encouraged to "exert themselves to obtain for such societies a single locale." In the aftermath of the Great Exhibition in 1851 there was some suggestion that the Royal Society and the various disciplinary bodies should be relocated to Kensington. When the Royal Society itself was removed from Somerset House to a new location in Burlington House, there was indeed some measure of juxtaposition as the Linnean and Chemical Societies moved with it. As late as 1872, giving evidence to the Devonshire Commission, Grove was still insisting that juxtaposition "would save much useless expense, lead to more co-operation, and promote more discipline in scientific workers, so that work in common for definite purposes might take the place of disjointed efforts."[19]

LECTURE ROOM EXPERIMENTS

The London Institution where Grove performed his lectures on correlation had originally been established as a competitor to the Royal Institution. That had been established in 1799 by Benjamin Thompson

(1753–1814) with the aim of placing natural philosophy at the service of agricultural improvement. The Royal Institution had originally been intended to cater to the scientific entertainment and economic betterment of the landed gentry. The London Institution, on the other hand, was meant to serve a more commercial clientele, though by the 1840s there would have been little to distinguish them. With Humphry Davy (1778–1829) at the helm, the Royal Institution had become a celebrated center for scientific entertainment and spectacle. Davy used experiments with the institution's large galvanic battery to great effect to establish his reputation.[20] Michael Faraday in turn learned the art of spectacular performance from Davy. He learned about the importance of meticulous preparation and the role of careful and disciplined choreography in making experiments work onstage. As David Gooding has noted, the translation of Faraday's tentative private experiments in the Royal Institution's basement laboratory into robust public performances in the lecture theater above required skilled work and practice.[21] Faraday also understood the importance of personal bodily discipline in making his performances work.

When John Tyndall inherited Faraday's mantle at the Royal Institution, he also inherited a particular view of how the discipline of physics should be performed and made visible.[22] Faraday, like his predecessor Davy, had worked hard at the business of scientific display. They had been highly successful in making the Royal Institution into what Thomas Carlyle (1795–1881) described as a "kind of sublime Mechanics' Institute for the upper classes." Both Davy and Faraday understood the importance of spectacle to their scientific performances. Their fashionable audiences wanted to see nature—and to see the performers as nature's masters. Tyndall's lectures, like those of his predecessors, were specifically and deliberately designed to appeal to his audience's senses and feelings.[23] They were appeals to emotion as much as reason. In the Royal Institution's lecture theater, images of authority were being constructed through public performances in which appeals to sensation were absolutely central. Bodies—those of performers and their audiences—were central to the construction of scientific authority. Just as Grove's performances and his politics of scientific reform can be understood as performances that instantiated a particular stance on disciplinarity, Tyndall's performances on the Royal Institution's stage also embodied an account of the discipline of physics and where it stood in relationship to other sorts of disciplinary performance.

Contemplating his predecessor and patron's success as a scientific performer, John Tyndall drew particular attention to Faraday's bodily discipline. His key quality, he said, "was his sense of order, which ran like a luminous beam through all the transactions of his life."[24] He elsewhere noted how "the habit of self-control became a second nature to him."[25] When Tyndall was appointed to his position at the Royal Institution, he was expected to continue in the tradition that Davy and Faraday had established. His lectures had to be "original, & such as can only be given by the original research of the Lecturer . . . illustrated by striking experiments, so as to present a beautiful outline-map of the subject, such as anyone, who would give continuous intelligent attention to the Lecture, would both apprehend and retain."[26] He was not always successful, in his early years at least. John Barlow, the Royal Institution's secretary, confessed to Faraday that "I dread the tendency of Tyndall's Lectures to become abstruse."[27] It is clear that Tyndall devoted a great deal of attention throughout his tenure at the Royal Institution to overcoming those kinds of problems. He had Faraday's model before him—and it was clearly a model that he took seriously and aimed to emulate.[28]

Tyndall's famous experiments with sensitive flames, first performed at the Royal Institution in 1865, offer an important example of the ways spectacles onstage were carefully choreographed and finely calculated to elicit particular kinds of responses from his audiences. They can be understood as instances of how Tyndall's disciplinary orientation with regard to other ways of knowing was made visible. The basic phenomenon was, to all appearances, relatively straightforward. Nevertheless, turning it into an effective performance required the marshaling of significant material and cultural resources. The effort and resources deployed should, if nothing else, remind us that these sorts of performances were not marginal or inconsequential affairs. What Tyndall's assistant William Barrett (1844–1925) had initially observed was that under particular circumstances the flames of a gas light were sensitive to sound. They reacted differently to different noises such as a low- or high-pitched note, a sharp clap, and so on. As Barrett described it: "While preparing the experiments for one of the Christmas lectures at the Royal Institution, I noticed that the higher harmonics of a brass plate (which I was sounding with a violin-bow in order to obtain Chladni's figures) had a remarkable effect on a tall and slender gas-flame that happened to be burning near. At the sound of any shrill note the flame shrank down several inches, at the same time spreading out sideways into a flat flame, which gave an increased amount of light from the more perfect combustion of the gas."[29]

Observing the phenomenon was not enough. The two experimenters had a great deal of work to do in order to turn the sensitive flame into a stable and exploitable demonstration that could reliably be performed onstage. Barrett and Tyndall experimented with different kinds of

sounds to see what effect they had on different kinds of flame. To make
the phenomenon work, the flame had to be carefully "tuned"—it only
worked at an optimum pressure, and experiment was needed to establish
what that pressure was. Barrett experimented with different shapes of
pipes and orifices to see what effect they had on the flame's sensitivity.
He made burners out of "glass tubing drawn out and the points broken
off so as to obtain orifices of various sizes." It was immediately clear that
"the *shape* as well as the *size* of the burner was an important element in
the production of the phenomenon." Barrett noted, for example, that
"the stem of a tobacco-pipe does not answer for a burner; the bore is
too small; but a gas-fitter's brass blowpipe, if straightened and filed to a
rather larger aperture, makes a very fair burner."[30] This kind of tinkering
was an essential aspect of the backroom preparation on which successful
spectacle depended.

At the heart of the sensitive flame demonstrations was a concerted
commitment to making the phenomena of sound visible, and the effort
at translating the aural to the visual underlines the key place that seeing
things occupied in this experimental culture. Tyndall's communications
to the *Philosophical Magazine* on the phenomena of sensitive flames often
read like instructions to prospective performers as much as they appear
as accounts of laboratory work.[31] This was a technology of display that
moved back and forth effortlessly between theater and laboratory, and
clearly a specific gestural knowledge needed to travel with the technology. Different kinds of flame of various shapes and sizes were cajoled into
responding to a variety of stimuli by expanding or shrinking. Tyndall
put the "most marvelous flame hitherto discovered" through its paces
in front of the Royal Institution by reading a passage from Spenser. He
explained to his audience how the "flame picks out certain sounds from
my utterance; it notices some by the slightest nod, to others it bows more
distinctly, to some its obeisance is very profound, while to many sounds
it turns an entirely deaf ear."[32] The *Intellectual Observer* described the
flames as "the most extraordinary and apparently magical things." They
described how "the creaking of boots, the rustle of a lady's dress, the patter of rain, and the tick of a watch—all influenced it in a striking way."[33]
According to the *Chemical News*, like "a living being, the flame trembles
and cowers down at a hiss—it crouches and shivers as if in agony at the
crisping of this metal foil, though the sound is so faint as scarcely to be
heard; it dances in tune to the waltz played by this musical box—and,
finally, it beats time to the ticking of my watch."[34]

The pervasive anthropomorphism implicit in these descriptions was
one of the most striking features of contemporary accounts of Tyndall's
sensitive flames. Barrett and Tyndall were complicit in this too. Barrett

described the flame's response to aural stimuli as "like that of a sensitive, nervous person uneasily starting and twitching at every little noise."[35] Tyndall described how his "most marvelous flame" responded to his presence: "When I shake this bunch of keys the flame is violently agitated, and emits a loud roar. The dropping of a sixpence into a hand already containing a coin at a distance of 20 yards, knocks the flame down. I cannot walk across the floor without agitating the flame. The creaking of my boots sets it in violent commotion. The crumpling or tearing of a piece of paper, or the rustle of a silk dress, does the same. It is startled by the patter of a raindrop. I hold a watch near the flame; nobody hears its ticks; but you all see their effects upon the flame."[36] The language emphasized the way in which the flame's presence and performance also drew attention to the presence and performance of the lecturer on the stage. The anthropomorphism and emphasis on symbiosis between phenomenon and demonstrator were presumably deliberate. They were part of a choreographed performance that was designed to appeal to the audience's aesthetic sense and appetite for cultivated wonder. It was a performance of disciplinarity.

Face to Face with Measurement

In March 1871 Maxwell told a correspondent that he had been busy "Tyndalising my imagination up to the lecture point."[37] The timing is not coincidental. Maxwell was in early 1871 considering the possibility of taking up the newly established chair of experimental physics at Cambridge. The university committee's report recommending the establishment was clear that this was to be an experimental position. The committee, chaired by the university's chancellor, William Cavendish (1808–1891), insisted that the "founding of a Professorship would be incomplete unless means were also supplied to render the Professor's teaching practical, and assistance given to him both in the Laboratory and Lecture-room."[38] Maxwell was therefore thinking hard about what would be involved in turning undergraduates into experimenters and how he might bridge that gap between laboratory and lecture. Lord Rayleigh had written to him a month or so earlier, suggesting that what was "wanted by most who know anything about it is not so much a lecturer as a mathematician who has actual experience in experimenting, and who might direct the energies of the younger Fellows and bachelors into a proper channel."[39] This is an interesting distinction that, implicitly at least, reveals a great deal about the broader cultural associations of lecturing at this juncture. A lecturer was someone like Tyndall who excelled at public experimental performances. It was a different kind of disciplined (and disciplinary) performance, and Rayleigh's implication was that this was not a skill that Maxwell himself possessed or that the university required. It put forward a different view of what kind of discipline physics at the Cavendish Laboratory would embody.

Maxwell himself, writing to Edward Blore, a leading member of the university and later vice master of Trinity College, mused that the "class of Physical Investigations, which might be undertaken with the help of men of Cambridge education, and which would be creditable to the University, demand, in general, a considerable amount of dull labor which may or may not be attractive to the pupils."[40] It seems that right from the beginning of Maxwell's Cavendish career he was thinking about ways of reconciling the necessary embodiedness (the "dull labor") of experimental life with the ideal mental life of Cambridge scholarship. Just as clearly, the kind of embodiedness that Maxwell was thinking about here was not the sort he associated with Tyndall and the Royal Institution. Following his appointment, Maxwell played a key role in the process of designing and equipping the new laboratory that would be built for him. He visited William Thomson's laboratory in Glasgow and Robert Clifton's (1836–1921) Oxford laboratory to familiarize himself with the requirements of a laboratory designed for university teaching. The resulting building embodied Maxwell's ambition to find a space for the values of precision experiment inside the university. On the outside, the impressive gateway featured "the arms of the Duke of Devonshire on the left, and the University arms on the right, the motto of the Cavendish family, 'Cavendo tutus,' occupying the center; and the whole is surmounted with a beautifully carved statue of the Duke in his robes as Chancellor of the University, and bearing in his hand the Cavendish laboratory." Inside, each room was designed for experiment.[41]

The concern with trying to figure out just how embodied experimental labor could be reconciled with the expectations of university culture was visible in Maxwell's inaugural Cambridge lecture only a few months after his appointment. It was clear that experimental physics, "while it requires us to maintain in action all those powers of attention and analysis which have been so long cultivated in the University, calls on us to exercise our senses in observation, and our hands in manipulation. The familiar apparatus of pen, ink, and paper will no longer be sufficient for us, and we shall require more room than that afforded by a seat at a desk, and a wider area than that of the black board."[42] Cultivated senses were essential to good experiment, he argued. Physics required "not only the trained attention of the student, and his familiarity with symbols, but the keenness of his eye, the quickness of his ear, the delicacy of his touch, and the adroitness of his fingers," and only by educating the senses could the experimenter "ensure the association of the doctrines of science with those elementary sensations which form the obscure background of all our conscious thoughts, and which lend a vividness and relief to ideas, which, when presented as mere abstract terms, are apt to fade entirely from the memory."[43]

Again, he drew the contrast between experiments of illustration that aimed "to present some phenomenon to the senses of the student in such a way that he may associate with it some appropriate scientific idea" and experiments of research in which "the ultimate object is to measure something which we have already seen—to obtain a numerical estimate of some magnitude." It was that kind of experiment: "those in which measurement of some kind is involved," that formed "the proper work of a Physical Laboratory."[44] This, from Maxwell's perspective, was exactly the difference between Tyndall's practice at the Royal Institution and the way he envisioned experiment at the Cavendish. Sensation was where physics started, not where it ended: "we have first to make our senses familiar with the phenomenon, but we must not stop here, we must find out which of its features are capable of measurement, and what measurements are required in order to make a complete specification of the phenomenon. We must then make these measurements, and deduce from them the result which we require to find."[45] Maxwell was trying here to adapt his perception of physics' sense of discipline to the broader context of the university's understanding of what discipline entailed. His understanding of measurement was that it was an embodied practice. It was work. This was clear in what he wanted to say that Cambridge physics should not be—it should not be only about measurement. If physics were just measurements then "our Laboratory may perhaps become celebrated as a place of conscientious labor and consummate skill; but it will be out of place in the University, and ought rather to be classed with the other great workshops of our country, where equal ability is directed to more useful ends."[46]

The notion of measurement as a laborious process—as something that needed work—has been noted by historians before, as has the emphasis on the discipline of measurement that underpinned the Cavendish regime under Maxwell and his successors.[47] What I want to speculate about here is how the recognition of experiment's laboriousness might be understood in terms of bodily practice and attention to sensation, and how this captures how the discipline of physics might be embodied. Maxwell worried that the similarity between factory and laboratory "may bring the whole university and all the parents about our ears" when they found out what was really going on at the Cavendish.[48] Maxwell's solution was to make the bodily discipline of experiment an adjunct of liberal education. If "the Devonshire Laboratory should be successful, we must endeavor to maintain it in living union with the other organs and faculties of our learned body." Disciplined experiment offered a complement to, rather than a replacement for, traditional mathematical work. Once students had "successfully bridged over the gulph between the abstract and the concrete, it is not a mere piece of knowledge that we have obtained: we have acquired the rudiment of a permanent mental endowment." It was all, he said, "a question of distribution of energy." Pushing the point, Maxwell joked that some "distributions of energy, we know, are more useful than others, because they are more available for those purposes which we desire to accomplish."[49] Energy physics offered a useful metaphor for how the student body should be put to work.

Not only Maxwell, but his successors at the Cavendish Laboratory understood that experimental training was bodily. Students needed to experience and embody the routines and disciplines that constituted metrology as much as its instruments did. Where Maxwell encouraged individuals to develop the self-discipline of experiment for themselves through engaging with iconic pieces of equipment such as the "great electro-dynamometer of the British Association" used to calibrate the ohm, Lord Rayleigh developed a more regimented and collaborative style of inculcating discipline.[50] Richard Glazebrook (1854–1935), later to be Rayleigh's demonstrator, reminisced that the "first experiments I can recollect related to the measurement of electrical resistance. I well remember Maxwell explaining the principle of Wheatstone's bridge and my own wish at the time that I had come to the laboratory before the Tripos, instead of afterwards." He recalled that under Maxwell, there "were no regular classes and no set drill of demonstrations arranged for examination purposes. . . . In Maxwell's time those who wished to work had the use of the laboratory and assistance and help from him but they were left pretty much to themselves to find out about the apparatus and the best methods of using it."[51]

Under Rayleigh's direction, space was organized to provide the "set drill of demonstrations" that both he and Glazebrook regarded as the key element in embodying the discipline of experiment in their students: "Each experiment was set out permanently on a table to itself, and written instructions were provided. The classes were at regular hours, and a demonstrator was in attendance, who assigned the experiment, and gave help in any difficulty, finally approving or disapproving the numerical result."[52] They modeled the regime on the recommendations of the American physicist Edward Charles Pickering (1846–1919) in his *Elements of Physical Manipulation* as a process for mass-producing disciplined student bodies: "Following this plan an instructor can readily superintend classes of about twenty at a time and is free to pass from one to another answering questions and seeing that no mistakes are made."[53] Glazebrook recalled his superior's practices as plain and unvarnished: "The apparatus throughout was rough and ready, except where nicety of workmanship or skill in construction was needed to obtain the result; but the methods of the experiments, the possible sources of error, and the conditions necessary to success were thought out in advance and every precaution taken to secure a high accuracy and a definite result."[54] Clearly there was nothing showman-like about Rayleigh's experiments, but nevertheless the proper use and deployment of his own and his assistants' bodies still mattered.

Reading Glazebrook and William Napier Shaw's (1854–1945) *Practical Physics* from this perspective is instructive. That book had its origins in "a series of MS. note-books" that had been compiled by Rayleigh's two demonstrators, "each dealing with one experiment, describing the apparatus actually in use in the Laboratory, the method of making the experiments, and in particular of entering up the results."[55] Experiment by the account offered here—which given its origins can be taken to represent the regime encountered by students at the Cavendish under Rayleigh—involved hard and frequently tedious work that needed a careful hand and eye. It required laborious and diligent effort. The textbook offered its readers extended descriptions of the repetitive work required to measure even relatively straightforward physical properties. Accurate measurement required not just care and diligence in performing the measurement, but laborious checking to ensure that all of the apparatuses were working correctly and performing a range of procedures to correct any errors in the equipment. Carrying out an experiment required constant attention and bodily engagement with the apparatus. Even as routine a business as checking the adjustment of a balance, for instance, required procedures that needed a dozen or so pages to describe.[56]

In one nice example Tyndall's sensitive flame apparatus was adopted as a detector to measure the wavelength of a high-pitched musical note. The method was based on a procedure that had originally been suggested by Lord Rayleigh. Rayleigh described investigations of the "beautiful phenomenon of sensitive flames" in which he established that under "the action of stationary sonorous waves a flame is excited at *loops* and *not* at *nodes*."[57] Glazebrook and Shaw in turn described how a "sensitive flame 'flares' when a note of sufficiently high pitch is sounded in its neighborhood; thus a hiss or the shaking of a bunch of keys is generally effective."[58] Drawing on Rayleigh's observation of where the flame was excited, the wavelength was found by setting up a standing wave by reflection and moving the flame along its line to determine the points where the effect of sound on the wave was at a minimum. The example is fascinating because it shows how what for Tyndall was a piece of spectacular showmanship on the Royal Institution stage could in another space be reconfigured as a novel technique for calibrating an apparatus. In different spaces and contexts, the sensitive flame experiments embodied different kinds of bodily performances.

The redeployment of Tyndall's sensitive flame and its reconfiguration as a piece of precision instrumentation is revealing because it underlines the ways in which experimental knowledge could follow a trajectory from stage to theater as much as the other way around. Sensational knowledge that had its origins in the genteel surroundings of a metropolitan lecture theater could travel and be accommodated to the disciplinary setting of a university laboratory where very different bodily regimes prevailed. The relationship between the two experiments is clear. Glazebrook and Shaw come close to quoting Tyndall's own description of the sensitive flame phenomenon with their reference to "a hiss or the shaking of a bunch of keys" as a suitable stimulus. They also offered an alternative method, leaving out the flame and using "an india-rubber tube leading to the ear" instead, listening directly for the positions of silence along the wavelength.[59] So the ear, properly disciplined, taught and calibrated could be a precision instrument as well as the eye. This raises interesting questions about what senses, under what circumstances and in what contexts might be considered appropriate generators of knowledge. Tyndall, on the stage, used sensitive flames to turn hearing into seeing. Glazebrook and Shaw, in their textbook, took the flame back out of the experimental ensemble, replacing the eye with the ear as a sensitive instrument for establishing facts. Rayleigh was borrowing the sensational practices in Tyndall's disciplinary performance and refashioning them in support of his own articulation of physics' discipline. It was explicitly interdisciplinary in that it offered itself as an exemplary instance of how bodies might be kept under control through experiment in any discipline.

TECHNOLOGIES OF ETHER

Oliver Lodge's *Modern Views of Electricity* (1889) articulated an account of the disciplinary ambit of physics that, while it owed a great deal to Maxwell's researches, contained a very different picture.[60] Trained at the London University and by 1881 professor of experimental physics at the newly established University College Liverpool, Lodge was a thorough advocate of Maxwell's theories of electromagnetism.[61] His approach to physics was nevertheless very different from the disciplinary regime of the Cavendish. For Lodge the ether was the defining feature of modern physics. As he expressed it, people "who are occupied with other branches of science or philosophy, or with literature, and who have therefore not kept quite abreast of physical science, may possibly be surprised to see the intimate way in which the ether is now spoken of by physicists, and the assuredness with which it is experimented on."[62] The task facing its proponents was to bring all of physics and the matter of disciplines beyond physics into its ambit, and Lodge was confident that the "next fifty years may witness these tremendous victories in part won."[63] Ether physics was meant to be more than a theoretical construct. It embodied a breadth of technologies and performances that defined the discipline of physics. Electricity—and Lodge meant the science of the ether—"has thus become an imperial science."[64]

Maxwell and his successors aimed to establish a distinction between two kinds of physics. Reviewing his friend Fleeming Jenkin's latest text-book in 1873, James Clerk Maxwell endorsed wholeheartedly the way in which the "author of this text-book tells us with great truth that at the present time there are two sciences of electricity—one that of the lecture-room and the popular treatise; the other that of the testing-office and the engineer's specification. The first deals with sparks and shocks which are seen and felt, the other with currents and resistances to be measured and calculated." He said much the same thing in the preface to his own *Treatise*, aimed at "those who have been brought face to face with quantities to be measured, and whose minds do not rest satisfied with lecture-room experiments."[65] This was not a disciplinary distinction that Lodge's ether physics recognized. Lodge's physics was built around technologies of display, particularly technologies of electrical discharge, as much as metrology and mathematical physics. Its iconic instruments were the induction coil and the Leyden jar. The focus of Lodge's physics on instruments like these that came from a tradition of sensuous technologies meant that the distinction Maxwell wanted to draw simply made no sense for Lodge's view of the discipline.

Lodge's ether physics drew on the work of experimenters like War-ren de la Rue (1815–1889) and Cromwell Varley (1832–1919), who carried out research on discharge phenomena during the 1860s and 1870s that can be seen to belong to the same tradition of sensational and sensuous practice as Gassiot's cascade and Pepper's monster coil.[66] William Crookes's investigations into the fourth state of matter belonged to this tradition too.[67] These were experiments in which highly skilled researchers worked to capture phenomena that were both spectacular and liminal. Crookes developed a variety of discharge tubes, such as his railway tube with "a little glass railway running along it from one end to the other," so that "the stream of radiant matter" from the poles caused a little paddle wheel to run along it.[68] Varley experimented with photography to try to overcome the problem of seeing through the spectacle and into its origins. In these sorts of experiments producing spectacle and making novel discoveries looked like the same thing, and both required the same ensemble of technologies and embodied skills. But as in the case of Rayleigh's appropriation of Tyndall's sensitive flames, technologies of display might be disciplined into other contexts. The experimental apparatus that Sir J. J. Thomson (1856–1940), for example, used in 1897 at the Cavendish Laboratory in Cambridge to carry out the experiments that led to the discovery of the electron was the direct descendant of discharge experiments such as these and the Ruhmkorff coils that powered them.[69]

In his autobiography Oliver Lodge offered an anecdote that illustrates nicely the dynamics through which instrumental and bodily skills crossed over (or sometimes failed to cross over) between different disciplinary spaces. He described an experiment by Thomson at the Physical Society "showing the electrodeless discharge in a vacuum bulb, induced by a coil surrounding the bulb, whenever a Leyden jar was discharged through the coil." In an experiment like this, the aim was to produce a visible discharge inside an evacuated tube or bulb that was not connected to any circuit, simply by induction. The Leyden jar provided the transient high potential needed. The experiment needed very specific conditions in order to work properly. In particular, all the components needed to be properly insulated to prevent charge leakage. However, the day on which the experiment was attempted "happened to be rather damp; everything was carefully insulated, with gas-heaters distributed about the lecture-table, so as to get things as dry as possible; and yet the experiment failed."[70] After several attempts to get the experiment to work, Thomson was obliged to halt the proceedings while his assistants tried to identify the problem.

Lodge related the anecdote because he "really knew what was the matter, and knew of a way to remedy it, but I lost the opportunity and held my tongue." Once proceedings were halted so that Thomson's assistants could try to adjust the apparatus, Lodge explained the solution to them, and when the audience returned "flashes were obtained without any difficulty, the electrodeless discharge in the bulb being brighter and more vigorous, to the evident astonishment of Thomson—it was probably stronger than he had seen before." Lodge "always regretted that I hadn't the prompt energy or, so to speak, impudence to get up and make this arrangement in front of the audience; for I was at that time very familiar with Leyden jar discharges, and it would have been an object lesson that would have attracted attention."[71] Lodge was "familiar with Leyden jar discharges" because he had been experimenting and performing with them from the late 1880s on. In a lecture at the Royal Institution in 1889, the "walls of the lecture-theatre, which was metallically coated, flashed and sparked, in sympathy with the waves which were being emitted by the oscillations on the lecture-table—an incident which must be remembered by many of those present."[72]

Lodge's insinuation was that he was better than Thomson at getting this kind of experiment to work because his sense of discipline embodied a more intimate connection with the experimental culture of "seeing and feeling." Instruments like the induction coil and its attendant apparatus needed to carry a bodily culture with them as they moved back and forth between different sorts of disciplinary spaces. As Lodge noted though, his performance at the Physical Society had its own afterlife. Thomson "must have examined the arrangement, and been struck by it, for he has a chapter, in his book on *The Electric Discharge through Gases*, giving the obvious theory of it and setting forth its advantages."[73] Thomson also discussed a similar arrangement in his *Notes on Recent Researches in Electricity and Magnetism.*[74] Robert Strutt remarked in his biography that Thomson "seems to have been originally excited by the beautiful experiments of Sir William Crookes." When Strutt "once made some remark to the effect that I liked the style of Crookes' papers, J.J. partially agreed, but said he thought they wanted editing by someone who would have cut out what he (Crookes) probably considered the finest passages!"[75] The remark reflects Thomson's own understanding that experiments like these needed to be differently embodied (rather than disembodied) in different performative settings.

BODY MATTERS

Conventional accounts of the disciplinary development of Victorian physics have tended to focus on a chronology that charts physics' development away from one kind of embodiment toward another. Physics might have started the nineteenth century as a science of sensation, but by the end of the period it was an exemplary instance of disciplined practice. My argument here is that there never was a settled and uncontested sense of discipline in Victorian physics. In different places and contexts, a variety of accounts of what physics was—or ought to be—as a disciplined body of knowledge and practitioners was articulated. Different accounts were offered of the space physics occupied with regard to other disciplines. Grove developed an account of physics' disciplinary place that emphasized its over-disciplinarity—it was a discipline designed to discipline other disciplines. Tyndall's performative discipline at the Royal Institution positioned his physics as a practice that worked with spectacle to create as exemplary aesthetic both for audiences and other disciplines. Maxwell worked to embody physics as a practice that integrated laboratory discipline with the mental discipline of liberal education, while Rayleigh's physics embodied a view of disciplined experiment that was explicitly designed to be exemplary across and between disciplines. Lodge's ether physics offered itself as a different kind of over-discipline built around technologies of display and sensation. These models of physics' discipline were complementary and competing, rather than chronological, and they all embodied different perceptions of where physics belonged in relation to the broader field of disciplinary knowledge.

So maybe the most fruitful way of thinking about the divisions among different disciplinary models of physics throughout the nineteenth century is to recognize them as clashes and accommodations among different sorts of bodily practices and performances as ways of making and presenting knowledge, rather than as attempts at displacement, or as chronological developments. The different disciplinary cultures of Victorian physics embodied different notions of how authority might be demonstrated through the body and inhabited competing centers of authority. They captured different notions of how physics' orientation to other disciplined ways of knowing might be understood. Their work was done in different spaces and for different audiences. But it is important to recognize that these different spaces and audiences were not entirely insulated from each other. It seems clear, as both coils and sensitive flames indicate, that skills and instruments moved around. These movements were not the conventionally understood ones of dissemination from laboratory to theater, either. They were artful borrowings and transformations between cultures of experiment that occupied different spaces and spoke to different audiences. It might be worth thinking more, as well, about those audiences and how they, as well as participants and instruments, moved around the different, competing, and complementary spaces of Victorian physics.

IV

RELUCTANT COLLABORATIONS

Chapter 7

ANTHROPOLOGY, PREHISTORY, AND PALEONTOLOGY AS CROSS-DISCIPLINARY ENDEAVORS

CHRIS MANIAS

At the 1897 meeting of the British Association for the Advancement of Science, held that year in Toronto, the archaeological luminary John Evans (1823–1908) delivered the presidential address on "Archaeology and the Antiquity of Man" (something that he felt showed "recognition by this association of the value of archaeology as a science"[1]). Almost forty years before Evans had been one of the leading figures in the "establishment of human antiquity," the intellectual and scholarly drive that had cemented the idea of human "prehistory," placing early humans within geological timescales. Evans drew attention to how this process had not just depended on his own nascent specialism of archaeology, but on the collaboration of a range of different disciplinary approaches. As fossils, artifacts, and ethnographic materials were compared and aligned, scholars combined methods and sources to construct a long developmental past that linked human societies and the natural world. Considerations of the deep human past involved reflections on changes in the landscape and shifts in animal fauna that "the archaeologist pure and simple is incompetent to deal with," and so collaboration with "geology and paleontology becomes absolutely imperative."[2] Different forms of evidence required different disciplines with different forms of expertise, and a careful division of labor was required in order to check possible errors and misinterpretations: "If left to himself the archaeologist seems too prone to build up theories founded upon form alone, irrespective of geological conditions. The geologist, unaccustomed to archaeological details, may readily fail to see the difference between the results of the operations of Nature and those of art, and may be liable to trace the effects of man's handiwork in the chipping, bruising, and wearing which in all ages result from natural forces; but the united labors of the two, checked by those of the paleontologist, can not do otherwise than lead toward sound conclusions."[3] Further "aid" was sought from ethnology and anthropology, with the prehistoric peoples of Europe aligned through "the comparative method" with modern populations asserted as being at a "prehistoric" level of civilization. For understanding humanity's early condition, these fields were connected through the shared examination of these questions, and simultaneously marked by their own special skills but also defined by their own respective omissons and ignorance. Collaboration was crucial, but this collaboration required that these disciplines were themselves clearly marked and carried out by particular specialists.

The disciplines of anthropology, prehistoric archaeology, and vertebrate paleontology have often been traced to Victorian processes and debates around issues like the "discovery of human antiquity," the rise of racial and colonial science, and the formation of new scientific networks and institutions.[4] As Adelene Buckland has recently argued, these disciplinary consolidations often occurred "as means of organizing and making sense of the chaos of the past," and required constant choices and occlusions.[5] A standard narrative is that the Victorian period saw the decline of polymathic and amateur scientific work and the rise of disciplinary specialization, professionalization, and institutionalization. This can be seen particularly in the growth of increasingly specialist learned associations—such as the Geological Society of London, the various associations devoted to archaeological matters, and the formation of disciplines around the "science of man," such as the Ethnological and Anthropological Societies of London and their eventual coalescence into the Anthropological Institute of Great Britain and Ireland.[6] Museums devoted to natural historical and ethnological matters were established in increasing numbers throughout the century, and particularly toward its close with the rise of the "new museum movement."[7] Meanwhile, chairs and university professorships gave formal status to particular subjects, with history, archaeology, and anthropology all being established as professorships at numerous universities over the course of the century.

This increase in disciplinary specialization and professional institutionalization can certainly be identified in much nineteenth-century intellectual life. However, it was not necessarily detrimental to work across the developing disciplines. These new institutions and networks often drove cross-disciplinary work, as people and audiences were linked by common interests, loyalties, personal relationships, and agendas to combine nascent specialisms and engage with connected problems. As knowledge became increasingly divided into separate "branches," it was frequently asserted that collaborative work between specialists was essential for new understandings—and to give credence to controversial finds or projects.

However, it is important to remember that these processes of cross-disciplinary correlation and alliance building were not necessarily harmonious or easy. Often the aligning of material and evidence required demarcations of expertise, the finding of "neutral ground," and the willingness to accept where one form of authority ended and another began. Calls for cross-disciplinary collaboration could be undertaken with the rhetoric of inclusion and collaboration, but could also be underlaid by an annexationist desire to claim domains of knowledge and subjects of interest as under the authority of a particular figure or institution. Victorian claims to work across disciplines could therefore frequently be a mode of conflict and division, as much as a spur to unitary projects of collaborative work.

This chapter will examine how these dynamics operated in three instances, where relations between the deep past and the human sciences intersected in particularly dramatic ways and which illustrate some distinct manners through which cross-disciplinary work could simultaneously be productive but also lead to annexationist claims and demarcations of expertise. First, it will examine some of the key attempts to connect geological, archaeological, and paleontological approaches in the late 1850s and early 1860s, during the classic period of the "Establishment of Human Antiquity," examining how Joseph Prestwich (1812–1896) and John Evans presented their combined researches on ancient human stone tools to the Royal Society and the Society of Antiquaries. As they simultaneously crossed and conceived of the borders between geology and archaeology, they brought the human and natural pasts into dialogue, while nevertheless subjecting them to the expertise of distinct fields. Next, the chapter will move to a case of disciplinary rhetoric and institutionalization around the "Sciences of Man" in the 1860s, examining how James Hunt (1833–1869) and other figures associated with the Anthropological Society of London attempted to align other "branches of knowledge" behind their new anthropological project. This presents a contrasting instance of how cross-disciplinary work involved the mobilization of authority within a conflict that entailed the belittling of rivals, rather than an attempt to build cooperative links between acknowledged specialists. Finally, it will move into the latter part of the century, examining how William Boyd Dawkins (1837–1929), a key figure in mid- and late Victorian geology, prehistory, and paleontology, presented and related the deep natural and human past in his writings and museum arrangements. This case shows a further variation in cross-disciplinary work, with the human and natural sciences being linked into a single project as grading into one another in a way that humanized the natural world and naturalized many human societies, but that also faced difficulties from institutional interests in museum and scholarly work.

These three case studies enable this chapter to show some of the intellectual and methodological drives to cross-disciplinary work in the Victorian period and how it became a necessary part of promoting new knowledge about the past, while claiming authority for specific disciplines and institutions. The human and the deep-time sciences were particularly important in these processes, as these fields were often controversial and required multiple forms of evidence, while also being dependent on learned society and museum contexts, drawing on institutions that were often themselves new and fluid. The examples traced, ranging from the collaborative to the oppositional to the relational, show the different strategies through which cross-disciplinary work in the Victorian human and deep-time sciences could be pursued and how this simultaneously linked and reified different approaches.

ESTABLISHING HUMAN ANTIQUITY ACROSS THE DISCIPLINES

The establishment of human antiquity was one of the greatest conceptual changes in mid-Victorian society,[8] consisting of a range of projects in the mid-nineteenth century that asserted that human existence was not just confined to the six thousand years that could be deduced from scriptural chronologies, but in fact stretched deep into the geological and paleontological past that had been already established by the deep-time sciences. A series of finds in the late 1850s and 1860s drove this change forward, most notably the reevaluation of the collections of the French customs official Jacques Boucher de Crèvecoeur de Perthes (1788–1868), the excavations by the Brixham Cave Committee in the south of the United Kingdom, and the study of the Cro-Magnon site in southern France by Henry Christy (1810–1865) and Édouard Lartet (1801–1871). These all demonstrated that human stone tools (and in the case of Cro-Magnon, skeletal remains) could be found in ancient geological strata, alongside the fossils of extinct Pleistocene animals like reindeer, aurochs, mammoth, and hyena. This fundamentally reevaluated chronologies of human existence, placing humans within the deep processes of the natural world. Indeed, the significance of this shift is such that historians increasingly talk about a "mid-Victorian time revolution," where the linking together of geological and human chronologies defined new ways of discussing and understanding the human past and its relationship to nature, and potentially being as significant as the promotion of Darwinian evolution.[9]

What is important for the purposes of this chapter is that establishing human antiquity was dependent on the cooperation of a variety of different forms of evidence and developing disciplinary approaches. On the one hand, it required the alignment of evidence from a range of different fields, which were themselves often unstable and in the process of being defined during this period. These included geology, which had identified the layers of rock in the earth as corresponding to particular eras in earth's history; paleontology, which had identified the remains of extinct beasts, often adapted for quite different climates and conditions, in the earth's strata; archaeology, which had defined stone tools and handaxes as artifacts made by human hands; and physical studies of human diversity, which led to rare remains found in prehistoric sites to be classed as belonging to particular human "types" and "races," and driving the comparison of stone tools with objects produced by modern peoples asserted as existing in similar social conditions. As a result, the analysis of prehistoric sites depended on the coalescence of methods and approaches.

As well as linking specialists, the establishment of human antiquity also required recognition by the scientific communities attached to these different approaches, and so depended on cross-disciplinary collaboration. Indeed, Donald Grayson has drawn attention not only to the suddenness of the establishment of human antiquity, but also to how it took the form of a concerted publicity campaign among some of the leading London learned societies, describing it as having "the appearance of a coordinated onslaught" by the members of the cave committee attached to the Geological Society of London. The new prehistoric finds were presented to a number of leading British scientific associations, including ones with a general remit, like the Royal Institution and Royal Society, and more specific societies, like the Geological Society of London and the Society of Antiquaries. Comparing and contrasting some of these presentations can therefore show us attempts at and the limits of cross-disciplinary work, as a group of scholars aimed to explain this new field of research and the evidence required to their respective scholarly communities.

The longest account was given by Joseph Prestwich to the Royal Society—a notably broad scientific association—and was delivered on May 26, 1859, and published (in edited form) in the society's *Transactions* in

1860.[10] Prestwich was a wine merchant with strong interests in geology who had extensive contacts in France owing to his professional business.[11] He was also greatly involved in learned society life in London and farther afield, having been active in the Geological Society of London since the 1830s and being a member of the cave committee established to systematically excavate Pleistocene caves around Britain for evidence of human artifacts.[12] Prestwich's presentation went through the history of the human antiquity question, the work on the British caves, and his visits to some of the key sites in France, including Abbeville, St. Acheul, and Amiens. His argument was that finds from these sites could be correlated to provide all the necessary evidence to overturn the idea "that man did not exist until after the last of our geological changes and until after the dying out of the great extinct mammals, [which] had become almost a point of established belief."[13]

Evidence was drawn from a wide range of perspectives. Geological evidence, particularly the dating of the strata and caves in which human flint tools had been found, was brought in extensively. The tools themselves were divided into three types: flakes, points, and almond- or oval-shaped implements. The flints were described as all being worn in single directions, which implied intentionality and human working, not randomly chipped as might be expected from natural weathering. Knowledge was also drawn from technical manufacturing as well as gentlemanly science, with Prestwich noting how "as the gun-flint makers observe, 'flint has no grain,'"[14] which meant that "nothing surely but the human hand could have directed a series of blows not only parallel with one another on each surface, but also carefully directed along and maintained on one given plane."[15] Expertise was also invoked as to whether the flint implements could have been made later, and then driven into old layers of the ground: "To anybody accustomed to the examination of drift deposits, there is little difficulty in distinguishing between the fresh and uniform appearance of undisturbed beds, and the mixed and confused make of made ground, independently of the occurrence of any charred materials, pottery, &c., and of bones in a comparatively fresh condition. The lines of original stratification once broken cannot be so restored as not to show the break of continuity."[16] Cross-disciplinary and cross-associational networks were also crucial for the location of additional evidence. Prestwich wrote how he had been informed by John Evans of an earlier paper read at the Society of Antiquaries over fifty years before,[17] which "although known to antiquaries, its geological bearings had escaped notice."[18] In the 1790s the antiquarian John Frere (1740–1807), working in the village of Hoxne, had found flint remains alongside "some extraordinary bones, particularly a jaw-bone of enormous size of some unknown animal" that might date "even beyond that of the present world." Evans noted that Frere was "an antiquary unfettered by geological theories,"[19] which gave him the ability to reach conclusions on the age of the earth and human artifacts that the geologists of the time, who were still locked in a young chronology for human history, could not make—and which at the time went unrecognized. Continuing this work, Prestwich visited the rubbish heap of the still-worked brick pit at Hoxne, where the workmen showed him the bones of animals (including horse and elephant) and informed him that they had found more stone artefacts at the site in the course of their work (although Prestwich failed to find any himself).

Prestwich's article also included two appendixes, presenting letters from John Evans, representing the study of artefacts, and the paleontologist and natural historian Hugh Falconer (1808–1865), representing the study of fossils, to give further credence to the discoveries. Evans used ethnographic analogies to note how flint is "so extensively employed for the manufacture of weapons and implements by uncivilized man in all ages and in all countries where flints are to be found,"[20] raising implications on the grade of civilization of the ancient inhabitants of Britain and France, while corroborating the finds. However, his conclusion referred to both expertise and disciplinary demarcation and raised some of the potentials, but also some of the tensions, in this project:

> There appear to have been one or two other similar discoveries in England, and more would probably have been recorded, had not the rudeness of the workmanship of the weapons been such that they would hardly attract the eye of an ordinary observer, while those scientific persons who have been engaged in the investigation of the drift, have been more on the alert for fossil organisms than for traces of the hand of man. These diluvial beds will, however, now form a point of union on which both the geologist and the antiquary may prosecute their inquiries together; and on this neutral territory between Paleontology and Archaeology a wide field is opened for investigation, which must eventually lead to a great extension of our knowledge of the history of primeval Man.[21]

In this way, cross-disciplinary work was crucial to the study of these remains, but there were clear demarcations in this project, and "neutral ground" needed to be found. Both the geologist and the antiquary would combine their understandings to develop new knowledge of "primeval man," but they would be distinctly following their own research abilities and ceding interpretative authority to the other when analyzing objects under their purview.

John Evans developed these analogies even further in the lecture he delivered to the Society of Antiquaries, the oldest of the British societies devoted to the study of human artifacts, on June 2, 1859.[22] Evans was a paper manufacturer from the southeast of England who had been highly interested in antiquarian matters since the 1840s and in geology since the 1850s.[23] His friendship with Prestwich, however, derived from his professional work, with the two meeting in a legal case over water rights in 1855 (for which they were serving as expert witnesses, on opposite sides of the case). Following this, they worked closely on scientific matters, traveling to France in 1859 and corresponding on scientific work around prehistoric material, although with a self-conscious division of labor, knowledge and expertise.

Evans's lecture opened with a long discussion on the relationships between the two different disciplines, and how they had linked to deal with the question of human antiquity:

> The natural connection between Geology and Archaeology has at various times been pointed out by more than one writer on each subject; and it must, indeed, be apparent to all who consider that both sciences treat of time past as compared with time present. The one, indeed, merges by almost imperceptible degrees in the other; while the object of both is, from the examination of ancient remains, to recall into an ideal existence days long since passed away, to trace the conditions of a previous state of things, and, as it were, to repeople the earth with its former inhabitants.[24]

Geology and archaeology had important connections and similar agendas of understanding the ancient earth, and the boundaries between them were often very hazy. The two subjects were even described in fraternal (if hierarchical) relations: "Geology is, in fact, but an older brother of archaeology, and it is therefore by no means surprising to find that the one may occasionally lend the other brotherly assistance."[25]

There were, however, significant differences in methodology and in degrees of confidence with which Evans spoke of the material. He largely felt it unimportant to go into depth on the geological character of the ancient landscapes, noting that "Mr Prestwich, in the able Memoir upon this subject which he has communicated to the Royal Society, has gone so fully into the geological features of this part of the valley of the Somme, that any further details are needless."[26] However, when turning to the objects themselves, Evans spoke of the same three types of implements as Prestwich, but was much more confident as to what their purpose may have been, naming them:

1. Flint flakes, apparently intended for arrow-heads or knives.
2. Pointed weapons, some probably lance or spear-heads.
3. Oval or almond-shaped implements, presenting a cutting edge all round.[27]

The third were the more mysterious: "As to the use which this class of flint implements from the drift was originally intended to fulfil, it is hard to speculate. The workmen who find them usually consider them to have been sling-stones."[28] The human manufacture of the flints was highlighted not just through observation, but also experimentation. Evans noted how "the manner in which they have been fashioned appears to have been by blows from a rounded pebble mounted as a hammer, administered directly upon the edge of the implements, so as to strike off flakes on either side. At all events I have by this means reproduced some of the forms in flint, and the edges of the implements thus made present precisely the same character of fracture as those from the drift."[29] This again fed into the description and discussion of the material. Understanding it required the coalescence of a variety of forms of knowledge and expertise—both scholarly and practical—all bringing to light questions and issues of "primitive man."

These assertions by Evans and Prestwich of the establishment of human antiquity, of the different branches of knowledge coalescing around the same conclusions and developing a new conception of the human past, provided a common vision and also institutional credibility. This grew from material analysis, theoretical perspectives, and discussion with a wide range of individuals, such as the workmen who were the main diggers of material (and who often identified it long before the antiquarians and geologists themselves), or through experimentation in flint knapping. This meant that the "coordinated onslaught" of the initial promotion of human antiquity became a means of reconciling different claims to knowledge and forms of evidence, while ensuring that knowledge was demarcated according to acknowledged forms of expertise.

INTERDISCIPLINARITY AS CONFLICT IN MID-VICTORIAN ANTHROPOLOGY

Occurring at the same time and in a linked manner with the establishment of human antiquity, the division between the Ethnological Society of London and the Anthropological Society of London is one of the most iconic episodes in the history of Victorian science and ideas of human difference. The outlines of this are well known and have been told many times. The Ethnological Society of London, founded in 1843 out of the older Aboriginal Protection Society, developed as the leading society in Britain for studying the "sciences of man," with members and publications drawing on cross-disciplinary agendas linking philology, geography, anatomy, and antiquarianism as much as ethnographic observations. In 1863 however there was a dramatic split, as a faction within the society under Dr. James Hunt broke off to form the new Anthropological Society of London. The split heralded almost a decade of bitter conflict between the two groups over who had authority over the science of man, which was only eventually resolved in 1871 after the death of Hunt and the merger of the two societies into the Anthropological Institute of Great Britain and Ireland.

The split has been analyzed in many different ways. Older works by John Burrow and Ronald Rainger have emphasized the intense racialism of the new "Anthropologicals," often using them as an example of a hardening of racial discourse in the mid-nineteenth century.[30] More recent studies by Sadiah Qureshi and Efram Sera-Shriar have looked at the scientific, public, and institutional elements of the controversy, often seeing the conflict as being about rival claims to scientific authority and control over publications and institutions, with important areas of consensus as well as disagreement.[31] Meanwhile others, such as Peter Mandler, have drawn attention to the eventual merger and defeat of the Anthropologicals as an example to show that hard racial science failed to gain traction in Britain.[32]

This section will examine a different point in some of the rhetoric underlying the attempted establishment of "anthropology" by the Anthropological Society of London in the mid-nineteenth century. As well as claiming to be driving forward a new science (and opposing "old dogmas"), the Anthropologicals present a marked instance of how the language of linking disciplines was deployed to attempt to establish authority for a new institution. In this instance, claims to work across disciplines were not simply a means of promoting cross-associational cooperation or bringing distinct forms of knowledge to bear on a scholarly issue, but were about institutional power. This on the one hand showed a vision where cooperation and ordering were crucial to scientific work. But it also implied that this disciplinary ordering should be structured and hierarchical to allow anthropology to dominate as many aspects of the sciences of man as possible. While claims to bridge disciplines and use approaches from many lines of inquiry were very similar to those presented by many in the Ethnological Society (both in the 1860s and much earlier),[33] the starkness of the Anthropological Society's attempts provides an important instance of interdisciplinarity being mobilized for conflict as well as collaboration.

The drive of the society toward consolidating disciplines can be seen in some of Hunt's key pronouncements to the society. In his opening "Introductory Address on the Study of Anthropology," delivered on February 24, 1863, Hunt made a series of bold claims that disciplinary subordination was the key distinctive feature of the new anthropological approach. Anthropology was not just to be a study of "the relations of Man to the Mammalia,"[34] as he claimed it had been described by ethnologists like Robert Gordon Latham (1812–1888), but was to be much wider. Instead, he clearly asserted the cross-disciplinary nature of the field:

> I would not have it supposed that the science of Anthropology has any right to be confined to such limits. Anthropology is, on the contrary, the science of the whole nature of Man. With such a meaning it will include nearly the whole circle of sciences. Biology, anatomy, chemistry, natural philosophy, and physiology must all furnish the anthropologist with materials from which he may make his deductions. While Ethnology treats of the history or science of nations or races, we have to deal with the origin and development of humanity. So while Ethnography traces the position and arts of the different races of Man, it is our business to investigate the laws regulating the distribution of mankind.[35]

This passage is interesting for arguing that the project of anthropology brought disciplines and approaches together, and distinguished it from earlier approaches of ethnology and ethnography, which were called narrower and more limited. This is of course highly rhetorical and involves some degree of misrepresentation, particularly as earlier ethnographic writers, such as James Cowles Prichard (1786–1848) and Latham himself, also frequently mixed approaches from a whole host of fields, including philology, anatomy, physiology, and history.[36] Yet this was part of the Anthropological presentation of the relations between disciplines in a directly annexationist manner, seeking to subsume factional and institutional rivals.

The cross-disciplinary unification attempted by the Anthropologicals also involved ordering distinct specialisms into an organized framework for understanding humanity. This on the one hand took the form of a new mode of research, but on the other required cooperation within the learned society context. In his 1864 address Hunt described how "we are, indeed, trying to do something more than founding a new society; we are endeavoring to found a new science,"[37] and asserted:

> Anthropology includes every science which bears directly on the science of man or mankind, and includes Anatomy, Physiology, Psychology, Ethnography, Ethnology, Philology, History, Archaeology and

Paleontology as applied to man. Take either of these branches of science away, and we can no longer form a veritable science of man. But it may be asked, 'Is it proposed to do the duty that is now attempted by societies that are devoted to some of the objects?' Certainly not; we only make use of these sciences so far as they will throw light on the past, the present, and the probable future of the human family. The philologist has hitherto been working in ignorance of the results of the physiologist; the historian in ignorance of the deduction of the ethnologist; and archaeology and ethnography have hitherto been supposed to be two distinct sciences; while the psychologist and anatomist have had a mutual contempt for the deductions of each other. It is to remove these anomalies that we have formed ourselves together, with a determination of hearing all sides of the question, and examining the evidence of each special student in a perfectly unbiassed manner.[38]

This is in some respects similar to the mixture and collaboration of antiquarian and geological knowledge presented in the establishment of human antiquity, with specialists aligning their work to engage with a particular problem in new ways and overcome barriers, fixed ideas, and omissions. However, Hunt's agenda illustrates some of the tensions involved, as this project of cross-disciplinary ordering impinged on demarcations between already existing independent societies and networks.

This is borne out even further in additional statements by Hunt, which argued that this project of anthropological consolidation was as much a process of redistribution of methods and problems as it was one of unification. In his 1866 Anniversary Address Hunt attempted to simultaneously include the whole field of human prehistory under the rubric of anthropology (which followed the large number of prehistoric papers read at the society's meetings), while arguing that there was danger in the field becoming ill-defined and needing to balance its cross-methodological diversity with a common sense of purpose. Noting "how advisable it has become for all anthropologists to possess some clear conception and definition of the objects and limits of their science," he sketched a range of subdivisions for anthropology:

1. Archaic anthropology, or the past history of man, from his physical remains and works.
2. Historical anthropology, or the past history of mankind, as deduced from mythology, creeds, superstitions, language, traditions, etc.
3. Descriptive anthropology, or the description of man and mankind.
4. Comparative anthropology, or the comparison of different men and dif-

> ferent races of men with one another in the first place, and a comparison of man with the lower animals in the second.
>
> The questions then arise; do these subdivisions all go to make up one science which has a center within itself? Can any of these divisions be taken away and a veritable science yet remain? Do these divisions include the whole science of man?[39]

This was simultaneously an attempt to clarify the expertise required for the overall science of man, an attempt to subsume dramatic and important new developments, like prehistoric archaeology under the domain of anthropology, and ensure that what was called anthropology remained clear and consistent. Here archaeology was taken as a potential warning, with Hunt asserting that it had begun to take on a very vague meaning, referring to all studies of the material past without an overarching method or purview. He expressly noted that "in using the term archaic anthropology we must guard against giving it such a vague meaning as archaeology has now acquired,"[40] and contrasted this with the desired anthropological project, which was to create a shared set of methods to understand laws relating to human existence.

Rather than search for "neutral ground" between different experts, the rhetoric of the London Anthropologicals was predicated on merging different branches of knowledge within a single analytical framework. However, while the Anthropological Society itself managed to accommodate numerous positions and studies, it was only varyingly successful. Hunt's attempts, for example, to claim the whole field of human prehistory as "archaic anthropology" failed, with the definition of this field as "prehistory" or "prehistoric archaeology" becoming far more conventional. More broadly, following the unification of the Ethnological and Anthropological Societies in 1871, the new Anthropological Institute continued to mix different approaches, focuses, and specialisms. However, its rhetoric and goals were not so overtly annexationist, being more implied within its research projects. Cross-disciplinary work continued, but in the absence of such a stark conflict, the claims that one association should provide the node and ordering principles for all knowledge of humanity faded into the background.

UNDERSTANDING PLEISTOCENE BRITAIN

The above two sections have looked at some of the institutional collaborations and tensions within metropolitan learned societies over the deep past and human diversity. But the sciences of the past and humanity were also intermixed in individual careers. Here, the Manchester geologist and prehistorian William Boyd Dawkins, presents a highly illustrative case, showing how a single individual could link together a range of sciences across a career of field research, publication, learned society and university activity, and museum building. Dawkins was one of the foremost scientists in late nineteenth-century prehistory and geology. Initially trained as a geologist, he devoted much of his early career to tracing the implications of the mid-Victorian time revolution for the ancient British fauna and the assumed character of prehistoric British populations. Two important works—*Cave Hunting* (1874) and *Early Man in Britain and His Place in the Tertiary Era* (1880)—aimed to explain the prehistoric past to both scholarly and learned public audiences, and established Dawkins's reputation in this area. He also worked between economic geology and education in Manchester, where he promoted geological and natural historical education, becoming the first professor of geology at Owens College in 1874. In this way, Dawkins mixed studies and expertise in the various deep-time sciences across his professional work and leisure activity.

Dawkins's conception of disciplinary division was one informed by stratigraphic ideas from geology, which served as a way of conceptualizing the links among the different ways of knowing the ancient past. In the introduction to *Early Man in Britain*, he noted how, "of the many fields of inquiry opened out by the intense mental activity of this century, there is none which promises to be more fruitful than that which has been won by the joint labors of the geologist, the student of prehistoric archaeology, and the historian."[41] Dawkins then went on to poetically describe the contributions of these different fields. First, the geologist "tells of continents submerged, and of ocean bottoms lifted up to become mountains; and he points out to us that side by side with the ever-changing conditions of life there were corresponding changes in the living forms."[42] Next came the archaeologist, who has "raised the study of antiquities to the rank of a science by the use of a purely inductive method, and have accumulated materials which enable us to establish a tolerably complete sequence of events from the remote past in which man stands in the geological foreground down to the borders of history."[43] Finally, "the writers of history . . . have carefully sifted the true from the false, the certain from the uncertain, in the records of this country, and have consolidated, so to speak, their domain, so that it can be used by the archaeologist as a base for the conquest of what lies beyond."[44] These methodological approaches all linked together for Dawkins, providing sequential means of understanding the ancient past.

More importantly, though, these subjects were not just linked by a single object of study, but were also connected by a vision of progress that connected humanity and nature. Purely historical accounts were only of limited use for understanding this, as "in the pages of the historian, man appears in the high state of civilization marked by the use of letters, and the written record is silent as to his progress up to that point."[45] Other subjects enabled progress and development to be traced even further back. Archaeology showed "the steps by which man slowly freed himself from the bondage of the natural conditions under which all other creatures live,"[46] while paleontology and geology showed "group after group of animals and plants . . . each connected with that which preceded it, and each becoming more and more highly organized, until man appears the last born as well as the highest and the noblest creature."[47] In this way, the layering of time and the unitary idea of progress gave justification and a rationale to cross-disciplinary linkage.

Dawkins's works were structured around a layered concept similar to that of geological strata, with the fields devoted to understanding the past all grading into a single narrative (much as Evans had suggested how each of the two fields of geology and archaeology "merges by almost imperceptible degrees in the other"). There was just one past, but understanding the whole required distinct methods. Notably, though, Dawkins only considered himself capable of elucidating the geological and prehistoric pasts, with his books, lectures, and public speeches all focusing on deep time and ending usually with the neolithic (or in some rare cases, the Roman) period. Historic analysis required a division of labor: an often-cited story relating to his friendship with the (primarily Anglo-Saxonist) historian John Richard Green (1837–1883) states that while the two were students together at Oxford they resolved that Dawkins would study the deepest past, while Green would focus on the early periods of history.

Dawkins's major published works followed this line of layered time very closely. The 1874 work *Cave Hunting* started with "historic caves" and moved back into the deep past, and his *Early Man in Britain* (1880) began with the early Tertiary and moved forward into archaeological and human periods. The books linked together a wide range of evidence and methods, describing current and prehistoric landscapes, the fossils of animals and plants, the material artifacts of humans who may have lived in particular periods, and the rare human skeletal remains, which were analyzed according to conventions of racial anthropology. In this way disciplines and evidences were combined to present the development of prehistoric Britain and Europe from the depths of the past until the present.

Dawkins's treatment of humans shows how far these methods and concepts could be interrelated. His understanding of human prehistory in Europe was of a series of distinct racial groups, each related to particular environments but also bearing with them a particular set of technologies, cultural forms, physical features, and association with a specific fauna. First were the "river-drift" people, who were assessed from stone tools and the diverse animals found alongside them as being "a nomad hunter, poorly equipped for the struggle of life, without knowledge of metals, and ignorant of the art of grinding his stone tools to a sharp edge."[48] While Dawkins acknowledged that virtually no human skeletal remains had been found from the river-drift period, which made physical anthropological studies impossible, this did not prevent him from making ethnographic analogies. Comparison with similar tools found in North Africa, the Middle East, and India led him to conclude that this people had been spread across a wide area. More important though were the paleontological links, as they lived alongside animals "either of the temperate or southern fauna of Europe," including elephants, rhinos, lions, and hippos. This ensured that they were understood as a warm-adapted group, associated with these southern creatures, and that "we cannot refer them to any branch of the human race now alive, and they are as completely extinct."[49]

Next came the "cavemen," who possessed more sophisticated tools of bone, horn, and antler, and made artistic depictions of Pleistocene animals like reindeer, horse, and mammoth. Here racial comparisons were made much more explicitly, with both the subject matter and the style of artworks produced by the cavemen and the modern "Eskimo" judged as highly similar. Dawkins noted, "The most astonishing bond of union between the Cave-men and the Eskimos is the art of representing animals. Just as the former engraved bison, horses, mammoths, and other creatures familiar to them, so do the latter represent the animals upon which they depend for food."[50] The linking of cultural stages in human prehistoric development with current ethnology is a consistent theme in Victorian prehistory and anthropology. Dawkins presents an extreme variant of this: the prehistoric and modern peoples were not just at similar stages of development or living under similar conditions, but were exactly the same people, linked through climate and racial typology. The dynamics of human migration were understood in ways drawn from zoological geography and animal migration, and connected with ideas of climate, migration, and extinction. Thus the river-drift people were associated with the warm southern fauna of elephants, hippos, and lions, and receded from Europe once conditions began to change. The cavemen were associated with the Arctic fauna of reindeer and bison and followed their patterns of migration to persist in the modern Arctic, marking modern Arctic peoples as relics of the past.

Importantly, Dawkins's notion of layered time was not just confined to his writings, but also found physical embodiment at the end of the nineteenth century in the creation of the Manchester Museum. The museum, which has been described in depth in an excellent work by Sam Alberti as marked by conflicts between developing disciplines and areas of collection,[51] was formed within Owens College. The building (designed by Alfred Waterhouse, the architect of the British Museum of Natural History in London) was constructed through public subscription and opened to the public in 1888. The museum was free of charge, and by the early 1890s it was receiving on average seventy to eighty visitors per day, as well as hosting numerous lectures and study visits.[52] Dawkins's original plan for the museum echoed his ideas of the links among the geological, animal, and human worlds. The museum was organized according to a scale, with an initial section, Ancient History of the Earth, showing the earth's development through minerals, rocks, and the paleontology of Primary, Secondary, and Tertiary life, which then led to the galleries of the Modern History of the Earth, examining plants, animals, and humans.[53] The museum therefore moved through geological displays, to fossils, to extant creatures, and finally to human cultures, with both nature and humanity linked along a single developmental trajectory.

The core principle of the museum arrangement was that "there is no break of sufficient magnitude in the Tertiary fauna or flora to show the line where geology ends and history begins, and the continuity is so marked that the present face of nature may be viewed as the current but not the last of the stages of evolution during the Tertiary period."[54] The Tertiary exhibits were relatively few in number, but showed some of the most important specimens in the collection, including casts of Eocene mammals from the Paris basin, casts of proboscidean fossils from around the world (including "a series of Teeth, illustrating the ancestry of the elephant"[55]), and a large number of Pleistocene specimens, including elk, mammoth, reindeer, hyena, and cave bear remains. The galleries ended with human objects, showing casts of carved and engraved bones, tusks, and antlers from the Caves of Perigord, illustrating the "Dawn of Art" with "Modern Eskimo Implements for comparison."[56] These collections clearly showed Dawkins's stamp, very much following his notions of layered time and direct racial comparisons, and drawing on the dicta in his popular works that the different ways of knowing the prehistoric past graded into one another. Indeed, he described the arrangement as presenting "the only Museum in Britain in which the continuity of Geology and history is clearly set forth. You walk straight from the Tertiary collections to those of Crete and Egypt—the one into the other!"[57] In this way the museum realized the concept that all evidence and approaches to the past were linked and corresponded with one another, and together showed common development across the deep past. Yet like the geological strata of the earth, these approaches were still distinct in form and composition, despite forming a single structure.

The arrangement of the Manchester Museum did not persist. In the 1920s the museum was completely rearranged, being split up into more conventional departmental divisions based on particular fields of research. There were various reasons for this. The individual curators of the different departments were increasingly protective of their own collections and demanded space and autonomy for their displays. And new collections were organized and expanded—most notably the large ancient Egyptian holdings—which were popular and gained funding but did not fit so well into the progressionist scheme.[58] This reorganization was something that Dawkins intensely opposed:

> The proposal now under discussion to remove the Tertiary group from its place between Zoology and Egypt, will diminish the teaching value of the Museum to the University in Zoology and Paleontology and its attraction to the public. It will destroy the unique character of the Museum viz.—the proof of the continuity of Geology with History, and of the incoming of man in the Tertiary period—resulting from recent discoveries. It will also interfere with the arrangement of the section of Applied Geology, so much needed by Engineers and Architects.[59]

The single vision of a layered past therefore broke down in the face of changing institutional structures.

However, it would be wrong to say that the forms of cross-disciplinary work examined in this chapter were doomed in the face of even greater specialization in the twentieth century. Synthetic texts, like H. G. Wells's *The Outline of History* (1920), continued to reach large public audiences, and debates in learned societies continued to link the human and the natural pasts. The notion that understandings of the geological and prehistoric pasts provided a crucial grounding for later developments in human history and required a coalescence of methods to understand was a long-standing project. However, the urgency and widespread nature of these debates began to recede, particularly in more formally organized museums and universities. It was indeed more in the popular press and voluntary societies that these older nineteenth-century trends of disciplinary linkage continued.

The instances discussed above of Victorian cross-disciplinary work in human antiquity, anthropology, and long narratives of the deep past were essential aspects of elaborating prehistoric chronologies in the nineteenth century. They were also fraught and often based as much around conflict, division, and tense demarcation as they were around collaboration and mutual support. Cross-disciplinary work often operated through personal relationships, factional politics, and idiosyncratic attempts to correlate different ways of understanding the deep past. In this respect, later fractures should not be seen as a "failure" of interdisciplinarity, but as almost implicit in these processes. The different branches of knowledge needed to be connected to provide a unitary vision, but this required negotiation and agreement. Cross-disciplinary work depended on the disciplines themselves being well defined and maintaining their own autonomy and authority. The different disciplines were connected through debate and interaction, and the "neutral ground" where specialists could work together required a common sense of purpose, shared problems, and demarcation of expertise.

Chapter 8

Disciplining the Field

The British Arctic Expedition of 1875 and the Construction of Arctic Research Specialism

Nanna Katrine Lüders Kaalund

Lieutenant Weyprecht very truly remarks that the Polar regions are the most important portions of our globe for the study of the natural sciences. . . . There is hardly a branch of natural science which is not deeply interested in Polar research. In the unknown region lie the keys to the true theory of magnetism—to that riddle in physics, the cause and nature of the aurora; to the question whether the ice of the Polar regions is not the regulator of our climatic conditions. Here, too, the sciences of astronomy and geodesy, of geology, zoology, and botany will be materially advanced. We now know just enough of nature's doings in high northern and southern latitudes to show us how important a thorough scientific investigation of these regions is to natural philosophy in all its branches.

—*Geographical Magazine*, April 1, 1876

When the Association of German Natural Philosophers and Physicists met for their forty-eighth annual meeting in September 1875, one of the topics on the agenda was the future of Arctic exploration.[1] The Austro-Hungarian naval officer Karl Weyprecht (1838–1881) had recently returned from the Arctic, but when he addressed the large audience as the second presenter on the first day of the meeting, his aim was not simply to provide an account of the results from his polar travels. Rather, Weyprecht presented a research program for a new type of polar science, and proposed the framework for what would become the First International Polar Year (IPY), which took place between 1882 and 1883.[2] Weyprecht argued that it was necessary to decouple geographical discovery from scientific research in the polar regions. In particular, he pointed to the track record of British Arctic expeditions in undertaking scientific research, which he considered rather poor. Many British proponents of Arctic research took offense to this criticism, including Clements Markham (1830–1916), the secretary of the Royal Geographical Society (RGS) who, as the editor of the *Geographical Magazine*, admonished Weyprecht's proposal in the pages of the journal.[3] In effect, Weyprecht's proposal sought to develop polar science as a distinct discipline, guided by specific methodologies. It was a program of reform that contrasted sharply both with the traditional fieldwork practices of Arctic and Antarctic travelers as well as the scientists who used their data across disciplines.

The criticism directed to Weyprecht's scheme was in part related to the decision of the British government to send out their first Arctic expedition since 1857. Known as the British Arctic Expedition of 1875, the venture was under the command of Captain George Strong Nares (1831–1915) and sought to reach the geographical North Pole. This ambitious geographical goal was the primary aim of Nares's expedition, but as had been the case with all British government-sponsored voyages to explore the Arctic since the Napoleonic Wars, scientific research was the stated secondary concern. Because this was the first Arctic expedition organized by the British government in almost twenty years, the stakes were high, and the scientific societies that had lobbied for its organization, including the RGS and Markham, were anxious to ensure that it achieved as much as possible.

In the British tradition, Arctic explorations were usually large and costly ventures. The expense and the prioritization of geographical discovery meant that the science they undertook was opportunistic, as the travelers were tasked with collecting data relating to a broad range of fields whenever possible. The opportunism translated into a fieldwork practice that was in many ways interdisciplinary by necessity. This is still the case today. For example, the director of the Scott Polar Research Institute, Julian Dowdeswell described the recent Weddell Sea expedition of 2019 to the Antarctic as being "part of an interdisciplinary science program."[4] As a specialist research field and a discipline, polar studies is inherently interdisciplinary, understood in this chapter as the "integration of one or more academic disciplines."[5] It sits between and draws on multiple other specialisms, and is unified first and foremost by its

FIGURE 8.1. The British Arctic Expedition of 1875–1876, H.M. ships stopped off at Cape Prescott. *Source:* Image held by Science Museum Group, Creative Commons Attribution NonCommercial ShareAlike 4.0 License.

geographical focus on the Arctic and Antarctic. Interdisciplinarity in polar science is neither a contemporary buzzword nor an anachronistic description of past research, but rather an integral part of knowledge making in a field site that was, and is, difficult and costly for non-Arctic-based researchers to access.

What did it mean to study the Arctic regions, where did this re-search take place, and how did one establish Arctic authority? Such questions were continuously debated throughout the nineteenth century, in Britain and elsewhere. Often, being designated as an expert was achieved through the physical act of travel. Going to these regional sites, having observed Arctic natural phenomena in situ, was a powerful argument for both regional and scientific expertise in the European and Euro-American context, both in the nineteenth century and in the present day.[6] For example, the Prussian cartographer and Arctic enthusiast August Petermann (1822–1878) had to be content with being called a speculative geographer because he never set foot in the Arctic.[7] At the same time, simply traveling to the Arctic was not enough to construct scientific authority. By virtue of this presence, European whalers and fur traders, for example, embodied Arctic expertise, yet it was a form of expertise that was only infrequently considered scientific. Arctic Indigenous peoples were also not considered scientific knowledge-makers by the European and Euro-American colonizers and travelers, even when they drew on Indigenous knowledge and practices.

What this points to is how the development of Arctic research as a distinct field of study has historically been linked to the performance of a specific type of knowledge-maker, one who saw the Arctic first-hand while having links to the European and Euro-American centers of learning. And this is where we find a key similarity between the British Arctic Expedition of 1875 and the First IPY: the attempt to decouple Arctic science and authority from the field-worker. The projects differed in the way this decoupling was attempted. While one of the key tenants of the First IPY was to remove the exploratory aspect of Arctic research completely, the scientific instructions provided to the British Arctic Expedition prioritized geographical discovery.

The RGS and the Royal Society (RS) took a leading role in the composition of the scientific instructions for the expedition. When Nares and his crew departed for the Arctic in late May 1875, they were provided with two key sets of scientific instructions intended to direct their researches in the Arctic: *Manual and Instructions for the Arctic Expedition*, produced by the RS, and *Arctic Geography and Ethnology*, produced by the RGS.[8] These instructions, which were published in 1875, are particularly revealing as they expose the tenuous relationship between the field and the metropole, and the contested state of Arctic science at the time. In this chapter I am concerned with how the instructions provided to the British Arctic Expedition portrayed the relationship between the field and the armchair, the generalist naturalist and the scientific expert. The so-called scientific naturalists, in particular the X Club, were central in the attempts to shape the research produced by the expedition.[9] Six out of the nine X Club members actively contributed to the instructions provided to the expedition. The role of the scientific naturalists in determining the scientific aims of the British Arctic Expedition should be seen as an extension of their work to control the research field in Britain. It was about power and a vision of who should be considered a scientific expert and what type of knowledge-making practices were considered scientific. In this vision of Arctic research, the scientific expert was in the metropole, while the field-worker was a nonexpert working based on the experts' instructions.

In the first section I consider the influence of Thomas Henry Huxley

(1825–1895) and Joseph Dalton Hooker (1817–1911) in selecting the two naturalists for Nares's expedition: Henry Chichester Hart (1847–1908) and Henry Wemyss Feilden (1838–1921). The choice of Hart and Feilden, two naturalists with very unimpressive scientific backgrounds, takes on new meaning when considered within the broader aims of the scientific naturalists in reforming and controlling the scientific landscape. In the second section, I expand on the relationship between the composition of the scientific instruction provided to the expedition and the choice of Hart and Feilden for the venture. It has been well established, in particular by historians of science Ruth Barton and Bernard Lightman, that the extent to which the scientific naturalists actually leveraged this type of control has been overestimated in the historiography.[10]

There existed a diverse and active community of scientific knowledge makers and communicators in the period, both within and outside of the elite scientific establishment; attempting to exercise control did not guarantee that such efforts succeeded. Indeed, the story of British Arctic research specialism illustrates both the scientific naturalists' influence on the expedition's scientific program and the performativeness and limitations of this influence. In the final section, I consider how the working practices of the two naturalists, in particular Feilden, resisted the desired framework of Arctic specialism as being under the domain of metropolitan-based researchers. Feilden was an Arctic scientific expert in his own right, but one whose research practices involved a re-shifting of his lived experience in the Arctic. In asserting his own authority, Feilden erased the labor and knowledge of Inuit field-workers from his scientific results. Therefore not two but three groups of naturalists emerge as part of the British Arctic Expedition: the metropolitan scientific naturalists, the British traveling naturalists, and the Arctic Indigenous naturalists. Historian Mary Louise Pratt has highlighted how colonial encounters involve the "copresence, interaction, interlocking understandings and practices, often within radically asymmetrical relations of power."[11] The constructed relationship of expertise therefore pertains not simply to the relationship between Western science in the metropole and in the field, but also to the epistemic erasure of Arctic Indigenous knowledge. Rather than reflecting the actual landscape of scientific practice and cultural authority relating to Arctic research in this period, the scientific instructions and the work of Hart and Feilden reveal the efforts to moderate the role and function of fieldwork, to define, control, and discipline Arctic science.

Scientific Naturalism in the Arctic

George Nares and his large crew left England in 1875 in two ships, the HMS *Alert* and HMS *Discovery*.[12] The *Alert* and *Discovery* were fitted with enough stores to last multiple seasons in the Arctic, in addition to a significant amount of expensive scientific equipment to support their research. The plan was to travel north along the western coast of Greenland through the waterway known as Smith Sound. From there they intended to complete the journey to the North Pole by dogsled. Despite these plans and preparations, the expedition was, by all accounts, largely unsuccessful. They did not reach the North Pole, they suffered badly from scurvy and the effects of subzero temperatures, and there were serious allegations of disobedience among the crew—including the captain. However, the expedition still carried out extensive research, particularly while the *Alert* and *Discovery* were frozen in ice. They returned to England in October 1876, having spent only one winter in the high Arctic of the north-west coast of Greenland, but with notebooks full of measurements and observations.

An article in *Nature* concluded that the production of useful scientific research had increased when the ships had been frozen in ice, while the health of the crew had decreased when they had attempted to reach the North Pole. This inverse relationship between scientific results and travel, it noted, "point[s] to the adoption of Weyprecht's scheme" and that "we shall have to thank both the successes" in scientific investigations "and the failures" of geographical discovery and health "of this expedition for opening up a new era in Arctic exploration."[13] This new era, the article emphasized, was defined by separating geographical surveying from scientific investigations in the Arctic, thereby creating a new specialized form of polar research. However, not everyone agreed with this restructuring.

Travel was a central component of the design and function of British Arctic expeditions throughout the nineteenth century. Voyages were sent out in search of something, be it a geographical discovery or a lost expedition, and such searches were further intended to lead explorers to new field sites for scientific investigation. This emphasis on discovery and travel as underpinning scientific advancement in the region is key to understanding the research practices of nineteenth-century British Arctic explorers. Because expeditions were expensive to send out, and because they intended to travel through areas that were difficult to reach, explorers were tasked with undertaking as much research as possible. It was never a given that there would be another opportunity to revisit the location. For both travelers and expedition organizers, this link between travel and research posed a key problem: How do you ensure that the scientific results produced in a changeable field site were reliable and complete? As part of the attempts to manage this uncertainty, expeditions were provided with both official written instructions, unofficial recommendations from scientific societies, and formal and informal training to guide their movements and investigations in the Arctic. The official instructions to Nares in 1875, which were reprinted in his *Narrative of a Voyage to the Polar Sea* (1878), stated that the "scope and primary object" was to reach the highest northern latitude, and more specifically, the North Pole.[14] Although this was described as the "main feature of the expedition," the official instructions further noted that "the object of the expedition is for the advancement of science and natural knowledge."[15] As a mission statement, this phrasing of expedition aims had largely remain unchanged since the end of the Napoleonic Wars, some fifty years prior.[16]

In 1818 the British government sent out four ships to the Arctic. Their official instructions directed them to, first, search for the fabled Northwest Passage and, second, to "contribute to the advancement of science and natural knowledge."[17] In both 1818 and 1875 the primary aim was geographical, with a secondary aim to produce as much scientific data in as many areas as possible. However, the crew did not include experts in each scientific area, leading to a frequent point of controversy on whether it was possible for the crew to undertake research in all these areas. For example, the astronomer and physicist Edward Sabine (1788–1883) was engaged in a heated argument with the explorer John Ross (1777–1856) over the definition of *naturalist* and the scope of scientific knowledge making in the Arctic. Ross commanded one of the 1818 British Arctic expeditions, and he described Sabine as the expedition's naturalist.

Sabine, however, rejected this designation, as he claimed he was merely employed to assist Ross in fulfilling the scientific goals of the expedition. Therefore, Sabine argued, he could not be expected to focus on both natural philosophy and natural history.[18] Sabine was an army officer and a prolific independent scientific practitioner, and he would later draw on his experience with Arctic travel in the development of the so-called magnetic crusade. Rather than a generalist field-worker who was undertaking work outlined in the scientific instructions to Ross's expedition, Sabine was himself an expert and focused on his own areas of specialism. This was not unique to Sabine, and later Arctic travelers, such as the surgeon and naturalist John Richardson (1787–1865), also undertook fieldwork with an eye to advancing their own scientific research. Thus, while the British Arctic Expedition retained the exploratory format, the expedition differed from other British ventures in the selection of the two naturalists to the expedition: Henry Chichester Hart and Henry Wemyss Feilden.

Hart and Feilden were not specialist scientific researchers, and they had only limited formal scientific training and publishing records. This was highly unusual for naturalists attached to government-funded expeditions. Historian of science Trevor Levere has argued that the British navy pushed for an exclusion of non-naval personnel on the expedition and that the "navy's attitude to non-naval personnel" such as civilian scientific researchers had a big impact on the choice of naturalists for the expedition. Adding to this, though, is a further factor: the influence of the professionalizing strategies of the scientific naturalists, which is a central part of this story. Hart trained at Trinity College Dublin, and his privileged family background enabled him to pursue his scientific interests without the need to secure a paid position or specialize in an area. In this sense Hart was the more traditional choice for the expedition, as his broad generalist research focus mirrored that of the independently wealthy naturalists of the earlier part of the nineteenth century, such as Charles Darwin (1809–1882).

Feilden, however, had no formal scientific training. The second son of a second baronet, Feilden chose a career in the army. Though he had authored multiple papers on ornithology, Feilden was first and foremost an army officer, making him a particularly peculiar choice for the expedition. Indeed, as Levere has further noted, the choice of Feilden "was considered by some outlandish."[19] When Feilden and Hart were announced as the naturalists for the expedition, one anonymous article in the *York Herald* stated that "we may conclude that in the persons of these two gentlemen botany and zoology are provided for," though this was only one aspect of the scientific aims of the expedition. As the article further noted, the expedition had been requested to focus on "observations on magnetism or meteorology, on the tides which in the Arctic zone are peculiar, on geology and botany, on natural history . . . and last, though not least, ethnology."[20] Similarly, an anonymous article in the *Hampshire Telegraph* noted that Feilden and Hart were "two highly qualified naturalists" but that it was "to be regretted that Geology has been omitted from the programme, and that there will not be such a master of ice phenomena."[21] Thus there were concerns about the extent to which the scientific training of the two naturalists was comprehensive enough to fulfill the broad scientific aims of the expedition.

The responsibility of appointing the naturalists rested with the RS, and at this point both its president, botanist Joseph Dalton Hooker, and one of its secretaries, biologist Thomas Henry Huxley, were X Club members. In addition, the mathematician and physicist William Spottiswoode (1825–1883) was the treasurer of the society. Through their roles within the RS, Hooker and Huxley were particularly influential in shaping the direction of the instructions provided to the expedition and in selecting the two naturalists. For example, both Hooker and Huxley corresponded with the renowned naturalist Charles Darwin about the instructions and potential field-workers. In a letter to Darwin, Huxley wrote "on behalf of the Polar Committee of the Royal Society to ask for any suggestion you may be inclined to offer us, as instructions to the Naturalists who are to accompany the new Expedition. The task of drawing up detailed instructions is divided among a lot of us."[22] Similarly, Hooker wrote to Darwin: "Have you any Botanical suggestions for the Arctic Expedition, if so please let me have them at once—I recommend special attention to insect action & fertilization Hybrids &c sowing earth from Icebergs[.] Also to try experiments on germination of seeds exposed to various degrees of cold."[23] These two quotations illustrate the background preparations that went into putting together the scientific instructions. More significantly, they point to the role of the X Club in shaping the scientific instructions for the British Arctic Expedition of 1875.

As a group, the scientific naturalists were engaged in attempting to control the entirety of knowledge production in Britain. Scientific naturalism was a broad project that sought to apply standard theories and methods to all research areas. Historian of science Frank Turner and later Bernard Lightman have described the scientific naturalists as "naturalistic in the sense that they ruled out recourse to causes not present in empirically observed nature, and they were scientific in that they interpreted nature in accordance with three major scientific theories, the atomic theory of matter, the conservation of energy, and evolution."[24] With this definition in mind, the instructions provided to the British Arctic Expedition and the choice of two generalist naturalists for the venture reveal how specific British-based researchers sought to control Arctic research methodologies.

When considered within the broader efforts of the scientific naturalists, in particular the X Club, in shaping British science, the choice of Feilden and Hart as naturalists take on new significance for our understanding of the place of Arctic science in this period. It emerges as a clear statement of intent pertaining to the relationship between the armchair and the field, the professional and the amateur. The focus was not, as it was for Weyprecht, on changing the framework for Arctic research in the British tradition, but rather in shaping the way field-workers carried out their duties while traveling. Hart's and Feilden's main strengths as collectors and observers were in botany and zoology, respectively, as was the primary research focus of Hooker and Huxley. Yet Hooker and Huxley were not interested simply in influencing those research areas, but were involved in shaping boundaries between who was a professional and who was an amateur.

As scientific reformers, Huxley and John Lubbock (1834–1913) were not working toward creating a discipline of Arctic science. Their focus was on extending the methods of scientific naturalism to the Arctic within the preexisting structure of exploration. Selecting nonspecialists to carry out research, which had been outlined in extensive scientific instructions, was a powerful strategy for attempting to control data collecting in the field while retaining the specialism and professionalism of the scientific disciplines in the metropole. In this constructed vision of Arctic science, the scientific experts were in the metropole, while the amateurs were doing their bidding in the field. However, as I show below, this was a rhetorical and vocational strategy rather than a reflection of the working practices of the Arctic field-workers. This can be traced in the scientific instructions provided to the expedition, which were very different from what Weyprecht suggested for the future of polar research under the IPY.

The scientific instructions provided to the British Arctic Expedition form part of a long-standing tradition in the British Empire of preparing travelers for undertaking scientific research abroad, not just in the Arctic but around the world. Providing instructions to travelers was one way for individual researchers and scientific societies to attempt to exert control over the data produced in the field. For example, the British Association for the Advancement of Science's *Queries Respecting the Human Race, to Be Addressed to Travellers and Others* (1841) was designed by the society to formalize the collection of ethnographic data.[25] It is important to emphasize that such instructions, even when they specifically outlined methodologies, should not be taken as evidence of the actual research practices of the explorers. Rather, instructions to travelers were statements of intent that reflect the types of expectations placed on field-workers. It shows the rhetorical and vocational strategies of the metropolitan-based researchers as they sought to direct and control the field-workers to produce datasets that would not only fall within their areas of interest, but also adhere to specific criteria for ensuring their reliability. Reflecting the tradition of instructions to travelers and the broad scientific aims of British Arctic voyages, Nares and his crew were provided with several sets of instructions and manuals to guide their scientific research.

To fully appreciate the role of these materials and the influence of the scientific naturalists in shaping the expedition's findings, I turn to the instructions provided by the RS and the RGS, the *Manual and Instructions for the Arctic Expedition*, and the *Arctic Geography and Ethnology*. No less than six out of the nine X Club members actively contributed to these instructions: Huxley, Hooker, Spottiswoode, John Tyndall (1820–

1893), John Lubbock, and George Busk (1807–1886). The composition of the instructions produced by the RS and RGS reveal how the societies, in particular through the work of the scientific naturalists, sought to control the research areas and methodologies for a type of fieldwork that was interdisciplinary by necessity. They were written for generalists in the field who would work across discipline boundaries in their pursuit of advancing knowledge about the Arctic for use by expert researchers in the metropole.

ᴿESEARCHING THE ARCTIC FROM AFAR

Nineteenth-century British explorers and traveling naturalists were requested to undertake research in a broad range of areas, and this posed several difficulties for the crew, the organizers, and those who hoped to make use of the data later. First and foremost, consider the problem of reliability. How do you ensure that the findings and measurements are correct, when the person(s) who carried out the work is not an expert in all those areas? There is a certain level of performance to this. The perceived reliability of the scientific research undertaken while away rested to a large extent on the successful performance of the explorers and expedition organizers in the theater of exploration. The whole concept of scientific exploration, in the Arctic and elsewhere, was a highly curated construct. When the naturalists to the expedition, as was the case of Feilden and Hart, were generalist jacks-of-all-trades, how does that square with a construction of observational reliability and usability of that data in specialist research fields? Seemingly, the *Manual* and *Arctic Geography and Ethnology* aimed to guide the explorers, in particular Hart and Feilden, to a largely multi-disciplinary practice, with sections delineated into separate discipline-specific groupings. This served to signpost who was a disciplinary expert, that is, who was asked to contribute to specific research areas, while at the same time portraying the traveling naturalist as someone who was simply undertaking the research on behalf of the actual experts. In the instructions, Feilden and Hart were framed as data collectors rather than as experts in the fields they gathered data for.[26] This differentiation is significant as it points to the attempt to place the locus of scientific authority not in the Arctic, but in the metropole.

The relationship between the experts in the metropole and the nonspecialist field-workers was in part a rhetorical construction, which further sought to underpin the reliability of the collected data. That is, the reliability and usefulness of the results from Arctic expeditions were fundamentally linked to the perceptions of the explorers and traveling naturalists. In turn, these perceptions were constructed by specific portrayals of their field practices. It is useful to first consider the structural organization of the instructions provided by the RS and the RGS to fully unpack the relationship between the portrayed research specialism of the instructions and the working practices of the explorers. The RS's *Manual* was comprised of two parts: the instructions and the manual. The first was an outline of the key areas of research that Nares and his crew were asked to undertake, and the second was a curated reference manual with reproductions of several texts on Arctic matters. The two main sections were each subdivided into two parts: physical observations and biological and geological observations.

The physical part of the instructions was further divided into specialist subsections, including "Astronomy," "Terrestrial Magnetism," "Meteorology," "Atmospheric Electricity," "Optics," and a mixed bag titled "Miscellaneous Observations." The biological and geological part of the instructions was divided into "Zoology," "Botany," and "Geology and Mineralogy," clearly demarcated areas of specialist research. For example, Hooker contributed a section on "Instructions in Botany" to the *Manual*, where he directed the explorers to make observations on specific areas of Arctic botany.[27] Hooker further included methods for preserving and recording botanical specimens, which Huxley also referred to as the preferred methodology in his contribution to the instructions.[28]

Turning to physics, Tyndall contributed a section "Hints towards Observations in the Arctic Regions."[29] This included a point-form list of "subjects for observation," such as experiments on heat, sound, and ice and glaciers—research areas where Tyndall had positioned himself as the foremost expert. As with Hooker's methodology for collecting and observing botanical specimens, the other contributors to the *Manual* referred to Tyndall's framework for explaining glacial phenomena.[30] The RGS's *Arctic Geography and Ethnology* was not formalized along subdivided disciplinary lines in the same way as the RS's *Manual*; yet the influence of specific scientific reformers is still evident throughout. As the title indicates, the RGS's document was divided into two main sections: geography and ethnology. The editorial responsibility for the RGS's instructions rested with Clements Markham, its secretary. Markham was also elected as a fellow of the Ethnological Society in 1864 and joined the Royal Anthropological Institute (RAI) after its amalgamation with the Anthropological Society of London (ASL) in 1871. In addition to reports and questionnaires drawn up by the RAI, *Arctic Geography and Ethnology* included papers by ethnographers and geographers such as Robert Brown (1773–1858), Ferdinand von Wrangell (1796–1870), Hinrich Rink (1819–1893), Carl Ludvig Christian Irminger (1802–1888), and Richard Collinson (1811–1883).

As a direct instruction for Feilden and Hart, the questionnaires are particularly revealing of the preferred methodology and approach to the specialisms of the RAI and RGS. In the mid-nineteenth century British geographical research was fundamentally linked to ethnology and anthropology. As the study of "Man," ethnology considered how regional differences influenced the development of humanity, including physical characteristics and cultures. The close relationship between geography and ethnology was also reflected institutionally. For example, geography and ethnography had shared Section E at the annual meetings of the British Association for the Advancement of Science (BAAS) since 1851. When the ASL was formed in 1863, it defined itself in opposition to the older Ethnological Society of London (ESL, f. 1843). This fracture was related both to professionalizing and vocational ambitions, as well as differing views of the nature and scope of the discipline.[31] How researchers should approach and interpret other cultures was a central question debated between the two groups.

After much deliberation and negotiation, the two rival societies merged to become the Anthropological Institute of Great Britain (later the RAI) in 1871. Lubbock was the first president of the new society. He was later the society's president of the Arctic Committee, and through this role Lubbock exercised a great deal of influence, directly and indirectly, on *Arctic Geography and Ethnology*. After Lubbock, the society's president was Busk, another X Club member, who held the post while the organization of the British Arctic Expedition was underway. The third president during the time of the expedition was Augustus Pitt Rivers (1827–1900), who also sat on the RAI's Arctic Committee. Edward Burnett Tylor (1832–1917), another future president of the RAI, also contributed to *Arctic Geography and Ethnology*.[32]

Though Pitt Rivers and Tylor were not members of the X Club, their working practices were aligned with scientific naturalism, and they studied human history within an evolutionary framework.[33] Pitt Rivers's contributions to *Arctic Geography and Ethnology* included three questionnaires: "Enquiries into Customs Relating to War," "Enquiries Relating to Certain Arrow-Marks and Other Signs in Use amongst the Eskimos," and "Enquiries Relating to Drawing, Carving, and Ornamentation."[34] In all of these, the explorers were directed to make observations and collect specimens in a way that specifically considered cultural practices as an expression of evolutionary developments. For example, Pitt Rivers asked the explorers to consider if mark-making practices were "confined to particular tribes or common to the whole race," and if drawing is "more practiced in some tribes than others, and if so, does this arise from inclination or from traditional custom."[35] Historian George Stocking has written that Pitt Rivers viewed "human sociocultural development in the Darwinian context," or as historian Efram Sera-Shriar has further noted, he followed a version of evolutionary thinking that was indebted to Tylor and the ethnologist James Cowles Prichard (1786–1848).[36] That is not to say that there was a unanimous approach to ethnology among the members of the RAI, but it highlights the influence of those who favored the application of scientific naturalism to all areas of research.

Take, for example, the contribution of the physician and craniologist Joseph Barnard Davis (1801–1881), "Questions for Explorers (With Special Reference to Arctic Exploration)."[37] Davis was a former council member of the ASL before the amalgamation, and he directed the explorers to consider what he described as the "general" features of humanity, which was heavily focused on physical appearances and resonated with the particular interest of the ASL in physical anthropology over social or cultural anthropology. Davis further included a request for observations on the "systems of relationship," which he asked the explorers to undertake following a system outlined by Lubbock in the *Journal of the Anthropological Institute*.[38] In this context, "relationship" refers not only to social designators such as "parent" and "cousin," but also to divisions among and development of the so-called human races. Lubbock was an evolutionary anthropologist, in that he applied the concept of evolution to the study of all aspects of humanity. Together with Huxley and Tylor, another scientific naturalist and close ally of the X Club, Lubbock was a central figure in shaping the direction of ethnography and anthropology in the second half of the nineteenth century. From this perspective, *Arctic Geography and Ethnology*, as well as the *Manual*, appears to tell stories of discipline formation. Yet it was not an attempt to construct an Arctic science, or even a region-specific Arctic subdivision of the scientific fields. Instead, it was about demarcating who determined the standards of data for a particular aspect of Arctic field-based research. This was a rhetorical and vocational strategy, which positioned the generalist interdisciplinary work of Hart and Feilden as a nonexpert research practice.

There is a clear conflict between the portrayal of Hart and Feilden as generalists and the level of complexity in the scientific instructions provided to them. In considering the specialist subsections of the physical part of the *Manual*, the amount of requested experiments and observations seems overwhelming. Tyndall's contribution alone, which took up only two pages, had enough queries to keep Hart and Feilden very busy. Some questions were more straightforward; for example, question number one asked the explorers to consider the shapes and sizes of snow crystals in different atmospheric conditions. Other questions were more complex and would require that the naturalists be stationary over a long period of time. As Tyndall noted for question three, which related to the conduction of heat through ice, these observations were possible only if "there is plenty of time at the observer's disposal."[39] Question numbers four to ten concerned the formation of icebergs and the behavior of glaciers. Tyndall had carried out his own research on glacial motion and related issues in the Alps over several seasons, and he was aware of the difficulties involved in gathering reliable and complete data on the ice. Yet he was hoping that the naturalists could determine the mechanism by which glaciers break off to form icebergs and if it was possible to see the effects of ancient glacial action on ice-free landscapes. Question fifteen even proposed that the naturalists could fire a cannon "windward" in the Arctic to observe aerial echoes.

The specificity of these questions reveals how Tyndall was not simply suggesting methods for Hart and Feilden to undertake broad investigations on heat, sound, and ice and glaciers, but instead outlined requests that would advance Tyndall's research agenda. Indeed, question thirteen asked Hart and Feilden to repeat the French biologist Louis Pasteur's (1822–1895) experiments on airborne germs. As with Pasteur, Tyndall developed what would be known as the germ theory, thereby disproving the theory of spontaneous generation. In the instructions Tyndall referred simply to "experiments similar to those of Pasteur upon the Mer de Glace" and left the naturalists to determine the process for this work. The results of Tyndall's own experiments in the Alps in 1877 were a "striking confirmation of the experiments of Pasteur upon the Mer de Glace."[40]

Perhaps unsurprisingly, given their lack of experience with the physical sciences, Feilden and Hart did not produce the results Tyndall had hoped for, and their work was not cited in his later publications on the topic. Though Hart and Feilden had been recommended by X Club members and though the scientific naturalists had greatly shaped the content of the scientific instructions to the expedition, this did not easily translate into data production that could be implemented into desired non-Arctic-specific specialist research fields. Tyndall's set of questions to Hart and Feilden reflect an incredibly interdisciplinary practice, both for Tyndall and for the two naturalists. Tyndall's questions, which were part of the section on physics, ranged from germ theory, acoustics, atmospheric science, and glaciology. His physics alone was not something that could easily be defined as a single-discipline practice.[41] While Tyndall worked to frame his fieldwork in the Alps as a central component of his scientific expertise, this embodied form of specialism was not extended to Hart and Feilden.[42] That is not to say, however, that the scientific naturalists, in particular the X Club, dominated Arctic research in this period. As with all areas of nineteenth-century British science, there were plenty other actors striving for scientific authority—including Hart and Feilden themselves.

The two sets of scientific instructions provided to Nares and his crew illustrate an attempt to demarcate the boundaries of expertise in relation to the Arctic through the influence of the elite scientific societies. It was a circular argument: Arctic fieldwork was presented as non-specialist labor carried out under the direction of the scientific experts through the instructions, thereby decoupling the field-worker from the data they produced. Bernard Lightman has notably argued that the extent to which the scientific naturalists pushed out other scientific practitioners and other forms of knowledge-making than their own, was more performative than a reflection of the scientific landscape of the time.[43] In the context of Arctic science, this performative element to research specialization and control came to the fore both in the metropole and in the field. As the first British Arctic expedition in almost two decades, the instructions provide crucial insights into the portrayed and functional format of fieldwork, in the Arctic and elsewhere. The strategies for controlling the Arctic field site pushed a narrative of research specialism and expertise among the British-based researchers and, by extension, an orderly Arctic research practice carried out based on the instructions, which was greatly at odds with the actual interdisciplinary practices of the Arctic explorer.

ERASURE AS A STRATEGY OF SPECIALIZATION

The British Arctic Expedition returned to England in the autumn of 1876, after spending only one year in the Arctic. Despite their brief stay, the crew had carried out a significant amount of scientific research. In the years that followed, this research was utilized in many scientific articles across several disciplines and authored by both the contributors to the instructions as well as the expedition crew. Feilden wrote several papers based on his research, which were not simply descriptive accounts responding to the instructions for the expedition but analytical interpretations of the data. This is the first way in which we can see the complications in the relationship between the expectations of the metropolitan-based researchers and the working practices of the traveling naturalists, as portrayed in the *Manual* and *Arctic Geography and Ethnology*.

The second way has to do with the lived experiences of Arctic exploration and research, in comparison with the portrayed working practices of Feilden, Hart, and the other European officers. The key was creating observational reliability. In both instances, reliability in the data was linked to an erasure of what we can call multi-epistemology, which here refers to the plurality of forms of knowledge and of research methodologies along Eurocentric lines of inquiry. In other words, the European explorers erased the contributions of Arctic Indigenous peoples in the construction of their expedition results. Arctic fieldwork in the British context was disciplined by the construction of ideals of expert data collection, which was nonspecialist and nontheoretical and undertaken during the process of geographical discovery. This specific ethos of who was an authoritative Arctic field-worker was central to the perceived reliability of their data. The scientific societies worked under the assumption that the explorers would undertake research as instructed (rather than functioning as independent Arctic experts undertaking their own research), and the explorers also shaped the portrayal of their data collecting in a way that emphasized the reliability of their practices. In turn, the explorers presented their observations as undertaken through interdisciplinary practices within the discipline-specific research framework of the instructions, which involved an erasure of the way knowledge was produced in the Arctic.

Nares, the commander of the expedition, published his official narrative in 1878 as *Narrative of a Voyage*. This two-volume work included a large appendix, which contained much of the scientific results of the expedition divided into the following categories: "Ethnology," "Mammalia," "Ornithology," "Ichthyology," "Mollusea," "Insecta and Arachnida," "Crustacea," "Annelida," "Echinodermata," "Polyzoa," "Hydrozoa," "Spongida," "Rhizopoda reticularia," "Botany," "Geology," "Report on Petermann Glacier," "Game List," "Meteorological Abstract," and "Abstract of the Results Obtained from the Tidal Observations." Accordingly, it was divided into what we can recognize as specialist and subspecialist research areas similar to how the instructions were organized. Many of the papers that were included in the appendix, as well as the articles that were published elsewhere using the data from the expedition, were written by those who had contributed to the scientific instructions.

The appendix to *Narrative of a Voyage* is particularly interesting as it shows the portrayed division between how the data was collected on the one hand and analyzed on the other. For example, Hooker, who had authored the botanical instructions in the *Manual*, wrote a paper on the same subject that appeared in the appendix to *Narrative of a Voyage*, "Botany of the North Polar Expedition, 1875–76." Here Hooker referred to Daniel Oliver (1830–1916), professor of botany at University College London and keeper at the Royal Botanic Gardens in Kew as having "examined and named" the plants collected by Feilden and Hart.[44] By thus dividing the practical and analytical work, the transfer of the plants into scientific specimens was portrayed as having taken place in the universities of England, rather than in the Arctic.

Busk, another X Club member, contributed to the appendix of *Narrative of a Voyage* with the paper "Polyzoa, Arctic Expedition, 1875–76." Busk began his paper by noting that Feilden's collection "includes only about seventeen species" of Polyzoa (the phylum now known as Bryozoa), and of these, Busk was "able to ascertain" that three were "new to science." Though Busk noted that the collection was "interesting," he also emphasized that the three species he believed were previously unidentified were "represented unfortunately by such very scanty and imperfect specimens" that he could "only venture to propose it provisionally."[45] The implication from this was that Feilden had not analyzed the data on the spot, but merely collected the specimens—and had done a rather poor job of it. Busk was limited by the results provided to him by Feilden, and this points to the way the constructed relationship between the data collector and the scientific researcher shaped the usability of the data. Because the explorers had traveled to remote areas to gather data, it was difficult to attempt to redo or verify their observations. The instructions sought to bridge this gap between the naturalists' lack of expertise and the desire of the specialists for usable specimens.[46] Or that would have been the case had the naturalists in fact been working as suggested by the instructions, as an extension of scientific experts of the metropole.

Feilden also contributed several papers to the appendix. He wrote up the sections on ethnology, mammalogy, ornithology, and geology, the latter with C. E. De Rance (1847–1906) of the Geological Survey of England and Wales. The section on ethnology is particularly revealing for how the practices of the Arctic field-workers responded to the requests of the scientific instructions. The high Arctic was of considerable interest to those Europeans and Euro-Americans who were engaged in the study of human history, as they believed researching the local Indigenous peoples, Inughuit, could provide insights into the cultural and physical origin and development of humans. The conceptualization of the high Arctic as isolated, with its people living almost in time capsules, was a highly Eurocentric and racist understanding of Arctic societies, one that went hand in hand with imperialist and religious expansionism.

This view was by no means unique to those contributing to the *Manual* and *Arctic Geography and Ethnology*, and it continued to shape the research field throughout the twentieth century. The British Arctic Expedition had, in particular through *Arctic Geography and Ethnology*, been provided with extensive instructions for collecting data on Inughuit. The questionnaires directed Feilden and Hart to gather evidence that fit into the evolutionary anthropology of figures such as Lubbock,

Huxley, and Pitt Rivers. Note here that there were multiple evolutionary models at play in the second half of the nineteenth century, and that Darwin's theory of evolution by means of natural selection was only one of several explanatory models.[47] Nevertheless, the data gathered by the seemingly non-biased field-worker could be adapted into the theoretical models of either a Tylorian or Darwinian evolutionary model, or, indeed, as hybrids of either. This analytical work was placed firmly in the metropole.

Feilden's paper on ethnology was first published in the journal *Zoologist* and touched on key areas of research that had been highlighted in the instructions provided to him. The instructions had, for example, cited Lubbock's so-called systems of relationship to guide Hart and Feilden toward gathering information about kinship in the high Arctic, forming part of the broader aim of European ethnologists to construct a global system of human developmental relationships. Feilden's paper was not simply a descriptive chapter, but rather reflected the analytical work he had undertaken in the Arctic. "It has been assumed, somewhat too hastily" Feilden wrote, "that the 'Arctic Highlanders' are a race completely isolated from other human beings."[48] In making his argument, that Inughuit were not a separate group, Feilden drew on a variety of evidentiary materials. He referred to observations on abandoned settlements and ruins, the climatic conditions, dispersion of plants and animal life, excavated human remains, past accounts of previous expeditions and Danish settlers, in addition to local customs and the use of technologies.

Feilden further reflected that the traces of abandoned settlements at very high latitudes may point "to a change in the physical conditions of an extensive area lying within the Arctic zone."[49] As such, Feilden drew on and combined a broad range of scientific research areas to make the conclusion that Inughuit had continuous contact with Inuit farther south as well as on the eastern coast of Greenland, and he further extended his analysis to reflect on the changing environmental conditions of the Arctic. Similarly, in his paper on ornithology in the appendix of *Narrative of a Voyage*, Feilden combined detailed descriptions of birds, with observations on marine life, climatic conditions, botany, the physical landscape, skeletal remains in abandoned settlements, and the accounts of past explorers and Danish settlers.[50] The fieldwork practices of a generalist such as Feilden was highly interdisciplinary, and he was clearly attuned to the state of the sciences he was contributing to.

With each separately delineated field-specific instruction, the contributors to the *Manual* and *Arctic Geography and Ethnology* had attempted to extend their grasp of the research to the high Arctic. The intended users were Feilden and Hart, as well as the other officers working as part of the expedition. But the efforts to discipline the Arctic field site by putting together extensive manuals and guides were also a way of transforming the explorers from independent expert researchers to nonspecialist data gatherers. Yet, looking at Feilden's papers published following the expedition is illustrative of how this way of framing the traveling naturalists as nontheoretically biased observers who responded to an interdisciplinary research program was a performative vocational and professional strategy employed by both the explorers and the metropolitan-based researchers. Though the influence of the latter, as seen in the scientific naturalists, the instructions, and Feilden's later published papers, also shows the limitations of this control. What Feilden's articles do not reveal is how his lived experience traveling and researching in the Arctic influenced his research practice. This tells only part of the story of the construction of Arctic specialism, as the creation of the image of the traveling naturalist as an objective observer went hand in hand with the subjugation of Arctic Indigenous peoples and the erasure of Inuit knowledge.[51]

As with the majority of European and Euro-American Arctic expeditions before it, Nares's expedition employed Inuit in Greenland and relied on the support of the local communities in Smith Sound. Before venturing north, Nares took the two ships to Upernavik, a city on the western coast of Greenland, where he employed the experienced Inuk explorer Suersaq (ca. 1832–1889).[52] Suersaq had participated in the three previous American expeditions through Smith Sound and had lived in the region for an extended period of time after marrying a local woman, Mequ.[53] Suersaq was, accordingly, the expedition's expert on the area, yet he was portrayed as a manual laborer, assisting with the practical aspects of Arctic exploration. This was a deliberate rescripting of his expertise—one that went hand in hand with the portrayal of the collected data as reliable. This is clearer in the travel narratives, and even more so in the personal journals of the explorers, including Suersaq's 1878 memoir, which recounted his experiences as part of the four expeditions through Smith Sound.[54] Yet we can see glimpses of the plurality of knowledge systems in the published scientific results as well. In his paper on ethnology Feilden made references to conversations with local peoples: "From information derived from one of the natives resident at Etah" there "can be no doubt" that there was "casual, if not regular" contact between the settlements on each side of Smith Sound. Feilden's appropriation of Indigenous knowledge also allowed him to retrace the "route, by which the migration of the Eskimo from North America to Greenland" took place.[55] This was an important contribution to the research Feilden had been directed to undertake, and Suersaq was central in securing this information. Yet Feilden did not name Suersaq as his source.

The implicit assumption in the *Manual* and *Arctic Geography and Ethnology*, as well as in the research papers that were published following the expedition, was that the scientific research was undertaken by the European explorers and that any support from Inuit in knowledge making required assessment and verification by the explorers. As the explorers took Inuit knowledge and rescripted it into the disciplinary patterns of Western science, they also erased the multi-epistemology of the "contact zone," to use Mary Louise Pratt's term.[56] This reflects a systemic erasure of the contributions of Indigenous peoples to what has historically been referred to as "exploration science" throughout the world. The deliberate erasure and rescripting of Inuit knowledge was by no means unique to the British Arctic Expedition. Arctic explorations, as with exploratory missions elsewhere, were inherently imperialistic ventures. At the same time, it is important to acknowledge the diversity and fluidity of such contact zones, rather than replicating the portrayed and lived power structures of colonial science in our historiography. For example, historian of science Fa-ti Fan has shown how nineteenth-century British naturalism in China involved both instances of imperial violence as well as long-term and explicitly acknowledged cooperations between the foreign and Indigenous naturalists.[57]

In the context of the British Arctic Expedition, the erasure of Inuit knowledge and the role of individual Inuit explorers in shaping European and Euro-American science in the Arctic formed a central part of the explorer's tool kit. As a form of knowledge making separate from that of the European and Euro-American colonizers, Indigenous Greenlandic (Kaalaallit) science holds its own traditions and boundaries of expertise.[58] Though there was a plurality of knowledge forms in the Arctic contact zone, this multi-epistemology was not planned for in the scientific instructions, and it was erased in the Arctic field site. This takes on further significance when thinking about our understanding of fieldwork practice during a period of increased discipline formation and research specialization in Victorian Britain. Returning to the division between the metropolitan-based researchers and how the scientific instructions framed the traveling naturalists as nonspecialist data collectors, the division between who is a scientific authority and who is not becomes clearer. The relationship among the metropolitan scientific naturalists, the British traveling naturalists, and the Arctic Indigenous naturalists was constructed and upheld through imperialist structures.

Inuit explorers employed by the expeditions and the local communities in the Arctic were central in ensuring the expeditions achieved their desired goals. This was not limited to practical support with food and shelter, but also extended to the scientific aims of the expeditions. Yet, rather than revealing the collected data as having been cocreated through intercultural encounters, field-workers such as Hart and Feilden wrote themselves into their scientific results as the sole originators of their data. When considering the disciplinary history of polar research, as well as the implementation of data from the Arctic into other scientific research fields, this is a crucial point to remember. The creation of Arctic expertise, be it as field-workers or as metropolitan-based researchers, took place through the exploitation of Indigenous peoples and the erasure of Indigenous knowledge systems and expertise. Therefore the question of what counted as Arctic expertise, and who could be trusted as an Arctic expert, is particularly poignant in this context. The relationship between the professional and the amateur was continuously negotiated against the backdrop of the erasure of other knowledge systems and practitioners, both in the Arctic and in Britain.

TRAVEL, TRUST, AND A DISCIPLINARY HOME FOR ARCTIC SCIENCE

In considering the disciplinary developments of field-based sciences, the Arctic, or polar regions more broadly, is perhaps an unusual example. The relative remote location of the field site and the cost and difficulties associated with traveling there have historically created an imperative to undertake as much research as possible. As a result of this challenge, it was difficult to double-check any observations, thus fundamentally tying the usability and veracity of the gathered data to the researchers, the organizers, and the scientific societies that produced the instructions to the expeditions. In the 1870s British context, science in the Arctic was disciplined not through a specialized research focus, but rather through a formalized relationship between the researchers in the metropole and the field-workers and through a hidden erasure of Arctic Indigenous naturalists. On paper it may seem that the First IPY created a fundamentally new way of undertaking polar research. As Royal Navy lieutenant George T. Temple (1847–1935) argued at the annual meeting of the British Association in 1882, the IPY "marked a fresh point of departure in Polar investigation, which might now be considered as an accepted branch of study."[59] It would be a mistake, however, to conclude that the IPY clearly delineated the relationship between fieldwork and armchair (or laboratory) research, or that it created a specialist area of polar research. While contemporary polar science, encompassing Arctic and Antarctic research, has a "home," to use the language of historian Sarah Burton, this specialization did not happen with either the First IPY or the British Arctic Expedition of 1875.[60] By placing trust in numbers, to use the terminology of Theodore Porter, as the basis for scientific veracity and credibility in the performative decoupling of the subjectivity of the field-worker from their data, both the British Arctic Expedition and the First IPY sought to create a disciplinary framework for polar science that was fundamentally based on the exclusion of those who were not deemed objective observers.[61]

Attending to the relationship among the metropolitan scientific naturalists, the British traveling naturalists, and the Arctic Indigenous naturalists reveals how erasure and rescripting of research practice and expertise, in different ways, formed a central part of the tool kit of both the metropolitan-based and traveling British scientific knowledge makers. This was an important part of the rhetorical and vocational strategies for constructing research specialism and expertise. The metropolitan scientific naturalists sought to control the two traveling naturalists and used the performatively interdisciplinary scientific instructions to write them into a subordinate position. At the same time they positioned Hart and Feilden as the sole originators of the scientific results in the colonial field site, thereby making the erasure of Inuit expertise an a priori part of the knowledge-making process. The detailed scientific instructions provided to Nares and his crew reflect attempts to control the scope and methodologies of their scientific research as a way of navigating the uncertainties involved in travel for geographical surveying. In this scheme, Arctic scientific expertise could be established from the metropole, taking place through an assumed extension of preferred scientific paradigms and including a rhetorical division between data collection (in the field) and analysis (in the metropole). Disciplining polar science in the second half of the nineteenth century was not only about creating an intellectual home for Polar studies, but about reconsidering the way trust in data was constructed.

V

HYBRID FIELDS

Chapter 9

STRETCHING THE
‛BOUNDARIES OF
KNOWLEDGE

Victorian Spirit Investigations, Credible Witnessing,
and the Alleged Exposure of the Medium Henry Slade

‑EFRAM SERA‑SHRIAR

One of the more heated cultural debates of the late Victorian era was whether or not spirits and psychic forces were real. It was a topic that attracted researchers from all corners of the scholarly world. Yet, despite numerous attempts by believers and skeptics alike to resolve the matter once and for all, no single discipline seemed able to offer a definitive conclusion on whether there was any real weight to the spirit hypothesis—the idea that the phenomena produced by mediums at séances are caused by disembodied spirits. Any verdict regarding the veracity of spiritualism was informed by an "interdisciplinary" approach, though this is an anachronistic term to apply to the Victorian period.[1] Investigations into spiritualism embodied the Victorian ethos of knowledge construction.

When studying any topic, whether it be ocean barnacles, lexicons of Indigenous peoples, or chemical reactions, Victorians tended to move across disciplinary boundaries. As Mark Bevir has argued, "Victorian learning tended to pick particular objects [or topics], explore facts about them, and then make sense of these facts by locating them in a developmental narrative that covered the whole of human life, perhaps the natural world, and sometimes even the divine."[2] Spirit investigations were no different, and varied quite broadly in approach. This chapter will explore the interdisciplinary nature of spirit investigations by examining in detail the American medium Henry Slade (1835–1905) and his supposed exposure as a fraud, which occurred in London during the autumn of 1876.[3] This event is a good case study for exploring how spirit investigations drew knowledge from many fields. As a result of the supposed exposure, figures possessing an array of disciplinary backgrounds and expertise weighed in on the revelation; all of them providing important testimonies both in support of or against the legitimacy of Slade's mediumship.

Since the 1980s scholarship on the history of spirit investigations during the late Victorian period has flourished. Important seminal works, such as Janet Oppenheim's *The Other World* (1985), Logie Barrow's *Independent Spirits* (1986), and Alex Owen's *The Darkened Room* (1989), laid the groundwork for more sophisticated studies of the modern spiritualist movement.[4] In the decades that have followed, historians continue to expand these narratives. One area in particular to receive considerable attention is scientific engagements with spirit and psychic phenomena. This is understandable given that so much of spiritualism's past has been about two central questions: (1) Did the observed phenomena actually happen? (2) How was it produced? In seeking answers to these questions, both believers and skeptics regularly appealed to scientific communities. Even the appropriation of the term *spirit hypothesis* indicates that spiritualists viewed their movement as possessing scientific and philosophical dimensions. While there has been some important and ground-breaking research on the intersection of science and modern spiritualism, these works typically examine this history through two main disciplinary lenses—psychology and physics.[5] By approaching modern spiritualism's past in such a manner, scholars have produced a rather skewed and narrow historical picture. All investigations into spirit and psychic forces were inherently interdisciplinary, and thus it is only by stretching across a much broader range of disciplinary boundaries that we truly come to appreciate the nuances and complexity of the movement's past.

To become a credible witness of spirit and psychic phenomena, one had to employ a specific kind of visual epistemology that incorporated a wide range of observational techniques and rigorous training. Researchers working to legitimize spirit investigations developed all sorts of discriminating practices, which sought to identify those characteristics that were of importance to studies of spiritualism and those to ignore. As the Henry Slade affair demonstrates, debates over the genuineness of spirit and psychic forces were hotly contested during the late Victorian era. No single type of disciplinary specialism could claim authority over spirit investigations, and cases were strengthened considerably by incorporating a broad range of perspectives from science and beyond when analyzing the evidence.

That is not to say that specialist training could not be strategically used to promote the veracity of a researcher's findings, but one's background alone was insufficient for gaining authority as a credible witness of spirit and psychic phenomena. What often mattered most was a researcher's experiential knowledge of observing in situ spiritualist performances. A researcher's ability to demonstrate that they had substantial firsthand experience observing and interacting with spirits and psychic forces added substantial legitimacy to their accounts. Equally, a profound knowledge of the extant literature on the subject was also highly valued, and these materials are quite diverse, including a range of disciplinary perspectives from physics, psychology, and chemistry, to folklore, law, anthropology, and professional conjuring—to name a few examples. Those who did not demonstrate these practical and theoretical understandings of spiritualism were quickly dismissed, especially by believers.[6]

The chapter is divided as follows. In the first section I will give some historical background on the development of the modern spiritualist movement in Britain during the second half of the nineteenth century, followed by two short biographical sketches of the two main combatants in this story—the medium, Henry Slade, and the investigator, Edwin Ray Lankester (1847–1929). In the second section I will examine the initial exposure of Slade as a so-called fraud. After visiting Slade on two occasions in September 1876, Lankester, along with his friend and corroborating witness, Horatio Bryan Donkin (1845–1927), wrote letters to *The Times* announcing that Slade was a charlatan. This was soon followed by a series of letters by multiple spirit investigators of different backgrounds, either supporting or rejecting the veracity of Lankester's and Donkin's testimonies. Much of these discussions centered on what it meant to be a credible and skilled witness of spirit and psychic phenomena. In the conclusion I will briefly discuss what transpired after Slade's supposed exposure in *The Times*, before reflecting more broadly on spirit investigations as an important example of Victorian interdisciplinarity.

THE ORIGINS OF THE SPIRITUALIST MOVEMENT, HENRY SLADE, AND EDWIN RAY LANKESTER

Modern spiritualism is typically traced back to 1848 and the rise of the first American high-profile mediums known as the Fox Sisters: Leah (1831–1890), Margaret (1833–1893), and Kate (1837–1892). At their family home in Hydesville, New York, the two younger sisters Margaret and Kate allegedly began communicating with spirits through rapping.[7] News of these spirit communications quickly spread throughout the region, and within a few months much of western New York was gripped by the spiritualist movement. Historically the region was populated by various Christian evangelical groups. It had been described as the "Burned-Over District" and the place of "the second great awakening," because of the force by which religious revivalism had passed through it. In an environment swelling with spiritual fervor, it is easy to understand why local residents were quick to accept the early claims of mediums and their followers as true. It was not long before news of spiritualism spread farther across the United States, before eventually traveling overseas to Britain and continental Europe.[8]

The first modern medium to come to Britain was Maria B. Hayden from Boston, Massachusetts. She arrived in London in 1852 and was investigated by several high-profile researchers. This included the Scottish publisher, naturalist, and anonymous author of the *Vestiges of the Natural History of Creation* (1844), Robert Chambers (1802–1871); the Welsh manufacturer and social reformer Robert Owen (1771–1858); and the English mathematician Augustus de Morgan (1806–1871).[9] As this list of figures suggests, already during its formative years in Victorian Britain, investigations into spiritualism were attracting researchers from a range of disciplines and backgrounds. Known primarily for producing spirit rapping during her performances, Hayden achieved only moderate success during her short-lived stay in London, and she returned to the United States within a year, as rumors were emerging that she was a fraud. Another equally unimpressive medium followed suit in 1853— Mrs. Roberts. Unlike Hayden, her specialty was table-turning, a common phenomenon to be witnessed at séances where unseen spirits supposedly lift, jolt, or tilt tables. She too fled back to the United States once reports spread that she was a fake. [10]

It was not until the arrival of Daniel Dunglas Home (1833–1886) in 1855 that spiritualism's momentum began to truly take root and grow in Britain. Originally from Scotland, Home moved to the United States in the late 1830s to live with extended family in Connecticut. His return to Britain happened during a tumultuous period for the movement when prominent skeptics such as the chemist and physicist Michael Faraday (1791–1867) and the physician and zoologist William Benjamin Carpenter (1813–1885) were trying to disprove through naturalistic explanations the causes that produced spirit phenomena such as table turning. Using a specially devised apparatus to determine if sitters were applying muscle pressure to tables during séances, both men argued that the effect of table turning was the result of ideomotor responses (or unconscious reflexes). It is a classic example of the application of interdisciplinarity in spirit investigations, combining physics and physiology to study these extraordinary phenomena.[11]

While most early Victorian mediums were fairly clumsy in their performances, and easily detected as frauds, Home was of a different caliber.[12] His success was largely due to his ability to produce a broad range of extraordinary phenomena during séances in fairly good light. His most astonishing feats allegedly included physical manifestations such as musical instruments playing without any human interaction, spirit hands appearing and disappearing before sitters' eyes, and levitating himself several feet above the ground. Arguments for unconscious muscular movements alone were insufficient to discredit Home's alleged powers, and only a comprehensive and interdisciplinary model stood a reasonable chance of uncovering whether he was a charlatan or a genuine psychic. Spirit investigators from research fields as diverse as physics, chemistry, law, medicine, anthropology, and professional conjuring regularly scrutinized Home's incredible psychic feats, and yet he was never caught cheating. His reputation as a trustworthy medium, willing to be investigated by anyone who wished, did much to strengthen his legitimacy.[13] In many respects Home became the face of modern mediumship in Victorian Britain, and he did much to strengthen the veracity of the spiritualist movement's core beliefs. As the psychical researcher Frank Podmore (1856–1910) argued, "The main defenses of Spiritualism must stand or fall" with Home.[14]

Within a few years more mediums were coming to prominence in Britain through private and public performances, and proponents of spiritualism were clearly on the ascent. There was also a proliferation of spiritualist literature being published during the 1850s and 1860s, and many popular lecturers on the subject toured around the country.[15] Henry Slade entered the British spiritualist scene a bit later (figure 9.1). By the time he arrived in Britain in 1876, there was a vibrant spiritualist community in London, eager to attend the séances at his rooms at 8 Upper Bedford Place in Russell Square. Since the late 1840s Slade had traveled around Michigan and New York State honing his skills as a professional medium. He had developed a strong reputation among American spiritualists by specializing in spirit writing. These were messages that appeared on supposedly blank slates without any apparent human agency involved. Slade's backstory further strengthened his status among spiritualists. He grew up in Johnson Creek, New York, in the heart of the Burned-Over District, just a few miles away from the first séances led by the Fox sisters. Although he claimed to be from a family of mediums, it was not until the age of twelve that he was identified as having psychic powers. After the death of his wife Allie, sometime during the 1850s or 1860s, she became his main spirit control during his performances. With such strong credentials and connections to the

FIGURE 9.1. Woodcut of the medium Henry Slade. *Source:* Henry Ridgely Evans, *Hours with the Ghosts, or, Nineteenth Century Witchcraft* (Chicago: Laird and Lee, 1897), 47.

original core movement, Slade claimed membership among an elite group of pioneering mediums.[16]

Edwin Ray Lankester (1847–1929) was the eldest son of the surgeon and public health reformer, Edwin Lankester (1814–1874) (figure 9.2). Because of his father's leading position within the scientific and medical community, the young Lankester grew up attending dinner parties with eminent researchers. This included the naturalist Charles Darwin (1809–1882), the biologist Thomas Henry Huxley (1825–1895), and the geologist Edward Forbes (1815–1854). These early experiences had a profound impact on him, and once he was of age to attend university, he traveled to the European continent to study physiology at both Leipzig and Vienna, before eventually going to Jena to study comparative anatomy and embryology under the tutelage of the zoologist and physician Ernst Haeckel (1834–1919). He returned to Britain in the early 1870s and held several minor posts, which included demonstrator in Huxley's practical biology classes for schoolteachers in South Kensington. Finally, in 1875 he secured a major post as chair of zoology at University College London. Known for his staunch rationalism and adherence to the scientific naturalism of the day, it is unsurprising that Lankester was ea-

FIGURE 9.2. Color lithograph of E. Ray Lankester by Leslie Ward. *Source: Vanity Fair,* January 12, 1905. Reproduced with permission under the terms of Creative Commons from Wellcome Collections, Library no. 5252i.

ger to investigate, and hopefully expose, spiritualism as nonsense. Slade was a perfect target, as he represented the highest echelon of spiritualist performers. Exposing him would be a huge blow to the movement's credibility.[17] The altercation that ensued between Lankester and Slade in the autumn of 1876 was fundamentally a clash between two opposing philosophical positions—materialism and scientific naturalism on the one hand, and immaterialism and spiritualism on the other. Framed in this light, the Slade affair was part of the "contest for cultural authority" in Victorian Britain, and there was much at stake for both sides.[18]

LANKESTER'S INVESTIGATION OF SLADE AND THE CORRESPONDENCE IN *THE TIMES*

News first broke of Lankester's alleged exposure of Slade as a trickster in the correspondence section of *The Times*. His reason for choosing this newspaper was simple: He wanted to deliver a huge blow to the spiritualist movement by outing one of its most high-profile mediums. *The Times* was an ideal platform for doing so because by the middle of the nineteenth century it had a massive readership, selling over fifty thousand copies per issue.[19] His findings, however, did not go unchallenged, and a lively debate ensued after the publication of his first letter. In total seventeen letters about the Slade affair were published in *The Times* between Saturday, September 16, and Saturday, September 23, 1876. These correspondences are rather revealing and underline how spirit investigations often attracted a diverse range of specialists, all willing to comment on the legitimacy (or illegitimacy) of a medium's professed powers. Whether they were a physician, anthropologist, lawyer, professional conjurer, or a mixture of a few specialisms, every commentator weighing in on the Slade controversy claimed expertise as a skilled observer of psychic phenomena, highlighting the interdisciplinary nature of spirit investigations.

In his first letter to *The Times*, Lankester recounted his experiences attending Slade's spiritualist performances earlier in the week. He wrote that it was his friend, the lawyer, journalist, and publisher Sergeant Edward William Cox (1809–1879) who had convinced him to attend one of Slade's séances. Cox was impressed by the medium's abilities, and his opinions were significant to Lankester. Cox was an experienced spirit investigator, having examined numerous high-standing mediums, including Daniel Dunglas Home during the 1860s. His training as a lawyer provided him with a good foundation for scrutinizing a broad range of evidence. He was also establishing himself as a leading authority in the emerging discipline of psychology, a research field that was developing specialist interest in psychical research.[20] Cox cofounded the Psychological Society of Great Britain in 1875, and he was primarily interested in the psychological side of mediumship to determine whether the human mind played some important role in producing so-called psychic forces.[21] Thus his investigatory approach to studying mediumship was quite interdisciplinary. Although he was a self-professed skeptic, Cox was quite respectful toward spiritualist beliefs, and he employed a balanced approach in his investigations, with a genuine openness to the prospect of seeing evidence that could convince him of the validity of the spirit hypothesis.[22] In 1871 he published a book about his experiences investigating psychic phenomena, *Spiritualism Answered by Science.*[23] With such a strong and nuanced knowledge of the movement, Cox was seen as a trusted authority, and Lankester's earnestness to visit Slade was spurred on by his friend's endorsement. There was also a second motivation. According to Cox, the great opponent of spiritualism, William Benjamin Carpenter, was "very much shaken" after visiting and witnessing Slade's performance. If these claims were true, that both Cox and Carpenter struggled to make sense of how Slade was able to produce his extraordinary phenomena, then perhaps he really was a psychic and worth studying.[24]

On Monday, September 11, 1876, Lankester visited Slade's rooms. He believed that the key to uncovering the true cause of Slade's professed powers rested in his abilities as a trained scientific observer to skillfully watch the medium's performance. After all, Lankester had studied with some of the greatest scientific men of the day, establishing himself as one of the leading anatomists and physiologists in Britain. Lankester recorded the details of his investigation, and this process of outlining the particulars of his experience in the séance room transformed his readers into what Steven Shapin and Simon Schaffer have termed "virtual witnesses," allowing them to acquire an almost firsthand knowledge of Lankester's investigation.[25] It was a core aspect of Lankester's "visual epistemology," ensuring that he could claim to be a trustworthy observer of these extraordinary phenomena. It was also testament to the professed experimental practice he used in his scientific undertakings.[26]

Lankester noted that the room was well lit, providing good visibility throughout the séance. Although there were some supposed spirit knocks and raps during the early stages of the sitting, the "chief manifestation" was the so-called spirit writing. The setup for receiving these messages was simple in construction. Slade used two standard slates with basic wooden frames, a piece of chalk, and an unassuming, four-legged loo table. As Lankester recounted in his letter to *The Times*, he sat at the corner of the table facing the window. Slade sat across from him with his back to the light. He showed Lankester that the two slates were clean with no writing on them. After enclosing a small piece of chalk between the slates, the two men held them together under the table with one hand each, pressing them tightly against the table's underside. The pressure was to ensure that no hand or finger could squeeze between the slates. Their free hands were placed on top of the table in clear view. Soon scratching sounds were heard, and when the noises ceased, the slates were opened, revealing a handwritten message from a supposed spirit.

This process of receiving alleged spirit messages was repeated several times, and Lankester stated, "I watched Slade very closely during these proceedings . . . paying no attention to the raps, gentle kicks, and movements of table, of which I will say nothing further than they were all such as could be readily produced by the medium's legs and feet."[27] Lankester believed that the supposed spirit knocks and raps occurring during the performance were designed as a distraction to draw his attention away from the slates. Thus he took special care to watch the process several times, taking note of Slade's physical movements and mannerisms. He also did not want Slade to suspect that he was being tested. Therefore Lankester "simulated considerable agitation and an ardent belief in the mysterious nature" of what he saw and heard. This seemed to convince Slade that everything was completely aboveboard. They received several messages during the séance, some short and difficult to read, and others long and clearly legible.

More significant were the considerable delays between Slade preparing the slates for receiving spirit messages and inviting Lankester to take hold of them under the table. During these interruptions there was sufficient time for the medium to write some text on the slates; conceivably accounting for why the messages were often difficult to read. Slade's behavior was also rather irregular during these pauses, and Lankester recalled, "During the delay Slade made various excuses; took up the little piece of pencil and bit it, and also invariably made a peculiar grating noise by clearing his throat."[28] These attempts at misdirection were carefully designed to mask "a low but perfectly recognizable sound of a pencil traversing a slate." Lankester, however, quickly caught on and ignored the distractions. Moreover, drawing on his training as an anatomist and physiologist he also noted that "on looking at Slade's right arm, the elbow of which was visible . . . I saw movements from right to left," which Lankester assumed meant that the medium was writing on the slates. By the end of the séance Lankester reasoned that Slade was almost certainly cheating, and he determined that it was essential to further investigate the matter. Feigning amazement at what he had witnessed, Lankester promised to return later in the week for another sitting. This of course was all a ruse to develop some trust with the medium; and Lankester was already devising a plan to expose Slade's supposed trickery.[29]

On the morning of Friday, September 15, 1876, Lankester returned to Slade's apartment for a second sitting. This time he brought his friend Horatio Bryan Donkin to act as a corroborating witness. A key way of strengthening one's account of a spirit investigation was by producing corroborating reports from other credible witnesses who had attended the same event. As a Fellow of Queen's College, Oxford, and a physician at Westminster Hospital in London, Donkin was a respectable figure in the scientific community, making him an ideal corroborator for Lankester's investigation. His role was essential for establishing the scientific veracity of Lankester's own testimony. If two respected men of science claimed to have witnessed the same thing, their reports were most likely legitimate. This approach to producing credible testimonies had a long tradition in the human sciences, a kind of "collective empiricism," to borrow the term from Lorraine Daston and Peter Galison.[30]

The two gentlemen took their seats at the table in Slade's séance room. However, before trying to catch Slade in the act of cheating, they carefully observed his routine multiple times to ensure that their suspicions were justified. Lankester explained his plan in his letter to *The Times*: "I went with my friend Dr H.B. Donkin . . . to test my hypothesis by this crucial experiment:—I had determined to seize the slate at the critical moment—at the moment when Slade professed that it was entirely untouched."[31] While Slade was preparing his slates for another message by wiping them clean, Lankester decided that it was time to grab them. He recounted, "There had been the usual delay and fumbling on Slade's part when I put my hand and immediately seized the slate away, saying. 'You have already written on the slate.'" On opening the supposedly blank slates, Lankester revealed a prerecorded message. Seemingly dumbfounded by the sequence of events that had just transpired, Slade sat back silently in his chair while Lankester and Donkin stormed into the neighboring room, where other clients were waiting for their appointments with the medium. Triumphantly, the two men revealed their discovery to the room—that Slade was a total fraud.[32]

Donkin's account, in same issue of *The Times*, reiterated most of the details described by Lankester in his letter; strengthening the shared credibility of their testimonies. As with Lankester, Donkin drew on his authority as an expert in anatomy and physiology when observing Slade's performance. He remarked, for instance, how "some contraction of flexor tendons" on Slade's wrist were moving while the scratching noises were being heard.[33] Donkin also expanded on the details of the event in other important ways. For example, he noted that the writing on the slates was likely produced by "the agency of a minute piece of slate-pencil probably held under the nail in the middle finger." During the delays, when Slade was holding the slates underneath the table on his knees, there was ample opportunity for the medium to write messages with the concealed piece of chalk. Thus there were rational, material explanations for how the phenomenon might be produced.

Moreover, while Slade was interrupting his performance by clearing his throat with the "grating noises," as described by Lankester in his letter, Donkin observed that Slade's "right arm, now at some distance from the table, [was] moving exactly as though he were writing something on his knee." Yet, because of his position at the opposite side of the table, Donkin's view was obstructed, giving Slade the freedom to manipulate the slates rather easily without being spotted. The lack of visibility produced a subtle but crucial opportunity for Slade to cheat, suggesting that the arrangement of the room and seating was carefully designed by the medium. Finally, Donkin stated that just prior to Lankester snatching the slates from the medium, they received a message from an alleged spirit in which Lankester was addressed as "Samuel." Both Donkin and Lankester asked if perhaps the spirits meant "Edwin," and sure enough in the subsequent message the name was corrected, further suggesting that Slade was likely the one writing the messages. Both Lankester and Donkin concluded their letters with the same verdict—that Slade was a fraud and possessed no psychic powers. However, this was only the beginning of a much more complicated dialogue on the genuineness of Slade's mediumship, and the veracity of the spirit hypothesis more generally.

The publication of Lankester's and Donkin's letters in *The Times* created a sensation. It was not long before other spirit investigators weighed in on the subject. These researchers all possessed different sorts of qualifications and backgrounds, highlighting the range of disciplinary perspectives in spirit investigations. For instance, in the following issue of *The Times* from Monday, September 18, 1876, the physician and anthropologist Charles Carter Blake (1840–1887) criticized Lankester's investigation for being both brash and ungentlemanly. He wrote, "Let me, as an observer, who has very closely watched the abnormal physical phenomena which take place in the presence of Dr Slade, protest against the tone in which Professor Lankester has described the facts and investigated the subject. The adoption of a violent method towards a gentleman and foreigner may lead to misapprehension as to the fairness and scientific spirit of the majority of observers."[34] Just because investigators were scientifically accomplished did not mean that their investigations were credible. They were still required to act appropriately, as gentlemen, while participating in séances. If investigators broke social protocols and were impetuous or forceful in their examination of a medium, could they really be trusted as careful observers of spirit and psychic phenomena? In Blake's opinion, no; Lankester's aggressive investigatory approach was too belligerent and biased to be considered reputable.

Blake substantiated his stance by emphasizing his own experiences in the séance room with Slade. He was, after all, an experienced psychical researcher who understood the correct protocols for studying spiritualism. Since the late 1860s Blake had been actively investigating the movement. He was part of a special committee formed by the Anthropological Society of London (ASL) in 1868 to examine whether the famous American spiritualist performers Ira Erastus Davenport (1839–1911) and William Henry Davenport (1841–1877) were genuine psychics. Under the leadership of the ASL's president and cofounder, James Hunt (1833–1869), anthropologists in this period saw their discipline as the only one to examine scientifically all aspects of human life.[35] If the spirit hypothesis were true, anthropologists were also to study human afterlife. The Davenport brothers provided a significant opportunity for anthropologists to examine a high-profile case—and potentially verify the existence of spirits and psychic forces.

The Davenport brothers were famed for their spirit cabinet performance, where they were tied inside a specially designed wardrobe, containing musical instruments. Once the cabinet was closed, the instruments would begin to play, and upon opening the doors, the brothers were still tied in the positions in which they had started. It was argued that spirits were the true cause of the extraordinary phenomena, with the brothers serving as psychic conduits. However, things did not go to plan during their tour of Britain, and in 1864 during a performance in Liverpool, an argument erupted between them and some audience members over the use of a binding known as a "Tom Fool's Knot." They refused to perform using it, resulting in the audience calling them frauds.

Eventually, after several more altercations with audiences in different cities, the brothers left Britain for continental Europe. Their return to London was supposed to be an opportunity to absolve themselves of the scandal. Their friend and devout follower, a journalist named Robert Cooper, arranged for the ASL to investigate the brothers. The hope was that by gaining the support of a community of scientific researchers who positioned themselves as the main experts on all aspects of human culture, they could reestablish their credentials as genuine psychics. Ultimately the ASL's investigation of the Davenport brothers was a failure. The two sides were unable to agree to terms on what controls were allowed to be used during the séances. Yet it still underlines how anthropologists were committed to becoming experts in spiritualism. The details of the ASL's investigation of the Davenport brothers were published in the spiritualist periodical *Human Nature*.[36] These failures aside, however, Blake continued to study spiritualism over the next decade.[37]

Prior to reading Lankester's and Donkin's letters in *The Times*, Blake had already visited Slade and carefully observed his performance. Therefore his interpretations could be framed as being unaffected by Lankester's and Donkin's so-called exposure. He argued that his experience with the medium was "so at variance" with that of Lankester and Donkin, that he doubted the accuracy of their testimonies. Was this merely a case of two inexperienced researchers misinterpreting the evidence before them? Blake believed this to be the case, and he wrote, "If Dr. Slade plays tricks, his *modus operandi* is something very different from that which Professor Lankester would suggest." He continued by stressing that Lankester's and Donkin's accounts were wholly inconsistent with what every other investigator had reported, arguing, "The observers who have visited him, including some of the cleverest minds in science, have failed to detect any fraud."[38] Thus Lankester and Donkin were mistaken, and to prove it, Blake employed a comparative method.

There was a long tradition of using comparative methods in anthropology for verifying the trustworthiness of testimonies, and Blake was building on this tradition. Such an approach also aligned nicely with the practices of spirit investigations, further demonstrating how psychical research was a kind of anthropological pursuit. Spirit investigators regularly substantiated their observations by comparing them with other so-called credible accounts. If they could show that their report was similar to that of other credible witnesses who had examined the same medium, then their observations were deemed as reliable. If the report seemed anomalous, it was likely untrustworthy because it either contained some observational errors, or it was driven by some kind of confirmation bias. In Lankester's case, Blake suspected that it was the latter, especially given that Lankester was clearly determined to expose Slade as a fraud.[39] By contrast, he framed his own investigation as being far more candid and balanced: "Let me assure your readers that my opinions are not in any way influenced by any theory of what is called 'spiritualism,' . . . in the name of science and veracity, let me entreat enquirers to suspend their judgement till they have arrived at a *vera causa* of the facts, and weigh the facts as they alone stand."[40]

In the same issue of *The Times* from September 18, 1876, there was also a letter from the barrister Charles Carleton Massey (1838–1905), who was vouching for the genuineness of Slade's mediumship. Massey was an important figure among British spiritualists. He was member of the Ghost Club (f. 1862), an early group of trained spirit investigators; a cofounder of the British National Association of Spiritualists (f. 1873), which was a main body for the formation and dissemination of spiritualist knowledge; and a close friend and confidant of the renowned medium Rev. William Stainton Moses (1839–1892), who was also a respected spirit investigator. Massey also acted as Slade's lawyer, and he was highly critical of Lankester and Donkin's investigation. He argued that it had been hastily executed. Massey broke down the evidence with the finesse of a skilled barrister, finding inconsistencies in their testimonies.

A credible account by a witness required levelheadedness and frequent visits to ensure that all aspects of a medium's performance were judiciously observed. Unlike Slade's accusers, Massey had "taken considerable trouble to get at the truth by frequent investigation," and he argued that if Lankester and Donkin were correct in their estimations—that Slade was a fraud—why were their descriptions so at odds with the reports of other prominent and experienced investigators, who had been unable to detect any trickery?[41] In Massey's view, Slade was not being dishonest; Lankester and Donkin were at fault. They had not spent enough time with the medium to properly observe his powers. If they had done so, they would never have accused Slade of trickery.

Massey interrogated other aspects of their accounts, and stated that Lankester and Donkin could not be trusted and that their brash investigatory techniques compromised their findings. With proper controls in place and careful observation, Slade's performance could be seen in a more reputable light. He wrote:

> In all the accounts which have been thought worthy of publication or even of private mention, one fact is invariably found—the production of writing on a slate ascertained to be clean and which is never for a moment removed from above the table or out of sight of the investigator. Frequently, it is the investigator's own slate; sometimes a double folding slate just purchased. The writing, which is heard, comes on the under surface of the slate laid on the table, and, therefore, in a closed space. Sometimes Slade does not even touch the slate—never when it is not removed from view of the spectator, could he write upon it without instant detection.[42]

The majority of the other witnesses to have visited Slade saw spirit messages being produced, and they did not detect any trickery on the part of the medium. Moreover, none of them recalled losing sight of the slates while they were being prepared for spirit messages. In many cases, the investigators even supplied their own slates, ensuring that everything was completely aboveboard. The claim that Slade prepared the slates on his knees under the table was also inconsistent with other accounts, suggesting that Lankester's and Donkin's testimonies were erroneous. Massey asserted that the supposed exposure of Slade was nothing more than a misunderstanding and misrepresentation of the events witnessed in the séance room. Lankester's and Donkin's accounts should be rejected outright in favor of the more reliable ones produced by more skillful psychical researchers.

A third letter also appeared in that same issue of *The Times*. It was written by the agriculturalist and inventor John Algernon Clarke (1827–1887), who offered a different perspective to the conversation. He was a collaborator of the professional conjurer John Neville Maskelyne (1839–1917), and he had a detailed knowledge of how to produce a range of illusions designed to fool human perception. Clarke gained considerable public notoriety for cocreating the automaton "Psycho," a self-operating mechanical mannequin that played a predetermined sequence of card games as part of Maskelyne's regular show at the Egyptian Hall. Thus he could claim to be an expert in detecting trickery; as he stated, "I may claim to know something of the subtle mechanical illusions; and I am also familiar with the methods of substitution, concealment, and reproduction employed by professors of sleight of hand. Had there been any proofs of supernatural agency at work when I was with the medium, I could hardly have failed to perceive them."[43]

Clarke devised a fairly simple plan to catch Slade cheating. When the medium requested the name of a "departed soul" that Clarke wished to receive "direct communication" from, he gave a fictitious name. And yet the supposed spirit still wrote several messages to him. How was this possible? Clarke of course did not reveal this information to the medium, as he hoped to collect further damning evidence. For example, contrary to Massey's testimony, Clarke noted that Slade was preparing the slates under the table, with the arrangement of the seating designed in such a way that it was impossible for him to see what Slade was doing under there. Clarke recalled that "upon my proposing to sit on the floor so that I could watch the underside of the table, the medium's legs, and his right hand which holds the slate . . . I was told by Mr. Simmons (the manager) and Dr Slade that none of the phenomena can be produced unless all the persons present joined hands on the table." Even with his obstructed view, Clarke was not fooled, and he observed movements in Slade's "arm and wrist precisely such as would have been apparent if he were writing on the slate."[44] After witnessing Slade's performance, Clarke resolved that everything he observed was a sham and that the medium was obviously a fake.

In the following issue of *The Times* on Tuesday, September 19, 1876, the codiscoverer of evolution by natural selection, Alfred Russel Wallace (1823–1913), weighed in on the debate. As one of the most ardent supporters of spiritualism in Victorian Britain, it was hardly surprising that Wallace leapt to Slade's defense. He had comprehensive knowledge of the literature on spiritualism and had rigorously investigated and validated dozens of mediums' powers. His most significant work on the subject was his collection of essays *Miracles and Modern Spiritualism* (1875).[45] His spirit investigations were highly interdisciplinary, and he saw his research as "a new branch of anthropology" that combined methods from the emerging human sciences, experimental testing from the physical sciences, and the latest evolutionary theories from the natural sciences.[46] It was all outlined in his "theory of spiritualism."[47]

Wallace's response to Lankester's and Donkin's allegations appealed to a form of collective empiricism that aimed to delegitimize his opponents using a comparative method. He rejected Lankester's and Donkin's testimonies, stressing how they were "completely unlike what happened during [his] own visit, as well as the recorded experiences of Sergeant Cox, Mr. Carter Blake, and many others."[48] None of the "fumblings" and "maneuvers" that Lankester and Donkin described in their accounts were present during these other investigations. Thus Wallace positioned himself, Cox, and Blake as highly proficient spirit investigators, skilled in detecting trickery at séances, and Lankester and Donkin as novices, unaccustomed to observing the subtleties of psychic phenomena. Those who knew what they were doing would recognize Slade's mediumship as genuine, and those who did not would produce anomalous reports that should be ignored.

On Wednesday, September 20, 1876, *The Times* published a letter by Sergeant Cox, his first public statement about the affair. He stated that his visit to see Slade was part of his "official duties" as the current president of the Psychological Society of Great Britain. Cox had been given instructions to report on any "psychological phenomena" that he witnessed, which might be of interest to the society's members. Specifically, Cox and his colleagues were concerned with collecting evidence that might support the possibility of "psychic forces" originating in the minds of mediums.[49] He confirmed that much of what he witnessed during his sitting with Slade corresponded with the reports of other letter writers to *The Times*. Slade was able to produce various examples of alleged spirit messages with all sorts of controls in place. However, unlike these other commentators, Cox "carefully abstained from pronouncing any judgement as to the genuineness or otherwise" of the phenomena. His reason for remaining neutral was that he had yet to determine with any certainty their cause. It may very well be the case that all of the manifestations that he had observed in Slade's presence were fake, but because he had not caught Slade in the act of cheating, he could not outright reject the medium's professed powers. Further careful observation over multiple sittings was necessary.[50]

There was also a significant difference about Cox's séance with Slade compared to that described by Lankester and Donkin: "With me the slate was not placed under the table, but upon it, and the writing appeared upon the side next to the table, my eyes, as well as my hand, being upon it and found the whole side filled with writing end to end." Thus, in Cox's opinion, there was no opportunity for Slade to covertly write the message under the table in the way Lankester and Donkin argued in their testimonies. Cox did suggest, however, that it might have been possible for Slade to have used some sort of "sympathetic pencil, which becomes visible when the slate is warmed by the hands placed upon it." However, he contended that "Chymists [sic] will say if such a thing can be."[51]

Cox concluded his letter by restating that his experience as an investigator had taught him to form no judgment until a medium was observed "twice or thrice" and "under various conditions." Cox knew too well that "a clever conjurer can deceive the eye of a stranger." Thus it was essential to conduct the research over an extended period of time to ensure the full experience of a medium's séances were rigorously documented. What is striking about Cox's assessment of Slade's mediumship is that it is a rather classic example of the interdisciplinary nature of spirit investigations. Here we have a lawyer, positioning himself as a psychologist but appealing to chemistry and experimental testing, while also drawing on his knowledge of conjuring tricks.[52] A broad range of perspectives was therefore informing his examination.

Slade was one of the last people to comment on his supposed exposure, and his letter appeared in *The Times* on Thursday, September 21, 1876. This was Slade's first and only public statement about the affair. His account focused on what he saw as the "facts" of the case:

> On our sitting down to the table I held the slate against the under side of the table, when, after some delay, the sound of the pencil writing on the slate was heard. On withdrawing the slate there was found to be what might have been intended for a name very poorly written upon the upper surface. I then wiped this off the slate, saying, "I will hold it again: perhaps they will write plainer." Again a little delay ensued, when I said to Professor Lankester, "Perhaps if you will take hold of the slate with me they may be better able to write." He thereupon released his hand from where it was joined with my left and those of his friend upon the table, and instead of holding the slate with me, seized it, as he describes.[53]

Contrary to the version given by Lankester and Donkin in their letters to *The Times*—that the message included the name Samuel instead of Edwin—Slade claimed that the message was altogether illegible. There was no attempt to decipher it, and instead he explained that he cleaned the slate immediately so that they could attempt to receive a more legible message. There was also no mention of Lankester and Donkin correcting the name only to receive another message with the correct information. According to Slade, the group was struggling to receive a second message from the spirits, and when he attempted to strengthen the group's so-called psychic bond with the spirits, by readjusting the positioning of the slates, Lankester seized them. Moreover, unlike the version described by his accusers, Slade argued that the slates did not contain any complete sentences but instead "there were only two, or, at most three words on the upper surface of the slate."[54] The placement of the message on the upper side of the top slate, combined with its incompleteness, suggested that the supposed spirits were abruptly stopped while writing the sentences.

Most important, Slade asserted that had Lankester "listened as closely as he says he watched me" he would have heard him exclaim, "They are writing now." Slade contended that he stated this information right at the moment when Lankester seized the slates. He also claimed that during this sequence of exchanges "the sound of the pencil when writing" could be heard. Thus Lankester and Donkin's entire story was undermined by Slade's testimony.[55] His professed authority as a medium was key to strengthening his report, as were the endorsements he received from figures such as Wallace and Blake. Unlike his accusers, Slade could attest to how one could psychically communicate with spirits, and therefore he was able to explicate the reasons for the delays during the séance. These delays were not so that he could physically write the messages, but to request and encourage the spirits to do so. Thus Slade's letter added yet another kind specialist perspective to the overall debate—that of a psychic. The story of Slade's exposure did not end with the correspondence appearing in *The Times*, and during the weeks that followed the story escalated further. Throughout the affair, the legitimacy of Slade's mediumship and the veracity of the spirit hypothesis more generally took center stage.

SLADE GOES TO TRIAL AND THE AFTERMATH

Lankester was determined to destroy Slade's career as a professional medium. During the same period that he was publicly attempting to expose Slade as a fraud in *The Times*, he was also seeking to indict the medium under the terms of the Vagrancy Act of 1824, which essentially stated that it was illegal in Britain for a person to use any "subtle craft, means, or device by palmistry or otherwise" to deceive people.[56] Slade was summoned to court, and the trial was held throughout October 1876. During the buildup to the trial, there was a feverish effort by spiritualists to defend Slade against what they viewed as an unjust attack on spiritualism by some of London's scientific elite. The culmination of these efforts materialized in the form of a whole issue of the spiritualist periodical *The Medium and Daybreak*, being dedicated to the medium. It was called "The Slade Number," and proceeds from its sale were used to support the medium's legal fees. Much like the correspondences appearing in *The Times*, many of the entries in "The Slade Number" contained supportive testimonies by a diverse group of witnesses with different intellectual and practical specialisms; providing further examples of the interdisciplinary nature of spirit investigations.[57]

During the trial there was also a broad range of experts testifying on behalf of both the prosecution and defense. Some were scientific figures such as Lankester and Donkin, who were retelling their accounts. Other witnesses possessed different kinds of specialist knowledge. For example, the magician John Neville Maskelyne attempted to re-create for the court how Slade might have produced his so-called spirit messages. There were even testimonies from the carpenters who had made the table that Slade used during his performances. Many of the skeptics at the trial believed that the table played a key part in Slade's supposed deception, and therefore it became a major piece of evidence in the case. Ultimately, however, it was the law that decided whether or not Slade was guilty of trickery.

While the judge was willing to declare that he believed Slade to be charlatan, he refrained from making any formal judgment about the veracity of the spirit hypothesis more generally.[58] Despite the small triumph of a guilty verdict, it was a decisive blow to Lankester's larger aim to delegitimize the spiritualist movement in Britain. More important, it left the door open for believers to continue to assert the reality of spirit and psychic phenomena. After receiving the verdict, Slade sought an appeal and was eventually acquitted on a technicality—there was an error in the original paperwork submitted to the court by Lankester. While awaiting retrial Slade fled the country. Spiritualists claimed the acquittal as a victory for their movement. Slade meanwhile went on to have a mixed career, and most skeptics continued to believe that he was a fraud. His legacy therefore was greatly tarnished by the row with Lankester.[59]

By pausing and reflecting on the larger picture, these discussions bring the analysis back to the primary focus of the chapter: that despite innumerable attempts by believers and skeptics alike to resolve the matter once and for all, no single discipline seemed able to offer a definitive conclusion on whether there was any real weight to the spirit hypothesis. As this case study has shown, Lankester and Donkin's professed exposure of Slade as a fraud in September 1876 attracted the attention of a broad range of specialists interested in spirit and psychic phenomena. The correspondence in *The Times* is therefore a wonderful example of Victorian interdisciplinarity. It shows how there were many types of spirit investigators with diverse backgrounds and specialisms. Some figures such as Clarke approached their investigations by drawing on their expertise in one specific field, in his case professional conjuring. Others such as Sergeant Cox were more integrative in their approaches. These figures studied spirit and psychic phenomena by mixing together theories and practices from a broad range of disciplines. In Sergeant Cox's case it was law, psychology, and chemistry.

All of these examples show that it was only by drawing together diverse disciplinary perspectives that a more sophisticated and nuanced understanding of spiritualism could be formed. Both the sciences and humanities offered potentially valuable insights into the reality of spirits and psychic forces, and no single discipline was granted priority in these disputes. Even with the emergence of a more codified body of researchers specializing in spirit investigations, with the establishment of the Society for Psychical Research in 1882, studies of spirit and psychic phenomena remained interdisciplinary in scope. The society's founding members included the philosopher Henry Sidgwick (1838–1900), the classicist Frederic W. H. Myers (1843–1901), the physicist William Fletcher Barrett (1844–1925), and the politician Arthur Balfour (1848–1930), to name a few. This cast of figures exemplifies how spirit investigations continued to stretch the boundaries of knowledge.

Chapter 10

"ANIMALS CANNOT SUBSIST ON AIR"

Nutrition as a Hybrid Field in Early Victorian Science

JAMES F. STARK AND RICHARD T. BELLIS

On May 1, 1866, Thomas Henry Huxley (1825–1895) delivered a lecture at the prize giving at Saint Mary's Hospital Medical School, an annual event that accompanied the opening of teaching for the summer session. Reflecting on the relationship between physical and medical science, Huxley remarked on "the immense indigestible mass of information" that constituted the medical curriculum, ranging across "physics, natural philosophy, chemistry, botany, zoology, with comparative anatomy, human anatomy, histology, pathology, therapeutics, medicine, surgery, dietetics, jurisprudence."[1] "The thing was absurd," Huxley opined, according to a report published in the *British Medical Journal*.[2] There was virtually no field of scientific inquiry that Huxley considered to be beyond the purview of the contemporary medical curriculum, which had absorbed new areas as fast as they became codified during the nineteenth century. However, as well as established disciplinary specialisms such as physics, chemistry, and botany, fields that arguably sat outside the process of discipline formation—in particular, dietetics—were also included by Huxley, despite regular claims in the lay press that "roundly accused the medical profession of complete ignorance of the subject of dietetics."[3]

The coalescence of disciplines has long been considered a hallmark of nineteenth-century science. Indeed, historians have characterized this specialization as one of the primary factors behind the emergence of what we might term *modern science*. The decline of natural philosophy around 1800 and its replacement with a rather more fragmented approach to the creation of knowledge was at the root of Simon Schaffer's articulation of a new vision of how knowledge-making processes were understood and historicized in the early nineteenth century.[4] This model has since been refined by others, including Jonathan Topham, who located these dramatic changes in the means and processes of scientific communication as well as disciplinary change, and James Secord, who has been more critical of revolutionary change in science and of the so-called Scientific Revolution in particular.[5] For these scholars, disciplines become visible and concretized in large part because of modes of textual communication that functioned as expressions of shared practice, language, and identity.

Our engagement in this chapter is also with texts, the plurality of which belies the narrative of shared identity in the case of nutrition. Alongside those works produced by an emerging cadre of nutritional specialists, we broaden the scope of relevant texts to include manuscript recipes.[6] In doing so, we show that factors underpinning the organizational assembly of knowledge into disciplines cannot readily account for the plurality of approaches which gravitated around nutrition in the nineteenth century. In the case of nutrition, as Elizabeth Neswald, David F. Smith, and Ulrike Thoms have argued, we cannot look to "theoretical concepts of 'boundary work' and 'boundary objects'" to understand the "indistinct and variable borders of nutrition science."[7] We argue instead that nineteenth-century nutrition—in the form of both knowledge and practice—is best understood as a hybrid field. That is, it was a field that incorporated approaches from multiple nascent scientific disciplines, lay and unorthodox expertise, and long-standing experimental and domestic conventions. At a time of discipline formation, we also suggest that our understanding of other areas of Victorian scientific investigation that resist neat classification or definition might benefit from being viewed in this way.

Neswald, Smith, and Thoms argue that the field of nutrition science began to acquire those attributes so central to scientific disciplines in the late nineteenth century: specialized research institutes, professional societies, and dedicated journals. However, their perspective is entirely compatible with a process of field formation that occurred over a longer time period; implicit in their argument is the view that the early twentieth century marked a culmination rather than an initiation.[8] However, the claim that "modern" nutrition science was born when the "first vitamin was isolated and chemically defined in 1926" requires some revision and nuance.[9] Given that, in their words, "nutrition can hardly be accurately described as any 'younger' than most modern biomedical disciplines," any exploration of how nutrition emerged as a hybrid field in a significantly earlier period must necessarily take into consideration the broader scientific and sociocultural landscape.[10] This leads us to a central question: How did the far-reaching disciplinary and communicative formation of early nineteenth-century science impact a heterogeneous area of investigation such as nutrition and vice versa?

Drawing on nineteenth-century medical and scientific texts as well as Victorian manuscript recipes—textual forms central to professional and lay understandings and practices of nutrition—we argue that appreciating the heterogeneity of nutrition and its relationships with the practice of preparing food and medicaments in the home has two major consequences. First, we see that investigators of various stripes who attempted to understand human nutrition better drew heavily on a broad range of scientific disciplines that were themselves undergoing a process of differentiation. Second, while it might be a stretch to claim that there was a coherent scientific disciplinary identity for nutrition science in this period (and perhaps even thereafter), the example of nutrition provides a model for thinking about activities that functioned outside the confines of a nascent disciplinary framework that continues to dominate our understanding of the structure of Victorian science. For example, the importance of chemistry for nutritional thought in the nineteenth century owed much to the early investigations into caloric measurement pioneered by Antoine-Laurent Lavoisier (1743–1794). As we shall see, subsequent experimental work by Justus von Liebig (1803–1873) among others aimed to determine the heat-generating properties of foodstuffs, as well as their chemical composition.[11]

Given the breadth of interest in nutrition as it gradually and incompletely coalesced into a discipline, we begin not with the scientific rationale of healthy eating or undisturbed digestive processes but with recipes collated for domestic use, before moving on to consider how a wide range of scientific perspectives were brought to bear on questions of dietetics and nutrition. These forms of nutritional knowledge developed in parallel and yet shared many features. This reflects the centrality of everyday cookery practices in generating concepts of nutrition, largely independent of professional scientific concerns. We explore first the multiple ways in which health-related matters appeared within manuscript recipes, before considering how nutrition was bounded as a scientific undertaking, and finally revealing the central role of unorthodox and lay expertise in constructing nutritional knowledge.

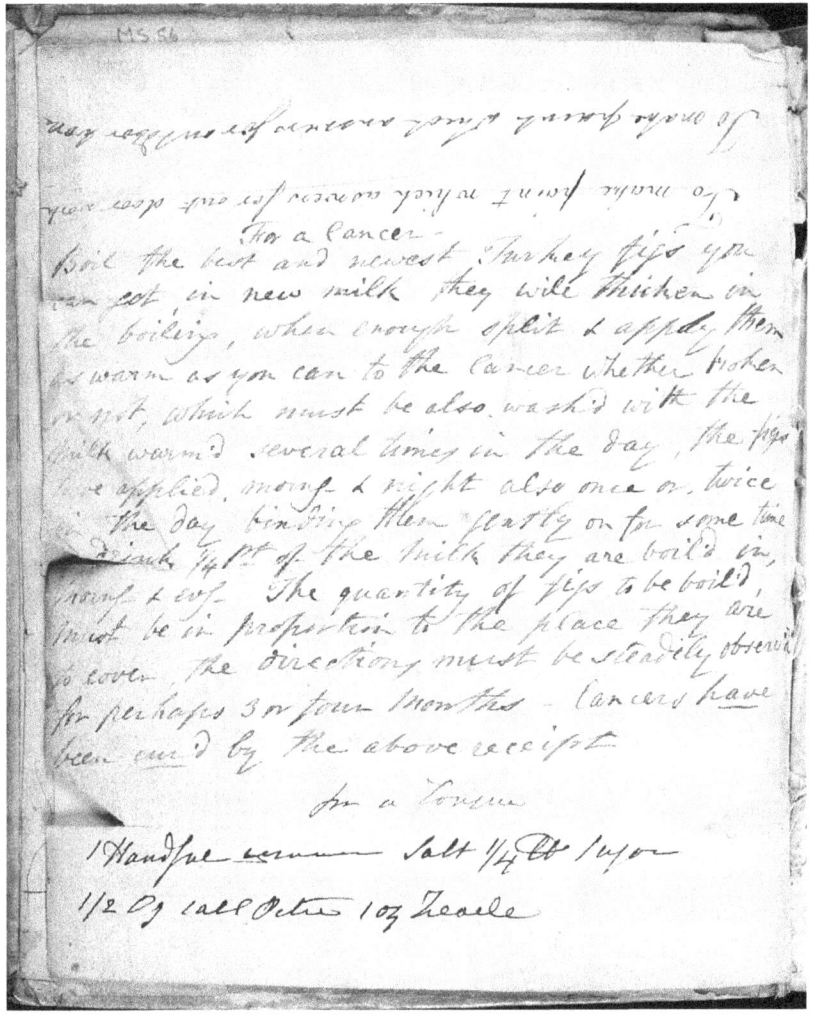

FIGURE 10.1. Recipe "For a Cancer," including more detailed discussions about how and when to apply the preparation. *Source: Cookery Book*, 1800[?], image 004, TD/006/001, Special Collections, University of Leeds.

MANIFESTATIONS OF HEALTH IN COOKERY

Our investigation centered on a series of seven lengthy manuscript recipe cookbooks compiled in the nineteenth century, and now held at the University of Leeds. Collated across the nineteenth century by a number of different, though usually anonymous, compositors, together these cookbooks emphasize the long-standing and widespread concern with running a well-fed and healthy household in the Victorian period. They reveal that there was not a neat divide between the preparation of foods and medicines in a domestic setting; remedies for both everyday ailments and more serious conditions coexisted with restorative recipes, emphasizing that household medicine and cookery and their systems of knowledge and practice were interdependent.

The households concerned were likely in comfortable financial circumstances—one recipe was titled "Cheap Soup for the Poor"—though the prevalence of home remedies for conditions such as rheumatism, measles, whooping cough, and "a cancer" (figure 10.1) suggests that preserving the household budget was at least as important as preserving meats, fruits, and vegetables.[12] Women's expertise features prominently in the cookbooks, as many recipes were credited as appropriate to their originator. This included medical recipes, which were given equal weight compared to those recipes that stemmed from medical practitioners. The health of the household thus rested on the judgment of the respective compilers and users of the cookbooks, who included whatever they viewed as useful for their own purposes. Together the cookbooks therefore provide an important insight into the concern with nutrition and health in the middle-class household in the Victorian period. They show that households both engaged with and were separate from wider scientific debates on nutrition and serve to highlight the importance of lay knowledge—generated independently from disciplinary structures—in the preparation of health-related foods and medicaments, as well as in defining bodily needs.

The recipes demonstrate the embeddedness of the compilers within the context of the home and its routines. Following advice like keeping tea biscuits "in [a] Tin Cannister [sic] to make them retain their Crispness" necessitated both a place in which to keep such an object, and the possibility of doing so for a long period of time.[13] More simply, the tools, such as cutlery, jars, pots, and pans, and processes, like fermentation, heating, and cooling, needed to make many of the recipes required access to a kitchen and perhaps a pantry to be able to use amenities such as the fire, the oven, the basin, and the shelf for periods of time ranging from a few minutes or hours, to days, weeks, and months. Timings for recipes were often discussed, either in terms of clock time ("bake it half an hour, in a quick oven" to make one custard pudding), or more empirical terms, engaging different senses.[14] For example, one recipe for apple puffs simply stated that they ought to be boiled "in a little water until they are dry."[15] Another required that "the vessel have some vent till you find by the noise ceasing that it has done fermenting,"[16] while a recipe for ginger wine advised "when it is blood warm put it into the cask."[17] Furthermore, these recipes were envisaged as suiting particular times of the day: a recipe for "Neat picking to your Tea" (an egg-based meal) was described as "the best thing for breakfast ever," an almond flummery "A genteel Dish for Supper."[18]

Those recipes more explicitly concerned with health might form part of a regular regimen. A "coffee cup full" of a "Diet Drink" was advised to be taken "three times a day"; a "strong herring soup" was to be stewed and then "taken every night and morning warm."[19] The "strengthening wash" of "Mock Sea Water" was advised to be had "fresh every morning"; a cure "For deafness" required a "drop or drops into the Ear morning + night."[20] More generally, beef tea was always "good for weak + sickly people," while a rhubarb cordial—somewhat cryptically—"will make a good medicine for those who are not very ill."[21] At the same time a number of recipes were intended to help keep the household clean and free from pests: alongside recipes for "Eye water" and Naples biscuits on one page were recipes "To Clean Tables" and "An infallible Recipe for destroying Rats."[22] It was not only individuals in the household who required a regular regimen to maintain good health, but the household itself too.

Nonetheless ill-health and disease were often irregular, with recipes providing relief to ailments that were more specific and short-term. One "remarkable good" gingerbread recipe was also "a very good receipt for any person who may have wind in their Stomach."[23] Recipes for coughs abounded, as did those for sore throat, mouth, and skin; indigestion and bowel problems; swallowed pins and bones; and toothache, cuts, and sprains. For such complaints the judgment of those in the household regarding the severity of the illness influenced the treatment. For example, one cough remedy containing laudanum was to be taken "2 or 3 Tablespoonful 2 or 3 times a day" as required.[24] Other recipes involved more specific sorts of judgment, incorporating diagnoses by members of the household of diseases such as ague (a type of fever or shivering fit that was also often called malaria), asthma, measles, rheumatism, and whooping cough. Such a judgment might see a member of the household specifically identified as sick and treated differently as a result. A syrup for "a Cough though a Consumptive one" was to be taken every morning, after which the "Patient" was "to lie in bed a little while afterwards."[25] On occasion, these judgments and the recipes of which they were a part required interaction with the medical profession of some kind. Mrs. Hind's recipe "For the Ague" was only to be taken after "an Emetic," likely procured from a druggist, had been taken first. Then the mixture was "to be taken every two hours when the fit is off."[26] Another recipe for weak or inflamed eyes was specifically "to be powdered at the Druggists."[27]

Compilers thereby located their works within a certain environment and place. The specific kinds of ailments and opportunities encountered, the ingredients available for use, and the local amenities imagined were all centered around the climate, trade, and infrastructure of Victorian Britain. Thus rhubarb wine was to be bottled in March, "the following June it will be fit for use"; ingredients were noted as foreign, such as "Lisbon wine" and "Jamaica Pepper"; and the druggist was assumed to be at hand.[28] The British regimen was thus located within a network of imagined and concrete relations that shaped what recipes were made and what they were made with. Ingredients emphasize that both national and international trade was vital to the contents of the Victorian pantry—staples included perishables like butter, as well as relatively recently introduced foreign spices such as cayenne pepper. National and international trade was fundamental to the scope of Victorian nutrition and an individual's dietary regimen. Nevertheless, for all of these recipes, the most important place was the home they were intended to serve.

Recipes enabled the household to treat issues as they saw fit. This central role extended to creating specific treatments intended to engender certain effects in the body. For example, the recipe for "Nitre" was to be used "when it is desirable to produce perspiration"—a specific and specialized purpose for which no explanation of usage was included.[29] The implication of this was that it was assumed by the compilers that other readers would know when such a recipe was appropriate. This authority pertained to how different members of the household were treated too. One recipe for "Hooping [sic] Cough," which contained extended instruction regarding its use ("Observe no other medicine must be given at the same time. If aperients should appear to be required, at first they must not be given, as the increasing doses of Alum will remove the necessity," and so on), listed the appropriate doses for different individuals in the household: "The dozes [sic] mentioned here are for children about 14, adults may increase the quantity to 26 grams—Infants may begin with 4 or 5 grains."[30] A necessary inclusion given the virulence of such diseases in the Victorian period, but one tempered by the recipe user's judgment, hedged terms such as *about* and *may* emphasizing where ultimate authority in determining dosage lay.

The overlap of ingredients between food and medical recipes was significant. For example, conventional foodstuffs such as sago, rice, and pearl barley, as well as Eringo root—a candied or pickled preparation of sea holly root—were core ingredients in a recipe for Restorative Jelly (figure 10.2).[31] Similarly, some recipes that were not self-evidently medicinal—such as one for a calf's foot cooked in milk and flavored with

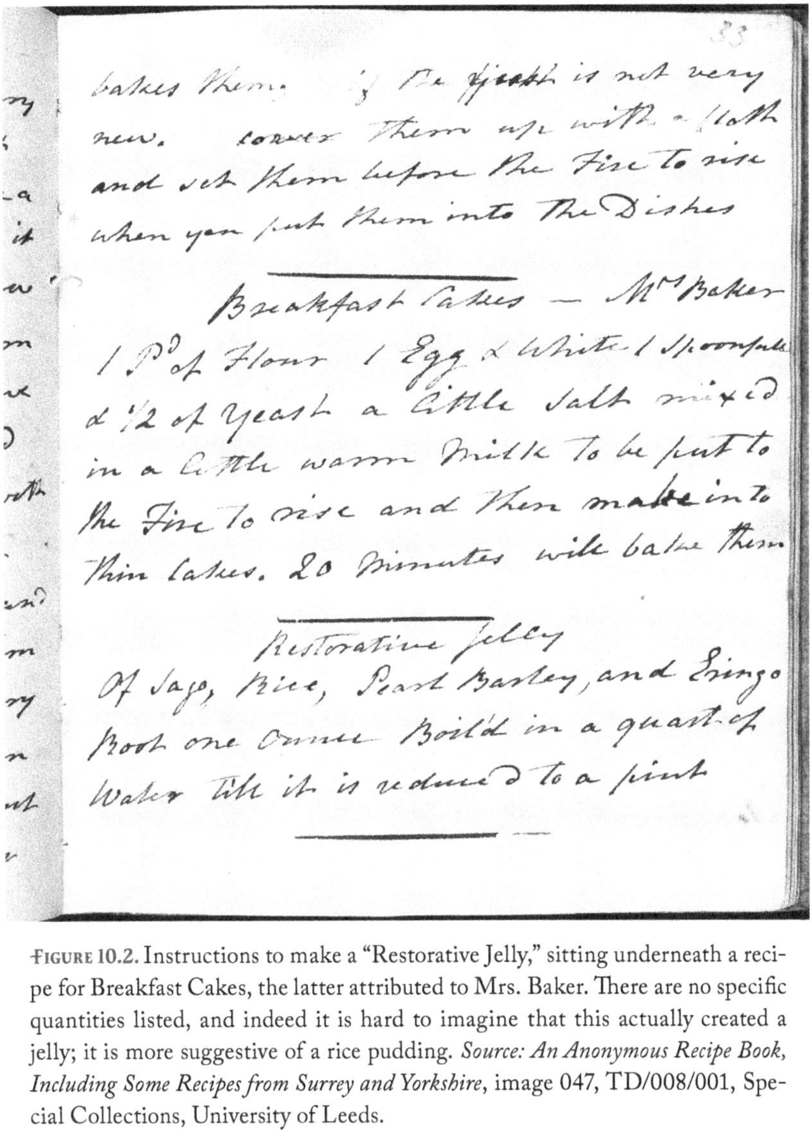
bakes them. if the yeast is not very
new. cover them up with a cloth
and set them before the Fire to rise
when you put them into the Dishes

Breakfast Cakes — Mrs Baker
1 Pd of Flour 1 Egg & White 1 Spoonfull
& 1/2 of Yeast a little Salt mixed
in a little warm Milk to be put to
the Fire to rise and then make into
thin Cakes. 20 Minutes will bake them

Restorative Jelly
Of Sago, Rice, Pearl Barley, and Eringo
Root one Ounce Boil'd in a quart of
Water till it is reduced to a pint

Figure 10.2. Instructions to make a "Restorative Jelly," sitting underneath a recipe for Breakfast Cakes, the latter attributed to Mrs. Baker. There are no specific quantities listed, and indeed it is hard to imagine that this actually created a jelly; it is more suggestive of a rice pudding. *Source: An Anonymous Recipe Book, Including Some Recipes from Surrey and Yorkshire*, image 047, TD/008/001, Special Collections, University of Leeds.

cinnamon and lemon peel—were also endowed with health-related potential, in this case as "an excellent strengthener."[32]

Medical recipes in the books were collected from a wide variety of sources, both lay and medical alike, in the same manner with which the other recipes were collected. That is, recipes from prominent physicians such as "Dr Baillie"—most likely Matthew Baillie (1761–1823), physician-in-ordinary to George III—sat alongside the rest with equal weighting, just as recipes for meals "from a famous cook in London" did.[33] The major source of expertise within these cookbooks, however, was that of women. This ranged from personal relations, such as "Mrs Willm Walker's receipt. got fr[o]m Mrs Bickerdick. her late housekeeper," to women in print, like the recipe for olives royal taken from "Mrs Lees English Cookery Book."[34] Local fairs, conventions, and meetings might have provided a source for new recipes too. The dating of a number of Mrs. Hind's recipes—June 22 and June 23, 1830—suggests that the cookbook's compiler attended a specific event where Mrs. Hind spoke.[35] Either way, the handwritten recipes presented the information contained within the cookbooks as being of equivalent authority, whatever the source. In that regard, the relation between the handwritten recipes and their sources is especially significant: How were recipes from other cooks and cookbooks collected, and what role did the assignment of credit to other authors have in what were primarily private works?

Perhaps the answer to those questions lies in the working nature of the books. These were both repositories of information and actively used within the household. To that end the cookbooks were variously organized through techniques such as numbering recipes or creating contents pages in the otherwise blank, or lightly ruled, notebooks. Moreover, all of the cookbooks we examined bore signs of use by at least one other individual, whether through the addition of further recipes, or through simply underlining important parts of a recipe in pencil. As repositories, these works would archive important procedures and sources of information however obtained; while in active use—perhaps propped open by the stove—specific parts of recipes, especially ingredients and their quantities, became of particular interest. Recipes circulated across society through the copying practices of cookbook compilers, but also down the years through the role of those works as a store of information. As a result ideas regarding health and nutrition, an appropriate regimen and treatment, circulated in the same way.

The virtue of practical experience was not simply restricted to the compilers and followers of recipes. As Charles Lane (1800–1870), for many years manager of the *London Mercantile Price Current* periodical, noted, "The dietician who continues to indulge in flesh knows as little of the merits of a bloodless diet as the fireside traveller understands of climes he has never visited."[36] Lane was a prominent advocate of vegetarianism, who in 1849 published a large text, *Dietetics*, which sought to identify the principles and discern the practical implications of ideal human nutrition.[37] In a chapter dedicated to anatomy, Lane devoted considerable time to discussing organic chemistry, "a science yet in its earliest infancy, though of deeper importance to human welfare than any other department of analytical knowledge."[38] His argument about appropriate diet stemmed largely from the anatomical differences between humans and animals, and he bemoaned the absence of scientific consideration of cookery, which meant that "our eating habits have been left, as it were, to chance or the cooks, or, at best, to the capricious palate and the unenlightened taste."[39]

Throughout these recipes we see the importance of nutritional knowledge and its place in the context of both domestic cookery and everyday practices of health. If we take seriously the production, codifying, and use of these foods in generating and propagating knowledge about nutrition in the domain of the nonspecialist, then what we see is a key site of nutritional knowledge production that developed and functioned independently from organized disciplinary practices and organizations. However, even within specialist texts on the subject, nutrition was a hybrid field that flourished without the disciplinary superstructures so characteristic of other areas of Victorian scientific inquiry.

The Scope of Nutrition "Science"

The content and form of many nineteenth-century recipes show that the preparation of both food and home remedies relied on considerable practical skill. Extra-recipe comments also reinforce the centrality of lay expertise in administering these restorative or therapeutic preparations, but how was scientific information mobilized to determine the most appropriate diet for human bodies?[40] Such questions long predated the Victorian period, of course. The archetypal premodern natural philosopher, Francis Bacon (1561–1626), approached the topic with typical holistic breadth. In his *Essays, Civil and Moral*, Bacon noted in chapter 30, "Of Regiment of Health," that one should "beware of sudden change in any great point of diet" and "be free-minded and cheerfully disposed at hours of meat and of sleep and of exercise."[41] However, when we interrogate works concerned with nutrition and dietetics in the nineteenth century, we see a heterogeneous field that self-consciously eschewed alignment with any single emerging disciplinary specialism.

We know that myriad factors, including but not limited to age, overall health, race, and gender, were frequently considered by specialists and lay authors to be important factors when deciding how to fuel bodies, but how did the acknowledgment of bodily plurality and distinctiveness draw on multiple scientific disciplines, including physiology, chemistry, and physics? Dietetics was a central social, as well as a scientific, concern, and reformers variously advocated the benefits of specific dietary modifications, whether the vegetarianism so beloved of Romantic utopian fiction or the increased meat consumption favored by medico-social commentators.[42] In the search for determining an ideal human diet, a broad range of scientific actors pursued all possible lines of inquiry. As an updated 1866 edition of Andrew Ure's (1778–1857) celebrated *Dictionary of Arts, Manufactures and Mines* noted under the entry for nutrition, "in order to study the nature of the process of nutrition, we are obliged to take advantage of all the avenues to knowledge which present themselves."[43] Ure, who advocated a synthetic approach to nutrition, drawing on all the branches of science that considered the nature of food and its relationship with physiology, was himself the embodiment of the plurality of early nineteenth-century science. Having originally served as an army surgeon, he occupied the chair of natural philosophy at the University of Glasgow, where his investigations and publications spanned mechanical, astronomical, geological, and chemical phenomena.[44]

In a paper delivered to the Edinburgh Medico-Chirurgical Society on November 9, 1842, John Hughes Bennett (1812–1875) addressed the topic of "abnormal nutrition."[45] Bennett, an extra-academical lecturer in histology and physician to the Royal Dispensary in Edinburgh, mobilized not only the physiology of the circulatory system, but also the "chemical constituents" and "mechanical or any other causes acting more especially upon any part of the frame [of the body]" as key causative factors behind the imperfect distribution of nutriment through the human organism.[46] Uncovering the chemical composition of food was of course a critical step in determining its specific nutritional value and function within the body, but so too was the chemical composition of the digestive system central to a complete theory of nutrition. In the ninth edition of *The Physiology of Digestion*, published posthumously in 1849, the Scottish physician Andrew Combe (1797–1847) argued that "in the case of the gastric juice itself, its special properties are so subtle that an elevation of a few degrees of temperature is sufficient to annihilate them, although the ablest chemist cannot detect any change whatever in its chemical constitution."[47] Previous editions, such as the authoritative second edition (1836), made no such references, highlighting the increasing significance of chemical knowledge in understanding digestive and nutritive processes, alongside references to the "muscular fibres of the stomach" and highly detailed anatomical descriptions of its structures, which featured in the later tenth edition (1860).[48]

Nutritional concepts also reached beyond the purely alimentary and into the metaphorical and metaphysical. In *The Dietetics of the Soul*, Ernst von Feuchtersleben's (1806–1849) instructions in 1852 for ensuring that "the soul is preserved in a state of health," nutrition functioned as a crucial metaphorical device.[49] Feuchtersleben's approach rested on "weaving together ethics and dietetics . . . to give a practical demonstration of the power which the mind exercises over the body."[50] A surgeon by training,

Feuchtersleben practiced in his native Vienna in medical psychology and lectured at his alma mater, the university of the Theresian Academy.[51] In *The Dietetics*, which ran through numerous editions and translations in the mid-nineteenth century, Feuchtersleben argued that there were indispensable connections between dietetics and ethics, which he considered to be "a powerful demonstration of the power which the mind exercises over the body."[52]

In his address to the Hunterian Society of London on February 9, 1859, Alfred Smee (1818–1877) took as his subject *General Debility and Defective Nutrition*.[53] Although Smee was, like Feuchtersleben, first and foremost a surgeon, his wide-ranging interests included electrochemistry and metallurgy. He opened his address on what might seem a straightforwardly medical (or at least biological) topic with reflections on combustion, referencing "sooty naphtha compounds," "carbonic acid," and the resulting "mechanical power."[54] For Smee, unpacking "the economy of man" required nothing more than understanding the generation of heat in the body, a theory based on the interaction between, and immutability of, twelve chemical elements.[55] As if to reinforce the point, Smee explicitly advanced the argument that "continuous chemical research" was necessary to elucidate "the phenomena of organic bodies."[56] Combined with this appreciation for the role of chemistry in understanding nutrition was a characterization of the body itself as mechanical. As Smee wrote, "The stomach may be considered as a receptacle of matter, the heart as a distributor, whilst the lungs, skin, and kidneys, are eliminators of changed matter, to perform the same functions in man as the chimney does to the stove."[57] In a healthful body, in Smee's model, "the well-nourished man is neither thin nor is he fat; and the muscular substance is neither attenuated nor is it wasted," indicating that internal bodily balance—and balance with the external environment—was of primary importance.[58] In common with many of his contemporaries, Smee considered that "in all ages, and under all circumstances of debility, the food is the matter of primary importance."[59] He also outlined the process through which the physician should put this approach into practice, taking first a detailed description from the patient about their own "peculiar experiences as to food" given that "the food that is most proper for the majority of persons, is absolute poison to some."[60]

The centrality of the animal economy and the production of heat was also evident in other accounts of nutrition. James Henry Bennet (1816–1891), physician-accoucheur to the Royal Free Hospital in London who later treated both Queen Victoria (1837–1901) and Robert Louis Stevenson (1850–1890), set out in his 1858 text *Nutrition in Health and Disease* to persuade "medical brethren of the imperative necessity of studying dietetics in connexion with chemistry and physiology."[61] In his opening statement, Bennet referred to the object of his interest: "the various functions and operations through the agency of which the animal economy is developed, its waste is repaired, and its heat is maintained."[62] Through the course of his treatise it became clear that chemical processes—"nature's chemistry"—were the fundamental level at which the principles of nutrition became intelligible.[63] To take the case of the stomach, Bennet argued that "the chemical changes that take place in the stomach are important. The fibrin and albumen of animal tissues are chemically acted upon by the gastric juice"; in reality, for Bennet the physiology of digestion and nutrition made sense only in light of chemistry.[64]

Nor was Bennet alone in his reverence of bodily chemical processes. In 1861 Thomas Grainger Stewart (1837–1900), later president of the Section of Medicine of the British Medical Association, delivered a series of lectures at the University of Edinburgh on materia medica and dietetics, published the following year. Stewart highlighted the importance of "medicinal agents . . . [and] the principles of dietetics and of climatology" as important factors in both the prevention and cure of diseases.[65] Writing in the same year, William Brinton (1823–1867), physician and lecturer on physiology at Saint Thomas's Hospital, London, published *On Food and Its Digestion*. Here Brinton considered "the Natural History of the various alimentary substances in due subordination to their . . . bearing on life and health," highlighting the importance of the practical consequences of a closer knowledge of nutrition.[66] While Brinton was himself a physiologist and clinician, the opening salvo of this work resembled far more closely a chemical treatise as he invoked the researches of Justus von Liebig (1803–1873), Julius Vogel (1835–1899), and Carl Neubauer (1830–1879) on multiple occasions.[67] After a chapter on bodily waste, Brinton spent around thirty pages discussing the chemical nature of food and "alimentary constituents," only then moving to consider the organs of the digestive tract and, in the second half of the book, different forms of food and dietetics.[68] Most tellingly for our purposes, however, Brinton reserved an entire chapter to cookery. "The claims of cookery," asserted Brinton, "in a scientific point of view, have yet to be established."[69] Once a complete science of cookery was devised, it would, he argued, sit "somewhere between Chemistry and the Fine Arts," necessitating the investigation of specific foodstuffs using techniques from "chemistry and the microscope."[70] Brinton's vision for the field was therefore hybrid, even beyond the sciences.

The complexity of cookery methods, and in particular their relationship to different ingredients, was a source of bemusement for Brinton, who acknowledged the "great law of cookery: namely, that different substances require such different times and heats for their cooking, as baffle alike the clock and the thermometer to regulate."[71] The ambition to determine a universally healthy diet was further undermined by the fact that "the child requires, on every ground, a food which is purer, richer, simpler, and more digestible than that of an adult," while "moderation in the quantity of food, care in its preparation, and simplicity in its form, are the principal requirements in the Dietary of the aged."[72]

Some of the most revealing commentaries on diet in the mid-nineteenth century came not from medical texts in the hybrid field of nutrition science, which itself necessarily evades neat disciplinary definition, but from volumes focused primarily on cookery. Alexander Murray, a physician who turned his hand to such publications, outlined an approach that mirrored closely those found in manuscript recipe collections. In *The Domestic Oracle: or, A Complete System of Modern Cookery and Family Economy* (1826, reprinted in 1850) Murray outlined properties of different foods and extensive accompanying recipes, before detailing proper ways of managing domestic economy ("wastefulness of every description should be cautiously avoided") and around forty pages of medical recipes.[73] As we move to consider a third critical textual form—published but authored by the nonspecialist—it becomes clearer that nutritional knowledge and practices were generated and captured by a broad range of historical actors who did not self-identify with or draw on distinct disciplinary identities. More significant than that, perhaps, is that nutrition-related publications comfortably inhabited multiple disciplinary spaces, profitably and congruously.

Nutrition, Dietetics, and Unorthodox Expertise

Lay authors, such as the democratic temperance advocate Joel Pinney (ca.1790–1869), frequently bemoaned the failure of scientific investigations to uncover the fundamentals of the relationships between body and nutriment. As he wrote in his popular 1830 treatise on the reasons behind "the present deteriorated condition of health," "the mechanism of nutrition, however, cannot be completely explained, by reason of our want of knowledge how each organ operates upon the aliment presented to it."[74] Pinney acknowledged that he was transgressing on territory already well trodden by medical authorities. His approach to diet within a schema of longevity-inducing lifestyle suggestions was, however, arguably as rooted in the moral as it was the biological. Pinney's claim to authority was grounded in lived experience. He recalled his own adolescence, when he "fell prey to indolence and listlessness, the consequences of luxury and irregularity . . . by a departure from the temperance prescribed by nature."[75] Pinney published several other texts in similar veins over the following two decades, tracing an alignment between good character and healthful nutrition.[76] This was a prevalent theme in early nineteenth-century commentaries on diet and dietetics. Michael Donovan's (1790–1876) similarly wide-ranging *Domestic Economy* (1837), published in a series of 133 texts edited by Dionysius Lardner (1793–1859). *Lardner's Cabinet Cyclopaedia*, was a four-hundred-page treatise on solid foodstuffs. The genre in question—that of a "family library"—was a popular publication form in the first half of the nineteenth century, as the *Cabinet Cyclopaedia* was critical in establishing a system of knowledge calculated to "enforce the cultivation of religion and the practice of virtue."[77] Donovan argued that "excessive addiction to the pleasures of the table leads to the debasement of the mind, the injury of the body, and the waste of property."[78]

As well as explaining how "normal" physiological function influenced, and was influenced by, dietetics, the relationship between nutrition and pathological processes was also at stake. Indeed, defective nutrition was often mobilized as a cause or consequence of abnormal bodily states. As Peter Redfern (1821–1912), longtime lecturer in anatomy and physiology at King's College in Aberdeen, noted in his influential 1849 study *On Anormal Nutrition in Articular Cartilages* that "inflammation is a peculiar perversion of nutrition . . . [or] a process of anormal nutrition."[79] For these purposes, practitioners of both the scientific approaches observed in physiology and pathology, as well as clinical practice, had stakes in determining precisely how so-called defective nutrition manifested in both laboratory and clinic. Writing in 1839, Richard Baron Howard (1807–1848), a physician practicing in Ancoats through a dispensing pharmacy, reflected on his extensive experience of imperfect nutrition, informed by his former role as resident medical officer at the nearby Manchester Poor House. Howard noted that "many lives are lost . . . from the combined effects of cold and inadequate nutrition," charting throughout his treatise on "the morbid effects of deficiency of food" a close relationship between adequate nutrition and internal heat generation.[80] Notwithstanding this increasingly strong alignment between bodily heat and fuel, lay and unorthodox authors who trespassed on such territory were often dismissed, unless they engaged with the current science. For example, in his 1839 treatise *An Essay on Food*, the author and polymath William Grisenthwaite argued that the "almost sole use of food" was to generate heat as part of "the act of respiration."[81] In a review published in the *British and Foreign Medical Review*, Grisenthwaite was castigated for being "one of the many examples in the present day of persons determined to vindicate the Creator according to their own views,"

SINGULAR EFFECTS OF THE UNIVERSAL *VEGETABLE* PILLS ON A GREEN GROCER! *A FACT!*

Who Green'un like was order'd to live for the space of one Month upon
Vegetable Diet & to Take during that time 132 Boxes of Vegetable Pills
for the Cure of a Gangreen. & Being caught in a Shower of Rain in the
Green Fields in the evening of the 1ˢᵗ of April last was put to Bed 'midst
Shooting pains, & in the Morning presented the above Phenomenon of a
Moving Kitchen Garden !!!

Query —   Is he not one of the Productive Classes

FIGURE 10.3. A caricature of the results of taking an excess of Morison's Universal
Vegetable Pills by Charles Jameson Grant (fl. 1830–1852). *Source:* C. J. Grant,
"Singular Effects of the Universal Vegetable Pills on a Green Grocer," from a series Grant's Oddities, 1834, Wellcome Collection, https://wellcomecollection.
org/works/h8ppw33a.

while noting that his doctrine was not without merit and "defended by arguments drawn from chemistry."[82]

Understanding nutrition promised to unlock far-reaching questions about bodily control, such as those of the animal economy that had animated members of the Montpellier school of vitalism in the late eighteenth century.[83] As a whole-body process, nutrition was a lens through which one might access the fundamentals of the organ systems and connect proper physiological function with environment. Authors such as James Morison (1770–1840) in his book *Morisoniania* (1831), effectively an everyday manual of health, argued that because nutriment pervaded the body "there is a common source of nutrition for the whole body; a single centre of circulation; a common place of union for all sensations and volitions—for nervous energy of whatever kind."[84] The British College of Health, a grandly titled institution established in London by Morison in 1826, was designed to promote his understanding of disease, styled as "the hygeian theory of disease." Morison—"the Hygeist"—recounted in his publications an extended period of ill-health, impervious to the attentions of physicians and surgeons of every kind, before identifying the curative power of "Vegetable Universal Medicines," based on the fundamental premise that nutrition pervaded the whole body.[85] Such claims were ripe for satire, and cartoonists duly ridiculed the purported power of vegetables as the basis of universal health and a healthy diet (figure 10.3). While Morison might be most accurately characterized as unorthodox in relation to scientific practice, his views reflected a far more mainstream perception that achieving improved nutrition—and drawing on the intellectual responses of multiple disciplines to do so—could yield huge improvements in human health, productivity, and longevity. While this was frequently debated in medical, scientific, and popular texts, the practices associated with such views were almost universally located within the home.

Morison's British College of Health was by no means alone as an institution promoting a nutrition-based approach to disease management and treatment. James Wilson (1807–1867), a London- and Paris-trained physician who together with James Manby Gully (1808–1883) founded a hydrotherapy establishment near the spa town of Malvern in Worcestershire, appended large sections on "digestion, nutrition, regimen, and diet" to his expansive seven-hundred-page treatise on water cures.[86] Published in 1854, this text was structured, somewhat unusually, around a series of 113 conversations between doctor and patient. These were arranged into six parts, the third and fourth of which were dedicated to "a popular exposition of the scientific details of nutrition, digestion, and diet" and "the scientific principles of diet and regimen, embracing the applications of modern physiology and animal chemistry to human nutrition."[87] Wilson's textual mouthpiece, the Doctor, defined nutrition as "a series of complex processes, by which the crude pabulum of organised structures is gradually elaborated and perfected, circulated, and assimilated."[88] In Wilson's vision of the "economy of animated nature" (deploying again the concept of an animal economy), the maintenance of heat was again primary, allied to the building up of the organism and its repair.[89] Demonstrating the continuity between external and internal environments, Wilson also advanced a popular argument in mid-nineteenth-century nutrition and dietetics, namely that "the chemical constituents of blood and of food are the same."[90] This perhaps explains why the process of assimilation—the incorporation of the nutritional properties of food into the body—was such a central aspect of nutritional theory. It represented one of the most important aspects of body–environment interaction.

Along with the environmental influence, the anatomical structures of the body were also implicated in critical processes of digestive action. An anonymous pamphlet *Vital Nutrition: A Popular Application of the Principles of Modern Science to the Promotion of Health and Vital Energy* (1859), outlined a system of physiology that rested on the twin powers of vital fluid and vital power. Each was intimately connected with the body's alimentary intake and internal processes. The author contended that these were mutually interdependent and acted both chemically and physically on food. For example, "when we masticate a piece of bread, a stimulant effect is produced through which saliva necessary to the deglutition of the bread is made to flow; when we take food into the stomach, the stomach is induced . . . to commence a certain action necessary to the digestion of that food."[91] The emphasis on the reciprocal relationship between natural foods and proper digestive action belied the true intention of this pamphlet, however. The back cover contained a lengthy advertisement for two pills, to be taken in tandem, to promote and secure "vital nutrition," as well as a supplementary "cordial of vital nutrition."[92]

The transformation of aliment into materials suitable for bodily growth, repair, and maintenance was at the heart of nineteenth-century theories of digestion and nutrition. Medical practitioner Jonah Horner (1798–1869), who described these pathways as "the great chemico vital process, *the change of tissue*" in his popular 1855 text on the relationship between hydropathy and nutrition, was just one of many writers who argued that the full range of anatomical structures was implicated in ensuring well-regulated and efficient digestion.[93] For Horner this included a recategorization of ganglionic centers in the nervous system as "*nutritive nerves*, because of their serving the great purpose of nutrition."[94] The controlling function of the nervous system enabled the anatomical struc-tures of the digestive tract to elicit "real change in the food" through various chemical and physical processes such as "chymification," made possible by the introduction of secretions such as gastric juices, "oxygen in the saliva," and "free muriatic acid."[95]

NUTRITION AS HYBRID FIELD

So how are we to conceive of nutrition? In the preface to his 1844 prac-tical treatise on diet and longevity, Thomas Parry argued that "health, strength, sweetness, and beauty—development of intellect and long life are all dependent upon this science."[96] In Parry's scheme (and we can trace similar patterns in a swath of writing on nutrition from the mid-nineteenth century) nutrition spanned medicine, laboratory science, aesthetics, domestic economy, and lifestyle, his dietary system being based at least in part on biblical guidance.[97] At the same time, as we have seen, it is misleading to characterize investigations into the nineteenth century as being fundamentally different in character from the nutrition research that came to predominate in the early twentieth century. While the American food scientist Henry C. Sherman argued in 1950, in a line reiterated by nutritionist and writer E. Neige Todhunter in 1973, that "until late in the 1890s the word *nutrition* was not much used," this was clearly not the case.[98] Indeed, while dietetics has historically been more concerned with meal patterns and food habits and preparation, the nu-tritional value and virtues of particular foodstuffs were intimately bound up with such debates during the nineteenth century.

In many senses nutrition defies the conventional categorization on which rests the still-dominant narrative of nineteenth-century dis-ciplinary specialization. There is a strong temptation to characterize any treatment of nutrition as a disciplinary identity in the nineteenth century as simple anachronism. However, even as early as 1796, the American-born British cofounder of the Royal Institution, Benjamin Thompson (1753–1814), later Count Rumford, articulated a vision of "a very important subject, *the investigation of the science of nutrition*" in his *Essays, Political, Economical, and Philosophical*.[99] This is at odds with the characterization of an interdisciplinary modern nutrition science as be-ing located only from the early twentieth century, which has resulted in nutrition science often being conflated with vitamin science.[100] Never-theless, as we have shown, investigations into nutrition were not limited to orthodox science. Publications by the likes of Rumford sat alongside the frequently anonymous authors and compositors of manuscript reci-pes. There is limited evidence for interaction and exchange between such kinds of actors, but all were of significance in shaping contemporary understandings of what constituted a healthy diet. Put simply, the nature of investigations into nutrition demonstrate that beyond the confines of increasingly concrete disciplinary identities for prominent scientific subjects, with their specialized journals, societies, and institutional structures, hybrid fields such as nutrition flourished. The Victorian period, which witnessed the emergence of modern, disciplinary science as a dominant approach to knowledge creation and exchange, also saw the persistence of inquiry into areas that resisted such categorization. In addition to nutrition, we might also consider other areas such as climate and epidemiology suitable candidates for analysis in this way. Each of these drew centrally on distinct disciplines supposedly in the process of becoming concretized yet they were at the forefront of Victorian scientific enterprise. Indeed, given that the term *biology* did not feature in Darwin's publication of *On the Origin of Species* in 1859, but "natural history" was prominent throughout, in certain contexts we might even look on such hybrid scientific fields as of greater significance for knowledge production than individual disciplines.[101] The fact that, in the words of Neswald, Smith, and Thoms, "many of the problems and controversies faced by earlier nutrition scientists are similar to those faced by today's nutrition professionals" suggests that it is both possible, and potentially productive, to trace a more direct lineage from present to past in this case.[102] Indeed, it serves to highlight the persistence of hybrid fields that necessarily remained, and continue to remain, outside more formal disciplinary classification.

"WITHNESS" AND VICTORIAN SCIENTIFIC CONVERSATIONS

BENNETT ZON

What is an afterword? It first appears in William Bishop's *Truth of Times Vindicated* (1643) describing the location of a word found toward the end of a verse in the Bible: the word נתן, "to give," in this case.[1] By the 1660s, however, it acquired its new meaning of concluding remarks or reflections—a conversation, if you will, with previous material. The meaning of *afterword* had shifted from describing the physical location of a word toward the end of a sentence to an ideological position of thought expressed at the end of a book—from location to position, in as many words. The word *scientist* is not dissimilar. First used—ironically, admittedly—in 1834 by William Whewell (1794–1866) to self-caricature a neologism, the word *scientist* locates one type of work among others, like words in a sentence, such as *artist, economist,* or even *atheist*: "Some ingenious gentleman proposed that, by analogy with *artist,* they might form *scientist,* and added that there could be no scruple in making free with this termination when we have such words as *sciolist, economist,* and *atheist*—but this was not generally palatable."[2] Yet as a term, the meaning of *scientist,* like *afterword,* soon meant much more, even if its use was hotly disputed for reasons of philosophical and philological disagreement.[3] By 1840 *Blackwood's Edinburgh Magazine* could claim, for example, that "Leonardo was mentally a seeker after truth—a scientist;

Correggio was an assertor of truth—an artist."[4] As *Victorian Interdisciplinarity in the Sciences* so roundly proves, that distinction, between seeker and assertor of truth, would have consequences for Victorian science, not least in the way the sciences navigated and negotiated the increasingly specialized, yet manifestly interdisciplinary, nature of their quests.

Whewell captures this tension in the same review he used to coin the term *scientist*, of Mary Somerville's (1780–1872) *On the Connexion of the Physical Sciences* (1834), and it can be no coincidence that Bernard Lightman uses the beginning of that same polemic as his chapter's (and the volume's) opening gambit. Terrified that modern science was becoming like "a great empire falling to pieces," Whewell bemoans its "increasing proclivity for separation and dismemberment" and mourns the loss of unity previously embraced by the "learned."[5] Somerville, according to Whewell (and Lightman), provides an antidote "showing how detached branches, have, in the history of science, united by the discovery of general principles."[6] Whewell and Lightman challenge us to think differently about science and its "disciplinary imaginary," to use a term coined by literary theorist Elizabeth Goodstein,[7] prompting questions of an epistemic nature: What is a Victorian discipline, a Victorian science, a Victorian scientific discipline—and what therefore is a Victorian scientist?

As Lightman and I have shown recently in *Victorian Culture and the Origin of Disciplines* (2019), and previously in *Evolution and Victorian Culture* (2014), the Victorian (and today's) "disciplinary imaginary" is more than the sum of its parts; it is by its very nature intractably, limitlessly interdisciplinary because its subjects and objects—its people and their ideas—constantly interact, constantly converse.[8] *Victorian Interdisciplinarity in the Sciences* excavates more of these conversations—these "broad, interdisciplinary sensibilities"[9]—in scientific textbooks, for example, described by Lightman as "attempts to create or redefine the scientific disciplines";[10] in cookbooks, considered by Stark and Bellis as territories of "scientific debates on nutrition,"[11] or in history books like Buckle's *History of Civilization in England* (1857), described by Hesketh as "making history itself a science."[12] These books, and all interdisciplinary (and perhaps all disciplinary) projects, highlight an important set of conversations, not simply among scientific disciplines but between the people and ideas that coalesce to create disciplinary identities. Lightman and I have discussed many types of interdisciplinarity in our work: the epistemically fraught concept of the "interdiscipline," for example; the idea of metadisciplines as metapattern, or more broadly the organization of Victorian knowledge; and indeed there are many other perfectly legitimate formulations of interdisciplinarity. But we have never really considered the conversational nature of Victorian interdisciplinarity. This afterword is a first attempt, and one designed to replicate in structure the casual, spontaneous, and (hopefully not too meanderingly) improvised form of a conversation.

Withness

Any chapter of a book is an afterword, a conversation with previous material, and so are disciplines. Disciplines are all afterwords—conversations, improvisations—because they can converse only in a present of past disciplinary imaginaries. Despite the importance of their intrinsic "pastness," however, modern disciplinary imaginaries seem to be constructed more around concepts of space than time. In an editorial, education theorist Helen Nicholson, for instance, talks of the "interdisciplinary space between drama and education," and asks "questions about where knowledge is located, [and] where its boundaries lie."[13] Social theorist Christina Nadler follows suit, describing a discipline as "a boundless space without structure, only consisting of relations, movement, and affects,"[14] while discourse on networks, systems, domains, communities, and so on—and even "home"—continue to emphasize the structural nature of disciplinary space.[15]

But time is also a factor, if underplayed. Philosophy helps us make the connection. Henri Bergson, for example, suggests that to be fully human is to reclaim awareness of the "true duration" of "the real": "it is the flow of time," he maintains; "it is the very flux of the real that we should be trying to follow. . . . For, as soon as we are confronted, with true duration, we see that it means creation, and that if that which is being unmade endures, it can only be because it is inseparably bound to what is making itself. Thus, will appear the necessity of a continual growth of the universe, I should say of a life of the real."[16] Experiencing true duration is difficult, however, because the present (and everything, like disciplines, within in it) is constrained by habitually becoming its own past. Bergson prescribes a treatment designed to effectively elongate the experiential "inner becoming of things"—creative evolution—but its efficacy involves a behavioral change in our nature: "Instead of attaching ourselves to the inner becoming of things, we place ourselves outside them in order to recompose their becoming artificially."[17] Cultural theorist Raymond Williams espouses a not dissimilar, if politicizing, idea, deprecating the past's colonial oppression of the present; a "longer" present emancipates us from a cumulatively imperial past: "In most description and analysis, culture and society are expressed in an habitual past tense. . . . If the social is always past, in the sense that it is always formed, we have indeed to find new terms for the undeniable experience of the present: not only the temporal present, the realization of this and this instant, but the specificity of the present being, the inalienably physical, within which we may discern and acknowledge institutions, formations, positions, but not always as fixed products, defining products."[18]

Time is also important for what psychologist John Shotter calls "withness" or "withness-thinking," a term that has, inexplicably, escaped the theoretical attention of studies in disciplinary or interdisciplinary history:

> The interplay involved [in withness] gives rise, not to a visible seeing, for what is "sensed" is invisible; nor does it give rise to an interpretation (to a representation), for our responses occur spontaneously and directly in our living encounters with another's expressions. Neither is it merely a feeling, for it carries with it as it unfolds a bodily sense of the possibilities for responsive action in relation to one's momentary placement, position, or orientation in the present interaction. Instead, it gives rise to a shaped and vectored sense of our moment-by-moment changing involvement in our current surroundings—engendering in us both unique anticipations as to what-next might happen along with, so to speak, "action guiding advisories" as to what-next we might expect in relation to the actions we might take. In short, we can be spontaneously "moved" toward specific possibilities for action in such thinking.[19]

There is much to unpack in Shotter's definition of withness-thinking, not least as it applies to Victorian interdisciplinarity and the sciences, but among the highlights are the invisible sense of presence, the awareness of difference, and the urge to respond. He calls these collectively "noticings," and they prompt different types of "reversals," or responses. Noticings range anywhere from instances of being immediately "struck" by something particular in conversation (an event or happening, narrative or ideology) to moments of gradually sensing unity and difference. Reversals involve a systemic response through physical sense, emotional judgment, and eventually intellectual analysis. Together noticings and reversals effectively create what he calls "systemic-thinking."[20]

Like interdisciplinarity, withness-thinking is a dialogical, conversational process for sensing, understanding, and interacting with the "I" of another person—or in disciplinary terms, with the "I" of another discipline—in what is essentially an elongated present (arguably, like an afterword). Shotter describes this kind of conversation as being "from the inside"—of one "I" being *in relation to* another "I" (i.e., withness); its opposite, being "from the outside"—of being an "I" in relation to a "you" (i.e., aboutness).[21] Disciplinary "I"-ness, or professional identity, has always been treated with special privilege—and concern for surviving threats posed by the aboutness of "you"-ness. Joe Moran is not alone therefore when he describes the threat in terms of professional defensiveness: disciplines "defend their territory and reinforce their exclusivity through particular types of discourse."[22] But equally, a plethora of hyphenated discourses attests to the theoretical health of our current methodological "I"-ness, including a-, anti-, cross-, de-, hegemonic-, meta-, multi-, neo-, non-, pluri-, post-, re-, trans-, un-, and even deviant-disciplinarity.[23]

More simply perhaps, anthropologist Veronica Strang and physicist and natural philosopher Tom McLeish convert interdisciplinary withness into a form of knowledge exchange,[24] and their definition was adopted by the Interdisciplinary Advisory Panel for the United Kingdom's Research Excellence Framework for 2021: "Interdisciplinary research is understood to achieve outcomes (including new approaches) that could not be achieved by established disciplinary approaches alone. Interdisciplinary research *features significant interaction* between two or more disciplines and/or moves beyond established disciplinary foundations in applying or integrating research approaches from other disciplines."[25] Strang and McLeish echo Lightman echoing Whewell, especially when they describe interdisciplinarity as research that "features significant interaction." Withness is everything about significant interaction because it is the "I"-to-"I"-ness within a conversation whose feature is the most significantly interactive. It is the experience of "I"-to-"I"-ness that produces Bergson's "true duration" of "the real"; the "I"-to-"I"-ness that produces Williams's "undeniable experience of the present . . . the specificity of the present being."

Victorian Withness

"I"-to-"I"-ness—withness in the broadest sense—is also an inextricable feature in the development of Victorian interdisciplinarity, particularly because the sciences remain in close disciplinary proximity. Electrochemistry is a case in point because the very nature of its laws, phenomena, and practices necessitates the distinctly overlapping, disciplinary encounters of physics and chemistry. These unite, or converse, through the experiment itself, making the case of electrochemistry "recalcitrant"; indeed, all experiments make their disciplines recalcitrant because they are invariably a site of conversation—the true duration of the real, or the specificity of the present being, when and where the "I's" of disciplines creatively evolve. Victorian electrochemistry is not, however, unusual in this regard. Drawing on Mikhail Bakhtin's dialogical philosophy, social scientist Arthur Frank says that "no one person's voice is ever even his or her own; no one existence is ever clearly bounded. Instead, each voice is always permeated with the voices of others. Each voice resists and contests some voices, and it embraces others, but there is no one that could coincide with itself."[26]

What makes the disciplinarity of Victorian electrochemistry different is not the fact that, as a case, it is recalcitrant, but precisely the fact that its experiments, like the textbooks, cookbooks, and history books in *Victorian Interdisciplinarity in the Sciences*, are openly dialogical at a time when "I's" were increasingly meant to be developing into specialized "me's" (recalcitrantly opposed to specialized "you's"). Of course, withness theory is steeped in empathic dialogism—Bakhtin in particular, but also philosopher of conversation Hans-Georg Gadamer. Appropriately enough in regard to electrochemistry, Gadamer urges us to *fuse*, rather than *form*, our horizons into one: a conversation should be a form of active reciprocity that "evokes genuine understanding as not only being intersubjective, but also as being dialectical—a new meaning that is born out of the interplay that goes on continuously between the past and the present, and between different horizons."[27] As Cantor points out, Michael Faraday (1791–1867) hated the word *scientist*, and it was precisely for this reason: Faraday's "I" was not a "me."

What happened to the "I" of natural philosophy when interdisciplinary Victorian disciplines seemed to migrate ineluctably, unobstructedly, toward the specialized "me" of professional modern science? This question suggests a form of structural, advancing alienation—disciplinary alterity, perhaps—what, broadly speaking, Shott calls "aboutness" rather than "withness." Gadamer, as it so happens, is often criticized for underplaying the role alterity plays in conversation,[28] but *Victorian Interdisciplinarity in the Sciences* suggests we take it more seriously. Victorian alterity comes in many forms—colonial, religious, economic, industrial, technological, textual, for example—and in the broadest sense it is counteracted by empathy, or at the very least sympathy. Victorian literary theorist Rebecca Mitchell makes an important, if defeatist point, about the fixed nature of Victorian alterity in her work on realist fiction, for instance, when she claims that "these works [of realist fiction] teach us, as it were, not how to overcome the alienation of the radical alterity of the other, but rather that we cannot overcome that alterity."[29] Buoyed by hierarchical immobilities in the Great Chain of Being, realist fiction, and many other art forms,[30] may selectively teach Victorians resignation to alterity, but that view was not universally shared, even within literature of the time. Victorian literary theorist Audrey Jaffe suggests that Victorian characters often, in fact, sympathize by watching or spectating sympathetically: "Society becomes a field of visual cues and its members alternative selves: imaginary possibilities in a field of circulating social images, confounded and interdependent projection of identity."[31]

Williams and Jaffe may concentrate their attention on fiction but a similar range of attitudes to otherness applies to Victorian science as well, particularly in the way those attitudes shape the socio-professional structure of interdisciplinary sympathies. Nanna Katrine Lüders Kaalund highlights this in the story of two naturalists for the British Arctic Expedition, Henry Chichester Hart (1847–1908) and Henry Wemyss Feilden (1838–1921). Hart and Feilden were categorically "othered" as nonspecialist amateurs delegated with the responsibility for carrying out professional fieldwork according to the highly specialized instructions of the *Manual and Instructions for the Arctic Expedition*; indeed, the *Manual* is an effort "to discipline the Arctic field site by putting together extensive manuals and guides . . . transforming the explorers from independent expert researchers to nonspecialist data gatherers."[32]

Another good example can be found in Efram Sera-Shriar's chapter on spiritualists eventually "outed" by science. Spiritualists were, in some instances, relentlessly pursued—"othered"—by the arbiters of empirical science, for example, American medium Henry Slade (1835–1905) by second-generation scientist Edwin Ray Lankaster (1847–1929). Sera-Shriar describes their story as emblematic: "materialism and scientific naturalism on the one hand, and immaterialism and spiritualism on the other."[33] I wonder whether these disciplinary characterizations are that easily dichotomized between sympathetic conversation (withness) and unsympathetic argument (aboutness), however, and whether interdisciplinarity, that is, disciplinary sympathy, empathy, is a cause or effect of attitudinal change. In some respect I ask what zoologist and scientific popularizer (and polemicist) Richard Dawkins asks of altruism in *The Selfish Gene*: whether interdisciplinary altruism can arise from disciplinary "selfishness." As both Sera-Shriar and Kaalund suggest, however, interdisciplinarity complicates the question precisely because of its own genetically altruistic tendencies. Perhaps it would be more honest therefore to title this book *Victorian Science: The Selfish Discipline*. Perhaps all disciplines are "selfish," in a paradoxically altruistic way. Perhaps the first step in interdisciplinary *withness* is recognizing disciplinary *aboutness*.

Dawkins is at great pains to stress that he uses the word *selfish* metaphorically when describing the "motivation" of genes. But what are a discipline's genes exactly, and how do we know? What is the Victorian disciplinary gene that produces an altruistic, interdisciplinary, intersubjective "I" of withness, instead of a disciplinary, subjective "me" of aboutness? And what role does history play in the manufacture of those genes? In what way are those genes always and invariably "after-words"? Dawkins might suggest we consult memetics and the idea of a culture gene for an answer—this is what psychologist Susan Blackmore not unrelatedly proposes in *The Meme Machine* (1999)[34]—but the disciplinary gene (or meme) is elusive. Understandably, Elsa Richardson, for example, struggles to identify it in her chapter on sanitation science and the International Health Exhibition, something she calls "a riotously interdisciplinary space where different registers of knowledge and diverse forms of expertise circulated."[35]

Riotous interdisciplinarity suggests more, however, than particular conversations between select "I"s, but a more large-scale and even industrial meeting. It suggests a conference, or even congress, instead of a conversation, like the science and exhibition of sanitation itself. The level of interdisciplinary magnitude harkens back to Dawkins's own rhetorical question over the level or magnitude at which selfishness should be pitched—the groups, species, organism, or ecosystem—but the same rhetorical question asked of disciplines defies an easy answer. Shotter provides a way forward when he describes the "relationship dimension" of depth, when "dynamically intertwined" eyes focus on an object.[36] But riotous interdisciplinarity isn't necessarily deep; in fact, interdisciplinarity is often accused of the pitfall of shallowness.[37] Victorian interdisciplinary depth was just as prone to shallowness as its ancestor today and, just like today, often guarded against shallowness by disciplinarily self-selected watchmen. Chris Manias epitomizes their conversational struggles in anthropology, prehistory, and paleontology through meetings of learned societies like the Geological Society, the Ethnological and Anthropological Societies of London (eventually joining forces to become the Anthropological Institute of Great Britain and Ireland), the Royal Society, and the Society of Antiquaries.

Victorian Scientific Noticings

According to action and quality theorists Peter Reason and Brian Goodwin, meetings, like exhibitions, are places where "the human process of becoming" encounters that of the "other," where we can observe "complex emergent wholes."[38] Meetings are also sites of withness noticings. Victorian noticings are evident across a range of human activities, not least meetings and exhibitions, but also books, experiments, and ideas more broadly. Shotter defines eight kinds of noticings, including: (1) being "struck by" an event or happening; (2) slowly sensing of a qualitatively unique "unitary whole"; (3) sensing difference; (4) "incipient forms"; (5) community or polis; (6) "what is not being said" (the elephant in the room); (7) "telling moments" when "collective narratives or ideologies" begin to be revealed; and (8) "disquiets," or a feeling that there is still a "something more" than what has already been sensed.[39] What strikes me most about noticings is its collective emphasis on "sensings"—and sensing's own particular liminality. Sensing classically refers to sense perception as opposed to intellectual cognition. According to philosopher Michelle Montague, for example, "one particularly persistent reason for marking a distinction is the simple fact that perception and cognition just intuitively 'feel different.'"[40] Yet the distinction between them remains stubbornly unresolved, and questions over the defining role of consciousness persist.[41] Sense is, perhaps, more liminal than we thought, and it has inadvertent implications for withness noticings and the development of Victorian interdisciplinarity in the sciences.

For arts researchers Maggi Savin-Baden and Katherine Wimpenny, liminality is a sense for the in-between, as "a psychological or metaphysical subjective state of being at the threshold of two existential planes . . . a sense of in-betweenness . . . a sense of shifting and changing."[42] For psychologists Raffaele De Luca Picione and Jaan Valsiner, however, sensing, or what they call "sensemaking," is itself liminal: "the semiotic structure of borders generate a liminal space, which is characterized by instability, by a blurred space–time distinction and by ambiguities in the semantic and syntactic processes of sensemaking. The psychological processes that occur in liminal space are strongly affectively loaded, yet it is exactly the setting and activation of liminality processes that lead to novelty and creativity and enable the creation of new narrative forms."[43] Sensemaking, to use Picione and Valsiner's term, surfaces prolifically in the new narrative forms of the Victorian sciences, and a particularly good, if not paradigmatic, example is Charles Darwin (1809–1882). Darwin's greatest sensemaking is unarguably natural selection; perhaps we can even identify the very moment of sensemaking in time, when he scrawled "I think" above a hastily configured phylogenic tree in the pages of his *Beagle* Notebook B (1837–1838).[44]

Janet Browne brings Darwin's sensemaking into high relief in her chapter on his engagement with natural history. His books are especially important examples because they reveal how Darwin sensemakes wide-ranging disciplinary knowledge into "new narrative forms," that is, the "narrative" of natural selection. There is another dimension: at the same time Darwin follows the cue of scientific popularizers by using the genre of narrative literary forms to tell an otherwise specifically scientific story. Darwin, in other words, not only sensemakes Victorian scientific interdisciplinarity, he sensemakes literary interdisciplinarity as well; as Browne says, "Darwin's books were mostly accessible in the sense that he wrote engagingly (in the first person), did not use technical language, and that the structure of each was primarily narrative, following a story-telling trajectory almost as in a novel, rather than taking the format of a scientific disquisition."[45]

The case of Darwin raises another point about the liminal nature of sensemaking among the participants in a disciplinary conversation. Much has already been said about the spatial nature of interdisciplinary conversation, but the temporal nature of withness-thinking—its "pastpresentness"—is more problematical perhaps. Withness-thinking can, for example, give the impression of being paternalistic, even if it involves a benign presumption among interlocutors. Shotter refers to Merleau-Ponty asking whether one ever looks *at* a painting, or rather sees according to it—*with it*, in other words, "we begin to make sense of what we might call a *'withness-seeing.'*"[46] Shotter's use of the word *sense* is important here not only because it mirrors Picione and Valsiner's term *sensemaking* but because sensemaking is a temporal process with temporal implications—like an afterword, you might say. As he suggests, "after having seen one Cezanne, we can begin to look over other paintings with the image of a Cezanne picture in mind that shapes and instructs our looking; the Cezanne painting has 'taught' us, or we have 'learnt' from it, a certain 'way or style or genre of looking' that we can now apply to other Cezannes, to other paintings."[47] When we read Darwin we read him like a viewer "reads" Cezanne, but the evidence of Cezanne's paintings or Darwin's books shows they that they are also both sensemakers in time. Today we might call them "dialogue partners" because they help us arrive at an understanding not only of a conversation's "communicative function but also of the [dialogue partner's] communicative agenda."[48] Like Cezanne and Darwin, Shotter is himself an example; *Getting It* is what it represents: *withness-thinking* in theory and practice.

VICTORIAN SCIENTIFIC REVERSALS

Getting it—getting witness—means giving it too and responding to its noticings antiphonally, with what Shotter calls "reversals." So if, as suggested, noticings are about sensemaking—making sense of something—reversals are, in effect, about "makesensing"—making something of sense, creating something out of sense. According to Shotter these are the actions that reposition us as "'inside' rather than 'outside' thinkers," and not just *in* the world of our surroundings but *of* it.[49] Iwan Morus problematizes makesensing in his chapter on physics, explicitly challenging us to rethink the nature of disciplinary equality that interdisciplinary withness-thinking is purported to value. William Robert Grove (1811–1896) was, for example, perfectly happy to practice and preach disciplinary correlation, provided the discipline of physics he created led the sciences. Morus calls this *over-disciplinarity*—"a discipline designed to discipline other disciplines"[50]—and breaches the spirit of withness-thinking by abrogating its moral duty to fully understand its disciplinary interlocutors as "I's." When over-disciplinarity makesenses, it creates something hierarchical, something outside rather than inside; it creates something characterizing *aboutness* rather than *withness*.

Victorian sciences didn't always behave that badly, however, and Morus, and *Victorian Interdisciplinary and the Sciences* more generally, balances Grove's story with other, more salutary accounts depicting the liberal economy underpinning disciplinary reversals. Morus counterbalances Grove's overdisciplinarity with examples of greater cooperation—James Clerk's Maxwell's (1831–1879) laboratory practice, for example, and its accommodation of university educational frameworks. There are others of course, but often, frequently, cooperation appears to have been produced *unintentionally*, organically, with the resulting interdisciplinarity recognized only retrospectively. Intention is, arguably, a major component in reversals—intention is to reversals as unintention is to noticings—but it comes at the risk of limiting the creative diversity organicism brings to the evolution of any relationship, especially within a conversation between disciplines. Shotter says that "the coordinated execution of planned actions depends upon all concerned already sharing the set of existing concepts relevant to the formulation of the plan, thus all new plans depend on old concepts—the process results in the 'continual rediscovery of sameness.'"[51] The intention to create "a new way of looking" is, for example, fundamental to any creative disciplinary project, but the intention to create "a new way of interdisciplinary looking" raises the stakes, as Jerry Jacobs so ably maintains. His *In Defense of Disciplines: Interdisciplinarity and Specialization in the Research University* (2013), for example, shows that interdisciplinarity comes at a risk to disciplinary identity somewhere between self-harm and self-benefit.[52] What is Victorian interdisciplinary intention, then, and when did it begin? It would be easy to argue that as disciplinary specialization grew, interdisciplinary intention waned under the pressure as a consequence, but *Victorian Interdisciplinarity in the Sciences* disproves that theory emphatically.

The fact is quite the opposite: professional specialization and generalization grew hand in hand, but interdisciplinarity, as a concept and intention, had to reinvent itself to avoid becoming the "continual rediscovery of sameness." Victorian interdisciplinary intention—the expressed or manifest will to collaborate, to dialogue, to converse, with different, distinct disciplinary partners—is in many respects no different from modern interdisciplinary intention. What this book shows is how it happened, how it evolved, in the sciences, and how intention grew the pool of sameness rediscovered in conversation. One historical presumption suggests that Victorians never set out intentionally to "do" or "justify" interdisciplinary scientific research for its own sake, as we might today—the way someone like Tom McLeish does in *The Poetry and Music of Science: Comparing Creativity in Science and Art* (2019). McLeish claims, for instance, that we must "learn from the misunderstandings and the mutual pain of fragmented disciplines."[53] But in fact he bears a striking resemblance to Whewell decrying specialization, and to some extent, like *Victorian Interdisciplinarity in the Sciences*, he replicates the redemptive intention motivating Whewell's theory. Whewell says that "the separation of sympathies and intellectual habits has ended in a destruction, on each side, of that mental discipline which leads to success in the other province."[54]

The fact is that, like us, the Victorians "did" interdisciplinarity by necessity, with the intention of answering questions their own research posed, and developing or refining their own disciplinarity through dialogue and conversation. Stark and Bellis say, for instance, that "investigators of various stripes who attempted to understand better human nutrition drew heavily on a broad range of scientific disciplines that were themselves undergoing a process of differentiation."[55] The same could be said of any discipline today; moreover, disciplines are always undergoing a process of differentiation: to be a living discipline is to be always undergoing a process of evolutionary differentiation, a process of reversals. McLeish suggests this when he recommends that we pick up where Whewell and the Victorians left off, and continue to develop "a listening ear . . . and look without prejudice."[56]

AFTERWORD'S AFTERWORD

What is an afterword if not "a listening ear"? And what, more importantly, is "a listening ear," least of all one without prejudice? A listening ear is a philosophical question, with musicological implications—or is it a musicological question, with philosophical implications? Or is it a philosophical question with anatomical implications; an anatomical question with electrochemical implications; an electrochemical question with neurobiological implications; a neurobiological question with evolutionary implications; an evolutionary question with psychosocial implications; a psychosocial with philosophical implications . . . ? The answer lies not in the reply of any one discipline but in the interdisciplinary plurality underpinning the "a"-ness of the ear itself. Perhaps a listening ear is an interdisciplinary ear; *the* listening ear, a disciplinary one. Perhaps our two ears are always divided that way (perhaps to function they have to be)—one interdisciplinary (a listening ear); the other, disciplinary (the listening ear). Perhaps Shotter was wrong therefore to describe the perfect dialogue as withness-thinking; perhaps *withness-listening* would have been a better term. In the broadest sense withness is, after all, about *a listening* "I" (withness) rather than *the* just *hearing* "me" (aboutness). Gadamer might call this "openness."[57]

Openness gives conversation its "state of indeterminacy"[58]—its uncertainty and what Bruce Ellis Benson characterizes as "loose-play."[59] Being a musicologist, Benson applies his theory to musical improvisation, but the same could easily be said of disciplines, and Victorian disciplines are no different in that regard. Victorian disciplines may seem at times to pride themselves professionally, educationally, on an apparent state of determinacy, certainty, and "tight-play," but that state is illusory. The fact of the matter is that Victorian disciplines, particularly those in closely-related sciences, were in a constant state of improvisational dialogue, with themselves and with other disciplines as well, constantly in a state of withness-thinking, or, better still, *withness-listening. Victorian Interdisciplinarity in the Sciences* proves it. It proves that the sciences improvised as they conversed, and conversed as they improvised; that their interrelationships were inextricable, inexhaustible, and unextinguishable—"open" (with) and "closed" (about), simultaneously. They were, in the very broadest sense, "liberated within constraint," to hazard invoking a theological term.[60] It may be paradoxical that a theological term summarizes the improvisatory character of Victorian scientific interdisciplinarity, but be that as it may, it somehow captures the meaningful exploratory purpose behind Victorian Britain's scientific enterprise to advance knowledge and improve our lives—to learn to listen interdisciplinarily, to do so with a listening ear, and without disciplinary prejudice.

NOTES

INTRODUCTION

1. There were three editions to Prichard's famous work, and by the third edition the book was extended into five volumes, each focusing on specific regions of the world and its peoples. See James Cowles Prichard, *Researches into the Physical History of Man* (London: John and Arthur Arch, 1813); Prichard, *Researches into the Physical History of Mankind*, 2nd ed., 2 vols. (London John and Arthur Arch, 1826); and Prichard, *Researches into the Physical History of Mankind*, 3rd ed., 5 vols. (London: Sherwood, Gilbert, and Piper, 1836–1847). For more on Prichard's ethnological research, see George Stocking, "From Chronology to Ethnology: James Cowles Prichard and British Anthropology 1800–1850," in Prichard, *Researches into the Physical History of Man*, ed. Stocking (Chicago: University of Chicago Press, 1973); Stocking, *Victorian Anthropology* (New York: Free Press, 1987), 48–62; Hannah Franziska Augstein, *James Cowles Prichard's Anthropology: Remaking the Science of Man in Early Nineteenth-Century Britain* (Amsterdam: Rodopi B.V., 1999); and Efram Sera-Shriar, *The Making of British Anthropology, 1813–1871* (London: Pickering & Chatto, 2013), 21–52.

2. James Cowles Prichard, "On the Relations of Ethnology to Other Branches of Knowledge," *Journal of the Ethnological Society of London* 1 (1848): 301.

3. Prichard, "On the Relations of Ethnology," 302.

4. Prichard, "On the Relations of Ethnology," 302, 310–11.

5. Thomas Henry Huxley, "On the Study of Biology," in Huxley, *Collected Essays: Science and Education*, 9 vols. (London: Macmillan, 1893), 3:270. For more on the special loan collection of scientific apparatuses to the South Kensington Museum, see Robert Bud, "Infected by the Bacillus of Science: The Explosion of South Kensington," in *Science for the Nation: Perspectives on the History of the Science Museum*, ed. Peter J. T. Morris (London: Palgrave Macmillan, 2013).

6. For more on the contest for cultural authority in Victorian Britain, see Frank Turner, "The Victorian Conflict between Science and Religion: A Professional Dimension," *Isis* 69 (1978); and Bernard Lightman, "Victorian Sciences and Religion: Discordant Harmonies," *Osiris* 16 (2001).

7. Jan Golinski, *Making Natural Knowledge: Constructivism and the History of Science* (Cambridge: Cambridge University Press, 1998), 67.

8. Jack Morrell and Arnold Thackray, *Gentlemen of Science: Early Years of the British Association for the Advancement of Science* (Oxford: Clarendon Press, 1981).

9. [William Whewell], "*On the Connexion of the Physical Sciences*, by Mrs. Somerville," *Quarterly Review* 51 (1834): 59.

10. John Herschel to William Whewell, February 7, 1835, a. 207, Whewell Papers, Trinity College, Cambridge; quoted in Morrell and Thackray, *Gentlemen of Science*, 424.

11. William Coleman, *Biology in the Nineteenth Century* (Hoboken, NJ: John Wiley and Sons, 1971); and George Stocking, *Victorian Anthropology* (New York: Free Press, 1987).

12. Morrell and Thackray, *Gentlemen of Science*, 451, 453–54.

13. Colin A. Russell, *Science and Social Change in Britain and Europe 1700–1900* (London: Macmillan Education, 1983), 193–219, 194 (quote).

14. Morrell tackles geology, discussing the making of British geology into a science from 1795 to 1830 against the background of the broader expansion of new scientific institutions between 1819 and 1841 in London that were "mainly voluntarily supported and specialist-oriented." He argues that the "new and old specialist societies in London usurped much of the ground previously held by the Royal Society; Jack Morrell, "London Institutions and Lyell's Career: 1820–41," *British Journal for the History of Science* 9 (1976): 135–36.

15. One exception is William R. Ashworth, "The Calculating Eye: Baily, Herschel, Babbage, and the Business of Astronomy," *British Journal for the History of Science* 27 (December 1994). Ashworth asserts that there were many competing astronomies in the early nineteenth century and that the Astronomical Society was meant to narrow and define the science's boundaries by removing astronomical speculation and placing astronomy on a solid mathematical base.

16. Frank M. Turner, "The Victorian Conflict between Science and Religion: A Professional Dimension," *Isis* 69 (September 1978): 359.

17. Ruth Barton, *The X Club: Power and Authority in Victorian Science* (Chicago: University of Chicago Press, 2018), 290.

18. Barton, *X Club*, 462.

19. Ruth Barton, "'Huxley, Lubbock, and Half a Dozen Others': Professionals and Gentlemen in the Formation of the X Club, 1851–1864," *Isis* 89 (1998).

20. Adrian Desmond, "Redefining the X Axis: 'Professionals,' 'Amateurs' and the Making of Mid-Victorian Biology—A Progress Report," *Journal of the History of Biology* 34 (2001): 5, 7, 11, 41; Paul White, *Thomas Henry Huxley: Making the 'Man of Science'* (Cambridge: Cambridge University Press, 2003), 170; Theodore Porter, *Karl Pearson: The Scientific Life in a Statistical Age* (Princeton, NJ: Princeton University Press, 2004); Jim Endersby, *Imperial Nature: Joseph Hooker and the Practices of Victorian Science* (Chicago: University of Chicago Press, 2008), 22–26. See also Bernard Lightman, "Introduction," in *Evolutionary Naturalism in Victorian Britain* (Farnham, UK: Ashgate, 2009), xii–xiii.

21. Jim Endersby, "Odd Man Out: Was Joseph Dalton Hooker an Evolutionary Naturalist?," in *Victorian Scientific Naturalism: Community, Identity, Continuity*, ed. Gowan Dawson and Bernard Lightman (Chicago: University of Chicago Press, 2014), 161.

22. One recent edited volume does examine the development of disciplines in the nineteenth century but does not focus on the scientific disciplines. This volume covers music, art history, history, classics, theology, English literature, and musicology in addition to zoology, anthropology, archaeology, psychology, mathematics, and the evolution of the scientific disciplines as a whole. See Bennett Zon and Bernard Lightman, ed., *Victorian Culture and the Origin of Disciplines* (Abingdon, UK: Routledge, 2019).

23. Bernard Lightman, "The Evolution of the Scientific Disciplines," in *Victorian Culture and the Origin of Disciplines*, ed. Bernard Lightman and Bennett Zon (New York: Routledge, 2020).

24. George Stocking, *The Ethnographer's Magic and Other Essays in the History of Anthropology* (Madison: University of Wisconsin Press, 1992), 3–4.

25. John Shotter, "More than Cool Reason: 'Withness-Thinking' or 'Systemic Thinking' and 'Thinking about Systems,'" *International Journal of Collaborative Practices* 3 (2012). See also John Shotter, "Understanding Process from Within: An Argument for 'Withness'-Thinking," *Organizational Studies* 27 (2006).

26. Charles Darwin, *The Origin of Species*, abridged by Philip Appleman (New York: W. W. Norton, 1975), 120.

CHAPTER 1: INTERDISCIPLINARITY IN MACMILLAN'S SCIENCE PRIMERS

1. [William Whewell], "*On the Connexion of the Physical Sciences*. By Mrs. Somerville," *Quarterly Review* 51 (1834): 58–60.

2. Mary Somerville, *On the Connexion of the Physical Sciences*, 7th ed. (London: John Murray, 1846; repr. New York: Arno Press, 1975), vii, 377, 432, 435.

3. Frank M. Turner, *Contesting Cultural Authority: Essays in Victorian Intellectual Life* (Cambridge: Cambridge University Press, 1993), 175–76; Theodore M. Porter, "The Fate of Scientific Naturalism: From Public Sphere to Professional Exclusivity," in *Victorian Scientific Naturalism: Community, Identity, Continuity*, ed. Gowan Dawson and Bernard Lightman (Chicago: University of Chicago Press, 2014), 276.

4. Thomas H. Huxley, *Science and Education* (New York: D. Appleton, 1897), 266–68.

5. Adrian Desmond, "Redefining the X Axis: 'Professionals,' 'Amateurs,' and the Making of Mid-Victorian Biology—A Progress Report," *Journal of the History of Biology* 34 (2001); Ruth Barton, "Huxley, Lubbock, and Half a Dozen Others," *Isis* 89 (1998); Jim Endersby, "Odd Man Out: Was Joseph Hooker an Evolutionary Naturalist?," in Dawson and Lightman, *Victorian Scientific Naturalism*.

6. Katey Ring, "The Popularization of Elementary Science through Popular Science Books, c.1870–c.1939," PhD diss., University of Kent at Canterbury, 1988, 80–81; Charles Morgan, *The House of Macmillan (1843–1943)* (London: Macmillan, 1944), 71.

7. Bernard Lightman, "Science, Scientists and the Public: The Contested Meanings of Science in Victorian Britain," in Lightman, *Evolutionary Naturalism in Victorian Britain: The 'Darwinians' and Their Critics* (Farnham, UK: Ashgate, 2009), 11–12. The definitive study of the X Club is Ruth Barton, *The X Club: Power and Authority in Victorian Science* (Chicago: University of Chicago Press, 2018).

8. Frank Miller Turner, *Between Science and Religion: The Reactions to Scientific Naturalism in Late Victorian England* (New Haven, CT: Yale University Press, 1974), chap. 2.

9. Bernard Lightman and Gowan Dawson, "Introduction," in *Victorian Scientific Naturalism: Community, Identity, Continuity*, ed. Dawson and Lightman (Chicago: University of Chicago Press, 2014), 20.

10. W. Tuckwell, "Science Primers," *Nature* 6 (May 2, 1872): 3–4.

11. Marga Vicedo, "Introduction: The Secret Lives of Textbooks," *Isis* 103, no. 1 (March 2012): 83.

12. Vicedo, "Introduction," 84.

13. Vicedo, "Introduction," 86.

14. Ring, "Popularization of Elementary Science," 80.

15. Bernard Lightman, *Victorian Popularizers of Science: Designing Nature for New Audiences* (Chicago: University of Chicago Press, 2007), 389–90.

16. Lightman, *Victorian Popularizers of Science*, 390. See also Ring, "Popularization of Elementary Science," 80–82, 355.

17. Ring maintains that eleven volumes were published by 1900, but I have been unable to find an eleventh volume. The Macmillan catalog of publications, which goes up to 1889, does not list an eleventh volume. It is unlikely that another volume appeared after the publication of Huxley's introductory volume. See Ring, "Popularization of Elementary Science," 355.

18. T. H. Huxley to Alexander Macmillan, January 13, 1877, f35, ADD 55210, Huxley notebook, Macmillan Archives, British Library, London.

19. Huxley to Macmillan, August 18, 1872, f22, ADD 55210, Huxley notebook, Macmillan Archives.

20. Efram Sera-Shriar, "Historicizing Belief: E. B. Tylor, *Primitive Culture*, and the Evolution of Religion," in *Historicizing Humans: Deep Time, Evolution, and Race in Nineteenth-Century British Sciences*, ed. Sera-Shriar (Pittsburgh: University of Pittsburgh Press, 2018).

21. Sera-Shriar, "Historicizing Belief."

22. Leonard Huxley, ed., *Life and Letters of Thomas Henry Huxley*, 2 vols. (New York: D. Appleton, 1902), 1:387.

23. Huxley to Henry Roscoe, September 12, 1879, 2:2, in Huxley, *Life and Letters of Thomas Henry Huxley*.

24. Macmillan to Huxley, April 23, 1880, 154, 22.154, The Huxley Papers, Imperial College London (hereafter Huxley Papers).

25. Professor Huxley, *Science Primers: Introductory* (Toronto: Canada Publishing, 1881), 16, 18.

26. Huxley, *Science Primers: Introductory*, 12.

27. Huxley, *Science Primers: Introductory*, 20.

28. Huxley, *Science Primers: Introductory*, 81.

29. Huxley, *Science Primers: Introductory*, 88.

30. Huxley, *Science Primers: Introductory*, 92.

31. Huxley, *Science Primers: Introductory*, 92.

32. Huxley, *Science Primers: Introductory*, 92.

33. Huxley, *Science Primers: Introductory*, 92.

34. Huxley, *Science Primers: Introductory*, 92.

35. Huxley, *Science Primers: Introductory*, 92.

36. Huxley, *Science Primers: Introductory*, 94.

37. Huxley, *Science Primers: Introductory*, 94.

38. Roscoe to Huxley, January 24, 1875, 22.144, Huxley Papers.

39. Roscoe to Huxley, January 24, 1875, 22.144, Huxley Papers.

40. ADD 55210, f30, British Library, Macmillan Archives.

41. Huxley, *Life and Letters of Thomas Henry Huxley*, 2:2.

42. Roscoe to Huxley, October 19, 1871, 25.273, Huxley Papers.

43. *Oxford Dictionary of National Biography*, s.v. "Roscoe, Sir Henry Enfield," by Robert Kargon, accessed February 9, 2020, https://doi.org/10.1093/ref:odnb/35827.

44. *Oxford Dictionary of National Biography*, s.v. "Stewart, Balfour," by P. J. Hartog, revised by Graeme J. N. Gooday, accessed February 9, 2020, https://doi.org/10.1093/ref:odnb/26463.

45. Crosbie Smith, *The Science of Energy: A Cultural History of Energy Physics in Victorian Britain* (Chicago: University of Chicago Press, 1998).

46. H. E. Roscoe and Balfour Stewart, preface to Roscoe, *Chemistry*, Science Primers (London: Macmillan, 1872), iv.

47. H. E. Roscoe, *The Life and Experiences of Sir Henry Enfield Roscoe* (London: Macmillan, 1906), 153. Roscoe believed that the chemistry primer had "met a want. It was written from the purely objective point of view, and was intended to serve as the first step in chemistry for young boys in schools. The experiments described were of an exceedingly simple nature, and of such a character that they could be made by the boys themselves." He proudly added that the total number sold up to 1906 was 335,000. Translations had appeared in Icelandic, Polish, German, Italian, Japanese, Bengali, Turkish, Malayalan, and Tamil (151).

48. Roscoe, *Chemistry*, iv.

49. Roscoe, *Chemistry*, 2, 8, 20, 29.

50. Balfour Stewart, *Physics*, Science Primers (London: Macmillan, 1872), 1–2, 36, 77, 127.

51. Sir Archibald Geikie, *A Long Life's Work: An Autobiography* (London: Macmillan, 1924), 163.

52. *Oxford Dictionary of National Biography*, s.v. "Geikie, Sir Archibald," by David Oldroyd, accessed February 9, 2020, https://doi.org/10.1093/ref:odnb/33364.

53. Archibald Geikie, *Physical Geography*, Science Primers (London: Macmillan, 1876), 17, 18, 19, 37, 47, 48, 49, 54, 63.

54. Geikie, *Physical Geography*, 7–8.

55. Matthew Stanley, *Huxley's Church and Maxwell's Demon* (Chicago: University of Chicago Press, 2015), 42.

56. Archibald Geikie, *Geology*, 2nd ed., Science Primers (London: Macmillan, 1874), 11, 25, 32, 35, 37, 44, 47, 53, 68, 74, 75, 78, 83, 84, 87, 89, 90, 99, 117.

57. Geikie, *Geology*, 4, 9.

58. Geikie, *Geology*, 128.

59. Barton later shows that a controversy involving Tyndall in the mid-1870s may have cooled the warm relationship between some scientific naturalists and Lockyer. See Ruth Barton, "Scientific Authority and Scientific Controversy in *Nature*: North Britain against the X Club," in *Culture and Science in the Nineteenth-Century Media*, ed. Louise Henson, Geoffrey Cantor, Gowan Dawson, Richard Noakes, Sally Shuttleworth, and Jonathan R. Topham (Aldershot, UK: Ashgate, 2004).

60. J. Norman Lockyer, *Astronomy*, Science Primers (London: Macmillan, 1879), 3, 40, 53.

61. Lockyer, *Astronomy*, 81, 88, 120.

62. M[ichael] Foster, *Physiology*, Science Primers (London: Macmillan, 1874), v, 4, 5, 6, 85, 92.

63. Leonard Huxley, *Life and Letters of Sir Joseph Dalton Hooker*, 2 vols. (London: John Murray, 1918), 2:151.

64. J[oseph] D[alton] Hooker, *Botany*, Science Primers (London: Macmillan, 1876), 1, 101.

65. Harriet A. Jevons, *Letters and Journal of W. Stanley Jevons* (London: Macmillan, 1886), 364.

66. W[illiam] Stanley Jevons, *Logic*, Science Primers (London: Macmillan, [1876] 1880), 11.

67. W[illiam] Stanley Jevons, *Political Economy*, 3rd ed., Science Primers (London: Macmillan, [1878] 1881), 7, 134.

68. Mario A. Di Gregorio, *T. H. Huxley's Place in Natural Science* (New Haven, CT: Yale University Press, 1984).

69. John Holmes and Sharon Ruston, eds., *The Routledge Research Companion to Nineteenth-Century British Literature and Science* (London: Routledge, 2017).

CHAPTER 2: THE RECALCITRANT CASE OF ELECTROCHEMISTRY

1. [William Whewell], *"On the Connexion of the Physical Sciences*. By Mrs. Somerville," *Quarterly Review* 51 (1834): 59–60.

2. Luigi Galvani, *De Viribus Electricitatis in Motu Musculari. Commentarius* (Bologna: Instituti Scientiarium, 1791).

3. "On the Electricity Excited by the Mere Contact of Conducting Substances of Different Kinds. In a Letter from Mr. Alexander Volta, F. R. S. Professor of Natural Philosophy in the University of Pavia, to the Rt. Hon. Sir Joseph Banks, Bart. K.B. P. R. S.," *Philosophical Transactions of the Royal Society of London* 90 (1800).

4. William Hyde Wollaston to Henry Hastead, August 30, 1801, Box 1, file 1, f. 20, Gilbert Papers, University College, London.

5. W[illiam] N[icholson], "Account of the New Electrical or Galvanic Apparatus of Sig. Alex. Volta, and Experiments Performed with the Same," *Journal of Natural Philosophy, Chemistry, and the Arts* 4 (1800).

6. Michael Faraday, *Experimental Researches in Electricity*, 3 vols. (London: John Murray, 1839–1855), 1:76–77.

7. Faraday, *Experimental Researches*, 1:107.

8. Faraday, *Experimental Researches*, 1:140.

9. Faraday, *Experimental Researches*, 1:145.

10. Faraday, *Experimental Researches*, 1:241–42.

11. Faraday, *Experimental Researches*, 1:195–98.

12. William Robert Grove, *On the Correlation of Physical Forces: Being the Substance of a Course of Lectures Delivered in the London Institution, in the Year 1843* (London: Samuel Highley, 1846). See also Iwan Morus's chapter in this volume.

13. Michael Faraday, "On the Conservation of Force," in Faraday, *Experimental Researches in Chemistry and Physics* (London: Richard Taylor & William Francis, 1859).

14. Jonathan Topham, "Beyond the 'Common Context': The Production and Reading of the Bridgewater Treatises," *Isis* 89 (1998).

15. William Nicholson, *An Introduction to Natural Philosophy*, 4th ed., 2 vols. (London: printed for J. Johnson, 1796), 1:112–294. In the 5th (1805) edition, the length of the chemistry section was reduced (1:112–259).

16. Nicholson, *An Introduction to Natural Philosophy*, 5th ed., 2 vols. (London, 1805), 2:336–49."

17. Neil Arnott, *Elements of Physics, or Natural Philosophy, General and Medical, Explained Independently of Technical Mathematics* (London: printed for Thomas and George Underwood, 1827), 1–4.

18. Arnott, *Elements*, xix, xvi.

19. W[illiam] T[homas] Brande, *A Dictionary of Science, Literature, and Art* (London: Longman, Brown, Green and Longmans, 1842), 926.

20. Michael Faraday to Christian Friedrich Schoenbein, May 15, 1854, in *The Correspondence of Michael Faraday*, ed. Frank A. J. L. James, 6 vols. (London: Institution of Electrical Engineers, 1991–2011), 4:679. There is a hint of this move toward electricity and away from chemistry in Faraday to Eilhard Mitscherlich, January 24, 1838, in *Correspondence of Michael Faraday*, 2:488.

21. Faraday to James Clark, April 8, 1848, in *Correspondence of Michael Faraday*, 3:684.

22. Brande, *Dictionary*, 217.

23. Joseph Agassi, *Faraday as a Natural Philosopher* (Chicago: University of Chicago Press, 1971), 3.

24. *Oxford English Dictionary*, s.v. "electrician," accessed March 27, 2019, https://doi.org/10.1093/OED/3893772288.

25. George William Francis, *Electrical Experiments . . .* (London: D. Francis and G. Berger, 1844), 90.

26. D. S. L. Cardwell, *The Organisation of Science in England: A Retrospect* (London: William Heinemann, 1957).

27. *Royal Commission on Scientific Instruction and the Advancement of Science*, 8 vols. (London: HMSO, 1872–1875); *Reports of the Royal Commissioners on Technical Instruction*, 2 vols. (London: HMSO, 1882–1884).

28. Thomas Thomson to Robert Jameson, September 9, 1817, quoted in J. B. Morrell, "The Chemist Breeders: The Research Schools of Liebig and Thomas Thomson," *Ambix* 19 (1972): 2.

29. William Thomson, "Scientific Laboratories," *Nature* 31 (1885): 410, quoted in Graeme Gooday, "Precision Measurement and the Genesis of Physics Teaching Laboratories in Victorian Britain," *British Journal for the History of Science* 23 (1990): 29.

30. Balfour Stewart, *Lessons in Elementary Physics*, new and enlarged 4th ed. (London: Macmillan, 1895), v.

31. Stewart, *Lessons*, vii.

32. Stewart, *Lessons*, 397–99, 403–8.

33. Stewart, *Lessons*, 431–38.

34. Stewart, *Lessons*, 440–44.

35. Stewart, *Lessons*, 454–56.

36. Stewart, *Lessons*, 462.

37. Stewart, *Lessons*, 461–68; Balfour Stewart, *The Conservation of Energy: Being an Elementary Treatise on Energy and Its Laws* (London: Henry S. King, 1873).

38. R[ichard] T[etley] Glazebrook and W[illiam] N[apier] Shaw, *Practical Physics*, 4th ed. (London: Longmans, Green, 1894), x.

39. Gooday, "Precision Measurement," 26.

40. Glazebrook and Shaw, *Practical Physics*, 491.

41. Glazebrook and Shaw, *Practical Physics*, 511–17.

42. Glazebrook and Shaw, *Practical Physics*, 517–22.

43. Henry Roscoe, *Lessons in Elementary Chemistry: Inorganic and Organic*, 6th rev. ed. (London: Macmillan, 1892); *Oxford Dictionary of National Biography* (hereafter *ODNB*), s.v. "Roscoe, Sir Henry Enfield," by Robert Kargon, accessed July 9, 2019, https://doi.org/10.1093/ref:odnb/35827.

44. Roscoe, *Lessons*, v.

45. Roscoe, *Lessons*, 33–34.

46. Roscoe, *Lessons*, 111–12. See also Henri Moissan, *Recherches sur l'Isolement du Fluor* (Paris: Gauthier-Villars, 1887).

47. William Ramsay, *A System of Inorganic Chemistry* (London: J. & A. Churchill, 1891), v–viii.

48. Ramsay, *System*, vii.

49. There was, however, a brief discussion of Jöns Jacob Berzelius's (1779–1848) theory of the decomposition of oxides, in which the oxygen component is described as electronegative and the other as electropositive. Ramsay rejected this theory (*System*, 206–7).

50. Ramsay, *System*, 73–74.

51. Ramsay, *System*, 62–63.

52. Ramsay, *System*, 584.

53. Robert Bud and Gerrylynn K. Roberts, *Science versus Practice: Chemistry in Victorian Britain* (Manchester: Manchester University Press, 1984), 158.

54. Bud and Roberts, *Science versus Practice*, 16.

55. "Technical Education," *London Evening Standard*, December 6, 1900, 5.

56. Jane Smeal Thompson and Helen G. Thompson, *Silvanus Phillips Thompson: His Life and Letters* (London: T. Fisher Unwin, 1920), 134–35.

57. Bertram Blount and A[rthur] G[eorge] Bloxam, *Chemistry for Engineers and Manufacturers: A Practical Text-Book*, vol. 1, *Chemistry of Engineering, Building and Metallurgy*. (London: Charles Griffin, 1896), 5.

58. "The Goldsmiths' Technical and Recreative Institute," *Kentish Mercury*, July 24, 1891, 4.

59. Blount and Bloxam, *Chemistry*, 1:83–95, 85 (quote).

60. Blount and Bloxam, *Chemistry*, 1:166–68, 220–21, 209.

61. Blount and Bloxam, *Chemistry for Engineers and Manufacturers: A Practical Text-Book*, vol. 2 (London: Charles Griffin, 1896), 51–53.

62. George Gore, *The Art of Electro-metallurgy; Including All Known Processes of Electro-deposition*, 4th ed. (London: Longmans, Green, 1890).

63. George Gore, "On Practical Scientific Instruction," *Quarterly Journal of Science* 7 (1870).

64. *ODNB*, s.v. "Gore, George," by Robert Sharp, accessed July 16, 2019, https://doi.org/10.1093/ref:odnb/33472; private communication from Robert Bud.

65. Gore, *Art of Electro-metallurgy*, vii.

66. August D. Waller, *An Introduction to Human Physiology*, 3rd ed. (London: Longmans, Green, 1896), 385–404.

67. James A. Secord, "Extraordinary Experiment: Electricity and the Creation of Life in Victorian England," in *The Uses of Experiment: Studies in the Natural Sciences*, ed. David Gooding, Trevor Pinch, and Simon Schaffer (Cambridge: Cambridge University Press, 1989).

68. Walter Larden, *Electricity for Public Schools and Colleges*, new ed. (London: Longmans, Green, 1893), 163–81.

69. Stewart, *Lessons*, 455; Ramsay's, *System*, 571. Neither Roscoe's *Lessons* nor Glazebrook and Shaw's *Practical Physics* mentioned the topic.

70. George Newth, *A Text-book of Inorganic Chemistry*, new ed. (London: Longmans, Green, 1896), 95.

71. Stewart, *Lessons*, 442–43; C. J. T[heodor von] Grotthuss, "Memoire upon the Decomposition of Water, and of the Bodies Which It Holds in Solution by Means of Galvanic Current," *Philosophical Magazine* 25 (1806). Grotthuss's hypothesis was also discussed in Larden's *Electricity*, 186–88.

72. Silvanus Thompson, *Elementary Lessons in Electricity & Magnetism* (London: Macmillan, [1881] 1892), 171–80, 394–97.

73. Blount and Bloxam, *Chemistry*, 1:87; Gore, *Art of Electro-metallurgy*, 42–47.

74. Gore, *Art of Electro-metallurgy*, 313–54.

75. Gore, *Art of Electro-metallurgy*, 354–67.

76. Compare Jürgen Renn, *The Evolution of Knowledge: Rethinking Science for the Anthropocene* (Princeton, NJ: Princeton University Press, 2020), 67-68.

CHAPTER 3: EXPERTISE ACROSS DISCIPLINES

1. Isaiah Berlin, *The Hedgehog and the Fox: An Essay on Tolstoy's View of History* (New York: Simon and Schuster, 1953), 1.

2. Ramin Jahnanbegloo, *Conversations with Isaiah Berlin* (London: Peter Halban, 1992), 188.

3. Most recently expressed by Oren Harman and Michael R. Dietrich, eds., *Outsider Scientists: Routes to Innovation in Biology* (Chicago: University of Chicago Press, 2013).

4. S. Andrew Inkpen and C. Tyler DesRoches, "Revamping the Image of Science for the Anthropocene," *Philosophy, Theory, and Practice in Biology* 11, no. 3 (2019).

5. John Pickstone, "Working Knowledges before and after circa 1800: Practices and Disciplines in the History of Science, Technology, and Medicine," *Isis* 98 (2007); and G. E. R. Lloyd, *Disciplines in the Making: Cross-Cultural Perspectives on Elites, Learning, and Innovation* (Oxford: Oxford University Press, 2009). See also Richard Whitley, Jochen Gläser, and Lars Engwall, eds., *Instituting Science: The Cultural Production of Scientific Disciplines* (New York: Oxford University Press, 2010); and Bernard Lightman and Bennett Zon, eds., *Victorian Culture and the Origin of Disciplines* (Abingdon, UK: Routledge, 2020). Timothy Lenoir focuses on Germany in *Instituting Science: The Cultural Production of Scientific Disciplines* (Stanford, CA: Stanford University Press, 1997).

6. Bernard Lightman, "The Evolution of the Scientific Disciplines," in Lightman and Zon, *Victorian Culture*.

7. Lightman and Zon, "Introduction," in Lightman and Zon, *Victorian Culture*, 1–17.

8. Sandra Herbert, *Charles Darwin, Geologist* (Ithaca, NY: Cornell University Press, 2005).

9. James R. Moore, "Darwin of Down: The Evolutionist as Squarson-Naturalist," in *The Darwinian Heritage*, ed. David Kohn (Princeton, NJ: Princeton University Press, 1985; repr. 2014).

10. George Henry Lewes, "Hereditary Influence, Animal and Human," *Westminster Review* 66, no. 129 (July 1856): 77.

11. Joseph Caron, "Biology in the Life Sciences: A Historiographical Contribution," *History of Science* 26 (1988).

12. Ernst Haeckel to Charles Darwin, October 26, 1864, in *The Correspondence of Charles Darwin*, ed. F. H. Burkhardt, Sydney Smith, and James R. Secord, 30 vols. (Cambridge: Cambridge University Press, 1985–2023), 12:379–82.

13. Thomas Henry Huxley, "On the Study of Biology," in Huxley, *Science and Education*, in *Collected Essays*, vol. 3 (New York: D. Appleton, 1893–1940).

14. Charles R. Darwin, "The Contractile Filaments of the Teasel," *Nature: A Weekly Illustrated Journal of Science* 16 (August 23, 1877): 339.

15. Robert M. Young, "Malthus and the Evolutionists: The Common Context of Biological and Social Theory," *Past & Present* 43, no. 1 (1969).

16. Francis Darwin, ed., *The Life and Letters of Charles Darwin, Including an Autobiographical Chapter*, 3 vols. (London: John Murray, 1887), 3:373–76.

17. James A. Secord, *Victorian Sensation: The Extraordinary Publication, Reception, and Secret Authorship of Vestiges of the Natural History of Creation* (Chicago: University of Chicago Press, 2000). For Humboldt, see Andrea Wulf, *The Invention of Nature: Alexander von Humboldt's New World* (New York: Alfred A. Knopf, 2015).

18. Menachem Fisch and Simon Schaffer, eds., *William Whewell: A Composite Portrait* (Oxford: Clarendon Press, 1991); and Richard Yeo, *Defining Science: William Whewell, Natural Knowledge, and Public Debate in Early Victorian Britain* (Cambridge: Cambridge University Press, 1993).

19. Pietro Corsi, *Science and Religion: Baden Powell and the Anglican Debate, 1800–1860* (Cambridge: Cambridge University Press, 1988).

20. A. R. Wallace to Darwin, July 6, 1870 (relating to publication of the *Descent of Man* [1871]), in Burkhardt, Smith, and Secord, *Correspondence of Charles Darwin*, 18:204; Charles Robert Darwin, *On the Origin of Species by Means of Natural Selection or the Preservation of Favored Races in the Struggle for Life* (London: John Murray, 1859), 459.

21. The philosophical groundings of Darwin's theory are discussed by David Hull, "Darwin's Science and Victorian Philosophy of Science," in *The Cambridge Companion to Darwin*, ed. Jonathan Hodge and Gregory Radick (Cambridge: Cambridge University Press, 2009); and in his earlier work, *Darwin and His Critics: The Reception of Darwin's Theory of Evolution by the Scientific Community* (Chicago: University of Chicago Press, 1983). For additional viewpoints, see Michael Ghiselin, *The Triumph of the Darwinian Method: With a New Preface* (Chicago: University of Chicago Press, 1984), 103–30; and Michael Ruse, "Darwin's Debt to Philosophy: An Examination of the Influence of the Philosophical Ideas of John W. Herschel and William Whewell on the Development of Charles Darwin's Theory of Evolution," *Studies in the History and Philosophy of Science A* 6 (1975).

22. William Paley, *Natural Theology; or, Evidences of the Existence and Attributes of the Deity, Collected from the Appearances of Nature* (London: printed for R. Faulder, 1802). Out of many historical studies on Paley and his influence, see particularly Jonathan Topham, "Biology in the Service of Natural Theology: Paley, Darwin, and the Bridgewater Treatises," in *Biology and Ideology from Descartes to Dawkins*, ed. Denis Alexander and R. Numbers (Chicago: Univer-

sity of Chicago Press, 2010); and Stuart Peterfreund, *Turning Points in Natural Theology from Bacon to Darwin: The Way of the Argument from Design* (New York: Palgrave Macmillan, 2012). An older analysis that touches on Darwin's substitution of natural selection for Paley's argument from design is given by Dov Ospovat, *The Development of Darwin's Theory: Natural History, Natural Theology, and Natural Selection, 1838–1859* (Cambridge: Cambridge University Press, 1981).

23. Roland Jackson, *The Ascent of John Tyndall: Victorian Scientist, Mountaineer, and Public Intellectual* (Oxford: Oxford University Press, 2018).

24. Soraya De Chadarevian, "Laboratory Science versus Country-House Experiments: The Controversy between Julius Sachs and Charles Darwin," *British Journal for the History of Science* 29, no. 1 (1996); and James Costa, *Darwin's Backyard: How Small Experiments Led to a Big Theory* (New York: W. W. Norton, 2017).

25. Janet Browne, *Charles Darwin: A Biography*, vol. 1, *Voyaging* (New York: Knopf, 1995), chaps. 7–14, esp. 338–39.

26. For the impact of Lyell, see Herbert, *Geologist*; and A. W. Sponsel, *Darwin's Evolving Identity: Adventure, Ambition, and the Sin of Speculation* (Chicago: University of Chicago Press, 2018), esp. 105–48. See also Browne, *Charles Darwin*, 1:186–90.

27. Francis Darwin, *Life and Letters*, 1:iii.

28. Darwin to John B. Innes, September 1, 1863, in Burkhardt, Smith, and Secord, *Correspondence of Charles Darwin*, 11:616–17.

29. Marsha Richmond, "Darwin's Study of the Cirripedia," in Burkhardt, Smith, and Secord, *Correspondence of Charles Darwin*, app. 2, 4:388–409.

30. Jonathan Hodge, "Darwin as a Lifelong Generation Theorist," in Kohn, *Darwinian Heritage*. For Grant, see A. J. Desmond, "Robert E. Grant: The Social Predicament of a Pre-Darwinian Transmutationist," *Journal of the History of Biology* 17 (1984).

31. Joseph Hooker to Darwin, September 4–9, 1845, and Darwin to Hooker, September 18, 1845, in Burkhardt, Smith, and Secord, *Correspondence of Charles Darwin*, 3:250–53.

32. Discussed by Roderick D. Buchanan and James Bradley in "'Darwin's Delay': A Reassessment of the Evidence," *Isis* 108, no. 3 (2017). For the comparative anatomical perspective, see Ghiselin, *Triumph of the Darwinian Method*, 103–30. See also Costas Mannouris, "Darwin's 'Beloved Barnacles': Tough Lessons in Variation," *History and Philosophy of the Life Sciences* 33 (2011); and Rebecca Stott, *Darwin and the Barnacle* (London: Faber, 2003).

33. Secord, *Victorian Sensation*; and A. J. Desmond, *The Politics of Evolution: Morphology, Medicine, and Reform in Radical London* (Chicago: University of Chicago Press, 1989).

34. Darwin to Richard Owen, November 25, 1846, in Burkhardt, Smith, and Secord, *Correspondence of Charles Darwin*, 3:372–73.

35. Charles Darwin, *The Autobiography of Charles Darwin 1809–1882: With the Original Omissions Restored*, ed. Nora Barlow (London: Collins, 1958), 118.

36. For a general account of Darwin's botanical work, see Peter Ayres, *The Aliveness of Plants: The Darwins at the Dawn of Plant Science* (London: Pickering & Chatto, 2008). For orchids, see Ayres, *Aliveness*, 71–77; Jim Endersby, *Orchid* (Chicago: University of Chicago Press, 2016), chap. 6; and Richard Bellon, "Inspiration in the Harness of Daily Labor: Darwin, Botany, and the Triumph of Evolution, 1859–1868," *Isis* 102 (2011).

37. Darwin to Hooker, June 19, [1861], in Burkhardt, Smith, and Secord, *Correspondence of Charles Darwin*, 10:330–31.

38. Francis Darwin, *Life and Letters*, 1:149; Darwin's working habits are described on 144–53.

39. Michael Reidy, *Tides of History: Ocean Science and Her Majesty's Navy* (Chicago: University of Chicago Press, 2008).

40. For example, E. C. Spary, *Utopia's Garden: French Natural History from Old Regime to Revolution* (Chicago: University of Chicago Press, 2000). See also Catherine G. Golden, *Posting It: The Victorian Revolution in Letter Writing* (Gainsville: University of Florida Press, 2010).

41. Janet Browne, "Corresponding Naturalists," in *The Age of Scientific Naturalism: Tyndall and His Contemporaries*, ed. Bernard Lightman and Michael Reidy (Chicago: University of Chicago Press, 2014).

42. Ruth Barton, *The X Club: Power and Authority in Victorian Science* (Chicago: University of Chicago Press, 2018).

43. James A. Secord, "Darwin and the Breeders: A Social History," in Kohn, *Darwinian Heritage*.

44. Evelleen Richards, "Redrawing the Boundaries: Darwinian Science and Victorian Women Intellectuals," in *Victorian Science in Context*, ed. Bernard Lightman (Chicago: University of Chicago Press, 1997); Joy Harvey, "Darwin's 'Angels': The Women Correspondents of Charles Darwin," *Intellectual History Review* 19 (2009); Tina Gianquitto, *Good Observers of Nature: American Women and the Scientific Study of the Natural World, 1820–1885* (Athens: University of Georgia Press, 2007).

45. Anne Secord, "Corresponding Interests: Artisans and Gentlemen in Nineteenth-Century Natural History," *British Journal for the History of Science* 27 (1994).

46. Darwin, *On the Origin of Species*, 224–35. For the correspondence with Miller, see Burkhardt, Smith, and Secord, *Correspondence of Charles Darwin*, 7:68–69, 93.

47. Kimberley A. Hamlin, *From Eve to Evolution: Darwin, Science, and Women's Rights in Gilded Age America* (Chicago: University of Chicago Press, 2014).

48. Richard H. Drayton, *Nature's Government: Science, Imperial Britain, and*

the "*Improvement*" of the World (New Haven, CT: Yale University Press, 2000); Jim Endersby, *Imperial Nature: Joseph Hooker and the Practices of Victorian Science* (Chicago: University of Chicago Press, 2008). Earlier work by Lucille Brockway also covers imperial Kew: *Science and Colonial Expansion: The Role of the British Royal Botanic Gardens* (New York: Academic Press, 1979).

49. John van Wyhe, *Charles Darwin's Shorter Publications, 1829–1883* (Cambridge: Cambridge University Press, 2009). This publication supersedes Paul Barrett, *The Collected Papers of Charles Darwin* (Chicago: University of Chicago Press, 1977).

50. Gillian Beer, *Darwin's Plots: Evolutionary Narrative in Darwin, George Eliot and Nineteenth-Century Fiction*, 3rd ed. (Cambridge: Cambridge University Press, 2009); and George Levine, *Darwin the Writer* (Oxford: Oxford University Press, 2011).

51. Darwin to T. H. Huxley, November 25, [1859], in Burkhardt, Smith, and Secord, *Correspondence of Charles Darwin*, 7:398.

52. Janet Browne, *Darwin's Origin of Species: A Biography* (London: Atlantic Books, 2006). See also Browne, *Charles Darwin*, vol. 2, *Power of Place* (New York: Knopf, 2002), 186–92.

53. Aileen Fyfe and Bernard Lightman, eds., *Science in the Marketplace: Nineteenth-Century Sites and Experiences* (Chicago: University of Chicago Press, 2007).

54. R. B. Freeman, *The Works of Charles Darwin: An Annotated Bibliographical Handlist*, 2nd ed. (Folkestone, UK.: Dawson, 1977). This work has been extensively revised by John van Wyhe for the website The Complete Work of Charles Darwin, http://darwin-online.org.uk/.

55. Innes M. Keighren, Charles Withers, and Bill Bell, *Travels into Print: Exploration, Writing, and Publishing with John Murray, 1773–1859* (Chicago: University of Chicago Press, 2015); Sylvia Nickerson, "Darwin's Publisher: John Murray III at the Intersection of Science and Religion," in *Rethinking History, Science, and Religion: An Exploration of Conflict and the Complexity Principle*, ed. Bernard Lightman (Pittsburgh: University of Pittsburgh Press, 2019); and Browne, *Charles Darwin*, 2:73–76, 96–97, 346–47.

56. Young, "Malthus and the Evolutionists."

57. Sponsel, *Darwin's Evolving Identity*, 125–44.

58. Elizabeth Platts, *In Celebration of the Ray Society, Established 1844, and Its Founder George Johnston (1797–1855)*, Publication no. 163 (London: Ray Society, 1994).

59. Burkhardt, Smith, and Secord, *Correspondence of Charles Darwin*, 5:7–8.

60. Burkhardt, Smith, and Secord, *Correspondence of Charles Darwin*, 5:2–6.

61. Browne, *Charles Darwin*, 1:335–36.

62. Sponsel, *Darwin's Evolving Identity*, 125–44, and Herbert, *Geologist*, 90–96.

63. M. J. S. Rudwick, "Darwin and Glen Roy: A 'Great Failure' in Scientific Method?," *Studies in History and Philosophy of Science A* 5, no. 2 (1974).

64. Jack Morrell and Arnold Thackray, *Gentlemen of Science: Early Years of the British Association for the Advancement of Science* (Oxford: Clarendon Press, 1981).

65. Bernard Lightman and Gowan Dawson, eds., *Victorian Scientific Naturalism: Community, Identity, Continuity* (Chicago: University of Chicago Press, 2014).

66. Burkhardt, Smith, and Secord, *Correspondence of Charles Darwin*, 7:146–47, and elsewhere. For the responses to Darwin and Wallace's paper, see I. Bernard Cohen, "Three Notes on the Reception of Darwin's Ideas on Natural Selection (Henry Baker Tristram, Alfred Newton, Samuel Wilberforce)," in Kohn, *Darwinian Heritage*; and Browne, *Charles Darwin*, 2:48–52.

67. Wyhe, *Shorter Publications*.

68. Geoffrey Cantor and Sally Shuttleworth, eds., *Science Serialized: Representations of the Sciences in Nineteenth-Century Periodicals* (Cambridge, MA: MIT Press, 2004); and Geoffrey Cantor et al., *Science in the Nineteenth-Century Periodical: Reading the Magazine of Nature* (Cambridge: Cambridge University Press, 2004).

69. Darwin, *On the Origin of Species*, 488.

70. Explored in Endersby, *Imperial Nature*. See also the special issue of *Journal of the History of Biology* 34, no. 1 (2001), on professionalization in the life sciences in the nineteenth century.

71. James Elwick, Bernard Lightman, and Michael Reidy, gen. eds., *The Correspondence of John Tyndall*, vol. 2., ed. Melinda Baldwin and Janet Browne (Pittsburgh: University of Pittsburgh Press, 2016).

CHAPTER 4: WITHOUT A DARWINIAN CLUE?

1. John Tyndall to Thomas Hirst, February 8, 1858, in *The Correspondence of John Tyndall*, vol. 6, *November 1856–February 1859*, ed. Michael D. Barton et al. (Pittsburgh: University of Pittsburgh Press, 2018), 328–29.

2. J. J. Sylvester to Tyndall, May 21, 1855, in *Correspondence of John Tyndall*, vol. 5, *January 1855–October 1856*, ed. William H. Brock and Geoffrey Cantor (Pittsburgh: University of Pittsburgh Press, 2018), 103.

3. Leslie Stephen, "An Attempted Philosophy of History," *Fortnightly Review* 27 (1880): 672.

4. [William Frederick Pollock], "Buckle's *History of Civilization*," *Quarterly Review* 104, no. 207 (July 1858): 38.

5. On Buckle's use of statistics, see Ian Hacking, *The Taming of Chance* (Cambridge: Cambridge University Press, 1990); and Theodore M. Porter, *The Rise of Statistical Thinking, 1820–1900* (Princeton, NJ: Princeton University Press, 1886), 60–65.

6. Ian Hesketh, *The Science of History in Victorian Britain* (London: Pick-

ering & Chatto, 2011), chaps. 1 and 2. See also Christopher Parker, "English Historians and the Opposition to Positivism," *History and Theory* 22 (1983); and T. W. Heyck, *The Transformation of Intellectual Life in Victorian England* (London: St. Martin's Press, 1982), chap. 5.

7. Leslie Stephen to C. F. Adams, June 2, 1899, in Frederic William Maitland, *The Life and Letters of Leslie Stephen* (London: Duckworth, 1906), 452. Note that Grant Allen makes a similar argument about Buckle being "a mere historical curiosity" in "Darwinism and Other Essays," *Academy*, August 16, 1879, 116. For a modern version of this argument see George W. Stocking Jr., *Victorian Anthropology* (New York: Free Press, 1987), 112–16.

8. Adrian Wilson, "Science's Imagined Pasts," *Isis* 108, no. 4 (2017); and Ian Hesketh, ed., *Imagining the Darwinian Revolution: Historical Narratives of Evolution from the Nineteenth Century to the Present* (Pittsburgh: University of Pittsburgh Press, 2022).

9. The secondary literature on "scientific naturalism" is extensive, but for a recent study that highlights both the utility and complexity of the term, see Gowan Dawson and Bernard Lightman, eds., *Victorian Scientific Naturalism: Community, Identity, Continuity* (Chicago: University of Chicago Press, 2014). For the larger cultural conflicts between scientific naturalists and the Anglican clergy, see Frank M. Turner, *Contesting Cultural Authority: Essays in Victorian Intellectual Life* (Cambridge: Cambridge University Press, 1993); and more recently, Ruth Barton, *The X Club: Power and Authority in Victorian Science* (Chicago: University of Chicago Press, 2018).

10. Henry Thomas Buckle, *History of Civilization in England*, 2 vols., 4th ed. (London: Longman, Green, Longman, Roberts, and Green, 1864), 1:3.

11. Buckle, *History of Civilization in England*, vol. 1, chap. 2.

12. Buckle, *History of Civilization in England*, vol. 1, chap. 4.

13. Buckle, *History of Civilization in England*, 1:239.

14. Buckle, *History of Civilization in England*, 1:292.

15. Buckle, *History of Civilization in England*, 1:559–60.

16. Buckle, *History of Civilization in England*, 1:318.

17. Buckle, *History of Civilization in England*, 1:462.

18. Buckle, *History of Civilization in England*, 1:318.

19. Buckle, *History of Civilization in England*, 1:585.

20. Buckle, *History of Civilization in England*, 1:337–42.

21. Buckle, *History of Civilization in England*, 1:462.

22. Buckle, *History of Civilization in England*, 1:736–822, 822 (quote).

23. Buckle, *History of Civilization in England*, 1:822.

24. "Buckle's History of Civilization," *Dublin University Magazine* 51, no. 301 (January 1858): 12.

25. [Robert Vaughan], "Buckle on Civilization: Destiny and Intellect," *British Quarterly Review* 28, no. 55 (July 1858): 3.

26. "Buckle's History of Civilization," *Gentleman's Magazine*, September 1857, 246.

27. [Mark Pattison], "The History of Civilization in England," *Westminster Review* 68, no. 134 (October 1857).

28. See, for instance, G. A. Simcox, "Henry Thomas Buckle," *Fortnightly Review* 27, no. 158 (February 1880): 277.

29. See, for instance, "Mr. Buckle and His Critics," *New Quarterly Review* 7, no. 28 (November 1858).

30. See note 6.

31. See, for instance, [Pollock], "Buckle's *History of Civilization*"; and [Richard Holt Hutton], "Civilization and Faith," *National Review* 6 (January 1858). For a discussion of the theologically motivated criticisms of Buckle, see Joshua Bennett, *God and Progress: Religion and History in British Intellectual Culture, 1845–1914* (Oxford: Oxford University Press, 2019), 208–10.

32. Darwin to Hooker, March 31, [1858], Darwin Correspondence Project, http://www.darwinproject.ac.uk/DCP-LETT-2248.

33. Darwin to Hooker, February 23, [1858], Darwin Correspondence Project, https://www.darwinproject.ac.uk/letter/DCP-LETT-2222.xml.

34. Darwin to Hooker, March 7, [1862], Darwin Correspondence Project, https://www.darwinproject.ac.uk/letter/DCP-LETT-3468.xml.

35. [Katharine M.] Lyell, ed., *Life, Letters, and Journals of Sir Charles Lyell*, 2 vols. (London: John Murray, 1881), 2:279–80.

36. Hooker and Lyell are quoted in Barton, *X Club*, 164. For other descriptions of Buckle's election to the Athenaeum, see Lyell, *Life, Letters, and Journals of Sir Charles Lyell*, 2:279–80; Alfred Henry Huth, *The Life and Writings of Henry Thomas Buckle* (New York: Appleton, 1880), 214–15; Adrian Desmond and James Moore, *Darwin: The Life of a Tormented Evolutionist* (New York: W. W. Norton, 1992), 464; and Desmond, *Huxley: From Devil's Disciple to Evolution's High Priest* (London: Penguin Books, 1997), 242.

37. Henry Thomas Buckle, "The Influence of Women on the Progress of Knowledge," *Fraser's Magazine* 57 (April 1858). Reviews of the lecture include "Influence of Buckle on the Position of Women," *Leader*, April 10, 1858; "The Over-education of Women," *Saturday Review*, May 8, 1858; and [William Caldwell Roscoe], "Woman," *National Review* 14 (October 1858).

38. John Tyndall, journal, March 19, 1858, MS RI JT/2/13c/1060–1061, Papers of John Tyndall, Royal Institution of Great Britain, London (hereafter Tyndall Papers).

39. Tyndall, journal, April 20, 1858, MS RI JT/2/13c/1071, Tyndall Papers; and Tyndall to Huxley, [April 21, 1858], in Barton et al., *Correspondence of John Tyndall*, 6:356.

40. Huxley to Tyndall, [April 22, 1858], in Barton et al., *Correspondence of John Tyndall*, 6:357.

41. Darwin to Hooker, February 23, [1858], Darwin Correspondence Project, https://www.darwinproject.ac.uk/letter/DCP-LETT-2222.xml; and Hooker to Darwin, February [25], 1858, Darwin Correspondence Project, https://www.darwinproject.ac.uk/letter/DCP-LETT-2225.xml.

42. Evelleen Richards, *Darwin and the Making of Sexual Selection* (Chicago: University of Chicago Press, 2017).

43. On the general turn toward large-scale histories of humans, see Efram Sera-Shriar, ed., *Historicizing Humans: Deep Time, Evolution, and Race in Nineteenth-Century British Sciences* (Pittsburgh: University of Pittsburgh Press, 2018).

44. Ian Hesketh, *Of Apes and Ancestors: Evolution, Christianity, and the Oxford Debate* (Toronto: University of Toronto Press, 2009), 77.

45. Joseph Hooker, quoted in Hesketh, *Of Apes and Ancestors*, 79.

46. John William Draper, *The History of the Intellectual Development of Europe*, 2 vols. (New York: Harper & Brothers, 1863), 1:1; the book was also published in London by Bell & Daldy in 1864.

47. "The Intellectual Development of Europe," *Saturday Review*, June 11, 1864, 726–27.

48. [Sheldon Amos], "The Intellectual Development of Europe," *Westminster Review* 27, no. 1 (January 1865): 102. See also "Review of *History of the Intellectual Development of Europe*," *Athenaeum*, July 2, 1864, 9.

49. [Henry Reeve], "Lecky's *Influence of Rationalism*," *Edinburgh Review* 121, no. 248 (April 1865): 426 ("family resemblance"); and "The History of Rationalism," *London Review*, June 17, 1865, 640.

50. John Lubbock, *Pre-historic Times, as Illustrated by Ancient Remains, and the Manners and Customs of Modern Savages* (London: Williams and Norgate, 1865), 465.

51. Lubbock, *Pre-historic Times*, 488–89.

52. Edward Burnet Tylor, *Researches into the Early History of Mankind and the Development of Civilization* (London: John Murray, 1865), 1.

53. Tylor, *Researches into the Early History of Mankind*, 4. For the parallels between Tylor's and Buckle's thought, see George W. Stocking Jr., *Race, Culture, and Evolution: Essays in the History of Anthropology* (New York: Free Press, 1968), 74–76; Stocking, *Victorian Anthropology*, 158; and J. W. Burrow, *Evolution and Society: A Study in Victorian Social Theory* (Cambridge: Cambridge University Press, 1966), 106, 253, 255.

54. Hooker to Darwin, July 13, 1865, Darwin Correspondence Project, https://www.darwinproject.ac.uk/letter/DCP-LETT-4873.xml.

55. Darwin to Hooker, [July 29, 1865], Darwin Correspondence Project, https://www.darwinproject.ac.uk/letter/DCP-LETT-4874.xml. The quote about Tylor's book is from Charles and Emma Darwin to Hooker, [July 10, 1865], Darwin Correspondence Project, https://www.darwinproject.ac.uk/letter/DCP-LETT-4868.xml.

56. Alfred Russel Wallace, "The Origin of the Human Races and the Antiquity of Man Deduced from the Theory of 'Natural Selection,'" *Journal of the Anthropological Society of London* 2 (1864).

57. Darwin to Wallace, September 22, [1865], Darwin Correspondence Project, https://www.darwinproject.ac.uk/letter/DCP-LETT-4896.xml.

58. Wallace to Darwin, October 2, 1865, Darwin Correspondence Project, https://www.darwinproject.ac.uk/letter/DCP-LETT-4906.xml.

59. Wallace to Darwin, October 2, 1865.

60. [James Macdonell], "Natural History of Morals," *North British Review* 47, no. 94 (December 1867): 377.

61. [Macdonell], "Natural History of Morals," 360.

62. For a recent discussion of the attempt to establish an evolutionary account of morality, see Ian Hesketh, "Evolution, Ethics, and the Metaphysical Society, 1869–1875," in *The Metaphysical Society (1869–1880): Intellectual Life in Mid-Victorian England*, ed. Catherine Marshall, Bernard Lightman, and Richard England (Oxford: Oxford University Press, 2019).

63. W. Robertson Nicoll, *James Macdonell, Journalist* (London: Hodder and Stoughton, 1890), 156.

64. John Morley, "A Fragment on the Genesis of Morals," *Fortnightly Review* 3, no. 15 (May 1865): 338.

65. Tyndall to Hirst, February 8, 1858, in Barton et al., *Correspondence of John Tyndall*, 6:328–29.

66. Ian Hesketh, "Technologies of the Scientific Self: John Tyndall and His Journal," *Isis* 110, no. 3 (September 2019).

67. Tyndall to Maria Goodwin MacKaye, June 2, 1870, in *The Correspondence of John Tyndall*, vol. 11, *The Correspondence, January 1869–February 1971*, ed. Adrian Kirwan and Elizabeth Neswald (Pittsburgh: University of Pittsburgh Press, 2022), 359.

68. John Tyndall, *Fragments of Science*, 7th ed. (London: Longmans, Green, 1889), 2:127; and Ian Hesketh, "The Making of John Tyndall's Darwinian Revolution," *Annals of Science* 77, no. 4 (2020).

69. For a compelling summary of Tyndall's address and its reception, see Bernard Lightman, "Scientists as Materialists in the Periodical Press: Tyndall's Belfast Address," in *Science Serialized: Representations of the Sciences in Nineteenth-Century Periodicals*, ed. Geoffrey Cantor and Sally Shuttleworth (Cambridge, MA: MIT Press, 2004).

70. For the distinction between Tyndall's "conflict thesis" and Draper's, see Bernard Lightman, "The Victorians: Tyndall and Draper," in *The Warfare between Science and Religion: The Idea That Wouldn't Die*, ed. Jeff Hardin, Ronald L. Numbers, and Ronald A. Binzley (Baltimore: Johns Hopkins University Press, 2018).

71. Tyndall, *Fragments of Science*, 2:200.

72. On the importance of cultivating a respectable persona with regard to scientific naturalism, see in particular Gowan Dawson, *Darwin, Literature and Victorian Respectability* (Cambridge: Cambridge University Press, 2007).

73. See, for instance, Buckle, *History of Civilization in England*, 2:328–29.

CHAPTER 5: GOOD HEALTH ON DISPLAY

1. "Opening of the International Health Exhibition," *Morning Post*, May 9, 1884, 3.

2. Julie K. Brown, *Health and Medicine on Display: International Expositions in the United States* (Cambridge, MA: MIT Press, 2009), 65.

3. Wilson Smith, "Old London, Old Edinburgh: Constructing Historic Cities," in *Cultures of International Exhibitions 1840–1940: Great Exhibition in the Margins*, ed. Marta Filipova (Surry, UK: Ashgate, 2015), 206.

4. Ernest Hart, "The International Health Exhibition: Its Influence and Possible Sequels," *Journal of the Society of Arts* (November 28, 1884): 5.

5. "The International Health Exhibition," *Western Mail*, May 9, 1884, 2.

6. "The International Health Exhibition," *Graphic*, May 17, 1884, 483.

7. Thomas Richards, *The Commodity Culture of Victorian England: Advertising and Spectacle, 1851–1914* (London: Verso, 1900), 4.

8. "The International Health Exhibition," *Graphic*, 483.

9. Annmarie Adams, *Architecture in the Family Way: Doctors, Houses and Women, 1870–1900* (Montreal: McGill-Queen's University Press, 1996), 13–14.

10. Thomas Curson Hansard, *Parliamentary Debates* (London: Hansard, 1884), 179.

11. Annmarie Adams, "The Healthy Victorian City: The Old London Street at the International Health Exhibition of 1884," in *Streets: Critical Perspectives on Public Space*, ed. Zeynep Celik, Diana Favro, and Richard Ingersoll (Berkeley: University of California Press, 1994); and Smith, "Old London, Old Edinburgh."

12. Adams, "Healthy Victorian City," 211.

13. Hart, "International Health Exhibition," 6.

14. Bernard V. Lightman and Bennett Zon, "Introduction," in *Evolution and Victorian Culture* (Cambridge: Cambridge University Press, 2014), 7.

15. Verity Hunt, "Narrativizing 'The World's Show': The Great Exhibition, Panoramic Views and Print Supplements," in *Popular Exhibitions, Science and Showmanship, 1840–1910*, ed. Joe Kember, John Plunkett, and Jill Sullivan (New York: Pickering & Chatto, 2012).

16. Lord Stanley, quoted in Lawrence Goldman, *Science, Reform and Politics in Victorian Britain: The Social Science Association 1857–1886* (Cambridge: Cambridge University Press, 2002), 198.

17. "The International Health Exhibition," *Lancet* (May 10, 1884): 853.

18. Tony Bennett, *The Birth of the Museum: History, Theory and Politics* (London: Routledge, 1995), 61.

19. Smith, "Old London, Old Edinburgh," 213.

20. George H. Birch, "Description of a Street Representing 'Old London' in the International Health Exhibition," in *International Health Exhibition: Official Catalogue* (London: William Clowes and Sons, 1884), xix (hereafter *Official Catalogue*).

21. Adams, *Architecture in the Family Way*, 16.

22. Henry W. Acland, "Preface," in *The Health Exhibition Literature: Health in the Dwelling*, ed. Acland (London: William Clowes and Sons, 1884), 1:3.

23. Acland, "Preface," 3.

24. *Official Catalogue*, xlviii.

25. *Official Catalogue*, v.

26. "Memoranda for the Guidance of Exhibitors," Acland, *Health Exhibition Literature*, 18:139.

27. "Health Laboratories as the Result of the Health Exhibition," *Nature* (December 11, 1884): 121–22.

28. "Memoranda for the Guidance of Exhibitors," 139.

29. Michael Worboys, *Spreading Germs: Disease Theories and Medical Practice in Britain, 1865–1900* (Cambridge: Sheffield Hallam University and Cambridge University Press, 2000), 170.

30. W. Watson Cheyne, W. H. Corfield, and Charles E. Cassal, *International Health Exhibition, London 1884: Public Health Laboratory Work* (London: William Clowes and Sons, 1884), 93 (hereafter *Public Health Laboratory Work*).

31. "Memoranda for the Guidance of Exhibitors," 140.

32. Iwan Morus, "More the Aspect of Magic than Anything Natural: The Philosophy of Demonstrations in Victorian Popular Science," in *Science in the Marketplace: Nineteenth-Century Sites and Experiences*, ed. Bernard Lightman and Aileen Fyfe (Chicago: University of Chicago Press, 2007), 337.

33. "Scientific Aspects of the Health Exhibition," *Lancet* (July 5, 1884): 24.

34. "The International Health Exhibition: First Notice," *Lancet* (July 19, 1884): 120.

35. Cheyne, "Part I.—Biological Laboratory," in *Public Health Laboratory Work*, 3.

36. See, for example, Gerald Geison, *Michael Foster and the Cambridge School of Physiology: The Scientific Enterprise in Late Victorian Society* (Princeton, NJ: Princeton University Press, 1978); and Stanley Joel Reiser, *Medicine and the Reign of Technology* (Cambridge: Cambridge University Press, 1978).

37. Ernest Hart, "Abstract of a Lecture on the International Health Exhibition of 1884: Its Influence and Possible Sequels," *British Medical Journal* 1249, no. 2 (1884): 1115.

38. "Scientific Aspects of the Health Exhibition," 24.

39. Charles DePaolo, *William Watson Cheyne and the Advancement of Bacteriology* (Jefferson, NC: McFarland, 2016), 82.

40. "The Fifth International Congress of Hygiene," *Lancet* (August 30, 1884): 384; and "The International Health Exhibition," *Times*, November 4, 1884, 2.

41. Adams, *Architecture in the Family Way*, 29.

42. Katherine Pandora, "The Permissive Precincts of Barnum's and Goodrich's Museums of Micellaneity: Lessons in Knowing Nature for New Learners," in *Science Museums in Transition: Anglo-American Cultures of Display in the Nineteenth Century*, ed. Carin Berkowitz and Bernard Lightman (Pittsburgh: University of Pittsburgh Press, 2017), 40.

43. Arthur a'Beckett, "Our Insane-itary Guide to the Health Exhibition," *Punch or the London Charivari*, August 30, 1884, 98.

44. Quoted in Anthony David Edwards, *The Role of International Exhibitions in Britain, 1850–1910: Perceptions of Economic Decline and the Technical Education Issue* (Amherst, NY: Cambria Press, 2008), 137.

45. a'Beckett, "Our Insane-itary Guide," 98.

46. "The International Health Exhibition," *London Journal*, June 7, 1884, 356; and "International Health Exhibition," *Graphic*, 483.

47. *The International Health Exhibition: Official Guide* (London: William Clowes and Sons, 1884), 11, 19 (hereafter *Official Guide*).

48. *Ward and Lock's Ready Guide to the Health Exhibition* (London: Ward, Lock, 1884), 39.

49. "International Health Exhibition," February 5, 1884, Exhibition International Health and Education 1883–1888 Ma/35/95, Victorian and Albert Museum Archive, London.

50. *International Health Exhibition Catalogue: Education Division* (London: William Clowes and Sons, 1884), xx–xxi.

51. *Official Guide*, 2.

52. W. Mattieu Williams, "The Science of Cookery"; and T. Spencer Cobbold, "The Parasites of Meat and Prepared Flesh Food," both in *Health Exhibition Literature: Health in Diet*, vol. 5 (London: William Clowes and Sons, 1884) (hereafter *Health in Diet*); *The Adulteration of Food: Conferences by the Institute of Chemistry* (London: William Clowes and Sons, 1884); and *Ward and Lock's Ready Guide*, 15.

53. George William Wigner, "Pure Milk," in *Health in Diet*, vol. 6; and John Jackson Manley "Salt and Other Condiments," in *Health in Diet*, vol. 4.

54. Alexander W. Blythe, *Diet in Relation to Health and Work*, in *Health in Diet*, 4:339.

55. Blyth, "Diet in Relation to Health and Work," 251.

56. Corinna Treitel, *Eating Nature in Modern Germany: Food, Agriculture and Environment c. 1870–2000* (Cambridge: Cambridge University Press, 2017), 95.

57. See Ian Miller, *A Modern History of the Stomach: Gastric Illness, Medicine and British Society, 1800–1950* (London: Pickering & Chatto, 2011), for a comprehensive account of the development of laboratory science and surgical interventions around the stomach from the mid-nineteenth century on.

58. *Official Guide*, 11.

59. "The Influence of the Vegetarian Dining Room at the International Health Exhibition," *Dietetic Reformer and Vegetarian Messenger* (August 3, 1884): 231.

60. See James Gregory, *Of Victorians and Vegetarians: The Vegetarian Movement in Victorian Britain* (London: Tauris, 2007), for an account of the movement's early history.

61. "The International Health Exhibition," *Dietetic Reformer and Vegetarian Messenger* (July 1, 1884): 197.

62. "Success at the International Health Exhibition," *Dietetic Reformer and Vegetarian Messenger* (September 1, 1884): 255.

63. "The Opening of the International Health Exhibition," *Lancet* (May 10, 1884): 353.

64. Blythe, *Diet in Relation to Health and Work*, 289; and Williams, "Science of Cookery," 11.

65. "A Converted Fruitarian," *British Medical Journal* (January 6, 1900): 37.

66. Thomas F. Gieryn, "Boundary-Work and the Demarcation of Science from Non-science: Strains and Interests in Profession Ideologies of Scientists," *American Sociological Review* 48, no. 6 (December 1983): 782.

67. "Success at the International Health Exhibition," 255; and "The International Health Exhibition," *Dietetic Reformer and Vegetarian Messenger* (June 1, 1884): 158.

68. Charles W. Forward, *Fifty Years of Food Reform: A History of the Vegetarian Movement in England* (London: Ideal Publishing Union, 1898), 100.

69. Gregory, *Of Victorians and Vegetarians*, 134–41.

70. Forward, *Fifty Years of Food Reform*, 101.

71. Blythe, "Diet in Relation to Health and Work," 267.

72. James Paget, "Reception by H.R.H. the Prince of Wales of the International Juries," *Health Exhibition Literature: Miscellaneous* (London: William Clowes and Sons, 1884), 18:4–5.

73. *Official Catalogue*, x.

CHAPTER 6: VICTORIAN PHYSICS AND ITS INTERDISCIPLINARITIES

1. There is an extensive literature here, but of particular importance are Simon Schaffer, "Late Victorian Metrology and Its Instrumentation: A Manufactory of Ohms," in *Invisible Connections: Instruments, Institutions and Science*, ed. Robert Bud and Susan Cozzens (Bellingham, WA: SPIE Optical Press, 1992); and Schaffer, "Accurate Measurement Is an English Science," in *The*

Values of Precision, ed. M. Norton Wise (Princeton, NJ: Princeton University Press, 1995). For an overview and critique of much of this material, see Graeme Gooday, *The Morals of Measurement: Accuracy, Irony and Trust in late Victorian Electrical Practice* (Cambridge: Cambridge University Press, 2004).

2. Sydney Ross, "Scientist: The Story of a Word," *Annals of Science* 18 (1962).

3. Michael Faraday to William Whewell, May 20, 1840, in *The Correspondence of Michael Faraday*, ed. Frank James (London: Institution of Electrical Engineers, 1993), 2:671.

4. William Robert Grove, "Physical Science in England," *Blackwood's Magazine*, October 1843, 524.

5. William Thomson, "On the Rigidity of the Earth," in his *Mathematical and Physical Papers* (London: C. J. Clay and Sons, 1890), 3:318.

6. The Physical Society of London (the precursor of the Institute of Physics) was established in 1874, with John Hall Gladstone (1827–1902) as its first president, but never achieved the status of the other disciplinary societies. See John L. Lewis, ed., *125 Years: The Physical Society and the Institute of Physics* (London: Institute of Physics Publishing, 1999).

7. David Gooding, "In Nature's School: Faraday as an Experimentalist," in *Faraday Rediscovered: Essays on the Life and Work of Michael Faraday*, ed. David Gooding and Frank James (London: Macmillan, 1985); Otto Sibum, "Reworking the Mechanical Value of Heat: Instruments of Precision and Gestures of Accuracy in Early Victorian England," *Studies in History and Philosophy of Science* 26 (1995).

8. Jimena Canales, *A Tenth of a Second: A History* (Chicago: University of Chicago Press, 2009); Chitra Ramalingam, "Natural History in the Dark: Seriality and the Electric Discharge in Victorian Physics," *History of Science* 48 (2010). See also Iwan Rhys Morus, "Worlds of Wonder: Sensation and the Victorian Scientific Performance," *Isis* 101 (2010).

9. Iwan Rhys Morus, *William Robert Grove: Victorian Gentleman of Science* (Cardiff: University of Wales Press, 2017).

10. William Robert Grove, *On the Progress of Physical Science since the Opening of the London Institution* (London: London Institution, 1842), 37.

11. Grove, "Physical Science," 518.

12. William Robert Grove, *On the Correlation of Physical Forces* (London: London Institution, 1846), 28.

13. Grove, *On the Correlation*, 50.

14. Bruno Latour, *Science in Action* (Milton Keynes, UK: Open University Press, 1987), elaborates the notion of obligatory passage points.

15. Iwan Rhys Morus, "Correlation and Control: William Robert Grove and the Construction of a New Philosophy of Scientific Reform," *Studies in History and Philosophy of Science* 22 (1991).

16. Grove, "Physical Science."

17. Edward Forbes to William Robert Grove, n.d., Grove Collection, Royal Institution, London.

18. Minute Book of the Philosophical Club, 17, Ms.721, Royal Society, London.

19. "Letter from the Honourable Mr. Justice Grove," in *British Parliamentary Papers: Education: Scientific and Technical*, vol. 3, *Sessions 1873–72* (Dublin: Irish University Press, 1979), 627.

20. Jan Golinski, *The Experimental Self: Humphry Davy and the Making of a Man of Science* (Chicago: University of Chicago Press, 2016).

21. Gooding, "In Nature's School."

22. Roland Jackson, *The Ascent of John Tyndall: Victorian Scientist, Mountaineer, and Public Intellectual* (Oxford: Oxford University Press, 2018).

23. Thomas Carlyle, quoted in Ursula DeYoung, *A Vision of Modern Life: John Tyndall and the Role of the Scientist in Victorian Culture* (London: Palgrave, 2011), 15.

24. John Tyndall, *Faraday as a Discoverer* (London: Longman, Green, 1870), 182.

25. John Tyndall, *Fragments of Science* (London: Longmans, Green, 1879), 1:452.

26. John Barlow to Michael Faraday, August 13, 1859, in *The Correspondence of Michael Faraday*, ed. Frank James (London: Institution of Engineering and Technology, 2008), 5:543.

27. Barlow to Faraday, August 13, 1859.

28. Jill Howard, "'Physics and Fashion': John Tyndall and His Audiences in Mid-Victorian Britain," *Studies in the History and Philosophy of Science* 35 (2004).

29. W[illiam] F. Barrett, "Note on Sensitive Flames," *Philosophical Magazine* 33 (1867): 216–17.

30. Barrett, "Note on Sensitive Flames," 218, 219.

31. John Tyndall, "On Sounding and Sensitive Flames," *Philosophical Magazine* 33 (1867).

32. John Tyndall, *A Course of Eight Lectures on Sound* (London, 1867), 241.

33. "Curiosities of Sound," *Intellectual Observer* 12 (1868): 48.

34. "On Musical and Sensitive Flames," *Chemical News* 3 (1868): 8.

35. Barrett, "Note on Sensitive Flames," 219.

36. Tyndall, *Course of Eight Lectures on Sound*, 240.

37. James Clerk Maxwell to C. J. Monro, March 15, 1871, in Lewis Campbell and William Garnett, *The Life of James Clerk Maxwell* (London: Macmillan, 1884), 289.

38. Quoted in Iwan Rhys Morus, *When Physics Became King* (Chicago: University of Chicago Press, 2005), 239.

39. John William Strutt to James Clerk Maxwell, February 14, 1871, in Campbell and Garnett, *Life of Maxwell*, 265.

40. Maxwell to Edward Blore, February 15, 1871, in Campbell and Garnett, *Life of Maxwell*, 265–66.

41. "The New Physical Laboratory of the University of Cambridge," *Nature* 10 (1874): 139. See also Isobel Falconer, "Building the Cavendish and Time at Cambridge," in *James Clerk Maxwell: Perspectives on His Life and Work*, ed. Raymond Flood, Mark McCartney, and Andrew Whitaker (Oxford: Oxford University Press, 2014).

42. James Clerk Maxwell, "Introductory Lecture on Experimental Physics," in *The Scientific Papers of James Clerk Maxwell*, 2 vols., ed. W. D. Niven (Cambridge: Cambridge University Press, 1890), 2:241.

43. Maxwell, "Introductory Lecture," 2:242.

44. Maxwell, "Introductory Lecture," 2:243–44.

45. Maxwell, "Introductory Lecture," 2:244.

46. Maxwell, "Introductory Lecture," 2:244.

47. Schaffer, "Late Victorian Metrology"; Sibum, "Reworking the Mechanical Value of Heat"; Graeme Gooday, "Spot-Watching, Bodily Postures and the 'Practised Eye': The Material Practice of Instrument Reading in Late Victorian Electrical Life," in *Bodies/Machines*, ed. Iwan Rhys Morus (Oxford: Berg, 2002).

48. Quoted in Simon Schaffer, "Metrology, Metrication and Victorian Values," in *Victorian Science in Context*, ed. Bernard Lightman (Chicago: University of Chicago Press, 1997), 33.

49. Maxwell, "Introductory Lecture," 248.

50. "New Physical Laboratory," 139; Falconer, "Building the Cavendish"; Schaffer, "Accurate Measurement."

51. Richard Glazebrook, *James Clerk Maxwell and Modern Physics* (London: Cassell, 1896), 76–77.

52. Robert John Strutt, *Life of Lord Rayleigh* (London: Edward Arnold, 1924), 105.

53. Edward Charles Pickering, *Elements of Physical Manipulation* (Cambridge: Cambridge University Press, 1873), quoted in Richard Glazebrook, "Lord Rayleigh's Professorship," in *A History of the Cavendish Laboratory, 1871–1910* (London: Longmans, Green, 1910), 45.

54. Richard Glazebrook, quoted in J. G. Crowther, *The Cavendish Laboratory, 1874–1974* (New York: Science History Publications, 1974), 98.

55. Glazebrook, "Lord Rayleigh's Professorship," 46.

56. Richard Glazebrook and Napier Shaw, *Practical Physics* (London: Longman, Green, 1889), 89–102.

57. Lord Rayleigh, "Acoustical Observations," *Philosophical Magazine* 7 (1879): 153.

58. Glazebrook and Shaw, *Practical Physics*, 181–82.

59. Glazebrook and Shaw, *Practical Physics*, 182.

60. Oliver Lodge, *Modern Views of Electricity* (London: Macmillan, 1889).

61. Bruce Hunt, *The Maxwellians* (Ithaca, NY: Cornell University Press, 1994).

62. Lodge, *Modern Views*, viii.

63. Lodge, *Modern Views*, xii.

64. Lodge, *Modern Views*, 336.

65. [James Clerk Maxwell], "Electricity and Magnetism," *Nature* 8 (1873): 42; Maxwell, *Treatise on Electricity and Magnetism*, 2 vols. (Oxford: Clarendon Press, 1873), 1:ix.

66. Warren de la Rue and Hugo Müller, "Experimental Researches on the Electric Discharge with the Chloride of Silver Battery," *Philosophical Transactions* 171 (1880): 109; Cromwell Varley, "Some Experiments on the Discharge of Electricity through Rarified Media and the Atmosphere," *Proceedings of the Royal Society* 19 (1871): 236.

67. William Crookes, "Molecular Physics in High Vacua," *Proceedings of the Royal Institution* 9 (1882): 140.

68. William Crookes, *On Radiant Matter* (London: E. J. Davey, 1879), 18.

69. Jaume Navarro, *A History of the Electron: J. J. And G. P. Thompson* (Cambridge: Cambridge University Press, 2012).

70. Oliver Lodge, *Past Lives: An Autobiography* (London: Hodder & Staughton, 1931), 187.

71. Lodge, *Past Lives*, 188.

72. Lodge, *Past Lives*, 185.

73. Lodge, *Past Lives*, 188; Lodge appears to be referring to the discussion in J. J. Thomson, *The Discharge of Electricity through Gases* (London: Archibald Constable, 1898), 173–74.

74. J. J. Thomson, *Notes on Recent Researches in Electricity and Magnetism* (Oxford: Clarendon Press, 1893), 93–94.

75. Robert Strutt, *The Life of Sir J. J. Thomson* (Cambridge: Cambridge University Press, 1942), 37.

CHAPTER 7: ANTHROPOLOGY, PREHISTORY, AND PALEONTOLOGY AS CROSS-DISCIPLINARY ENDEAVORS

1. John Evans, "Archaeology and the Antiquity of Man," *Popular Science Monthly*, November 1897, 95.

2. Evans, "Archaeology and the Antiquity of Man," 97.

3. Evans, "Archaeology and the Antiquity of Man," 98.

4. Martin Rudwick, *Earth's Deep History: How It Was Discovered and Why It Matters* (Chicago: University of Chicago Press, 2014); Ralph O'Connor, *The Earth on Show: Fossils and the Poetics of Popular Science, 1802–1856* (Chicago: University of Chicago Press, 2007); A. Bowdoin Van Riper, *Men among the Mammoths: Victorian Science and the Discovery of Human Prehistory* (Chicago:

University of Chicago Press, 1993); Donald K. Grayson, *The Establishment of Human Antiquity* (New York: Academic Press, 1983); George Stocking, *Victorian Anthropology* (New York: Free Press, 1991).

5. Adelene Buckland, "Introduction," in *Time Travelers: Victorian Encounters with Time and History*, ed. Buckland and Sadiah Qureshi (Chicago: University of Chicago Press, 2020), xvii.

6. Martin Rudwick, *The Great Devonian Controversy: The Shaping of Scientific Knowledge among Gentlemanly Specialists* (Chicago: University of Chicago Press, 1988); George Stocking, "What's in a Name? The Origins of the Royal Anthropological Institute (1837–71)," *Man* 6, no. 3 (1971); Efram Sera-Shriar, "Observing Human Difference: James Hunt, Thomas Huxley and Competing Disciplinary Strategies in the 1860s," *Annals of Science* 70, no. 4 (2013); Sadiah Qureshi, "Robert Gordon Latham, Displayed Peoples, and the Natural History of Race, 1854–1866," *Historical Journal* 54, no. 1 (2011); Philippa Levine, *The Amateur and the Professional: Antiquarians, Historians and Archaeologists in Victorian England 1838–1886* (Cambridge: Cambridge University Press, 2008).

7. Tony Bennett, *The Birth of the Museum: History, Theory, Politics* (London: Routledge, 2013); Tony Bennett, *Pasts beyond Memory: Evolution Museums Colonialism* (London: Routledge, 2004); Samuel Alberti, *Nature and Culture: Objects, Disciplines and the Manchester Museum* (Manchester: Manchester University Press, 2009).

8. Grayson, *Establishment*; Van Riper, *Men among the Mammoths*.

9. Clive Gamble and Theodora Moutsiou, "The Time Revolution of 1859 and the Stratification of the Primeval Mind," *Notes and Records of the Royal Society* 65, no. 1 (2011).

10. Joseph Prestwich, "On the Occurrence of Flint Implements, Associated with the Remains of Animals of Extinct Species," *Philosophical Transactions of the Royal Society of London* 150 (1860).

11. Jenny Bulstrode, "The Industrial Archaeology of Deep Time," *British Journal for the History of Science* 49, no. 1 (2016), has recently written a fascinating study on the links between Prestwich and Evans's business and economic interests, and their interests in human antiquity; "The Industrial Archaeology of Deep Time," *British Journal for the History of Science* 49, no. 1 (2016).

12. The work of the committee has been discussed extensively in Van Riper, *Men among the Mammoths*.

13. Prestwich, "On the Occurrence," 277.

14. Prestwich, "On the Occurrence," 295.

15. Prestwich, "On the Occurrence," 296.

16. Prestwich, "On the Occurrence," 298–99.

17. John Frere, "Account of Flint Weapons Discovered at Hoxne in Suffolk," *Archaeologia* 13 (1800).

18. Prestwich, "On the Occurrence," 304.

19. Prestwich, "On the Occurrence," 305.

20. "Appendix: Letter from John Evans to Joseph Prestwich, sent 25 May 1859," in Prestwich, "On the Occurrence," 310.

21. "Appendix," 312.

22. See Levine, *Amateur and the Professional*; and Van Riper, *Men among the Mammoths*, for the growth of archaeological associations.

23. H. Lamdin-Whymark, "Sir John Evans: Experimental Knapping and the Origins of Lithic Research," *Lithics—The Journal of the Lithic Studies Society* 30 (2009): 45; and Bulstrode, "Industrial Archaeology."

24. John Evans, "On the Occurrence of Flint Implements in Undisturbed Beds of Gravel, Sand, and Clay," *Archaeologia* 38 (1860): 280.

25. Evans, "On the Occurrence of Flint Implements," 280.

26. Evans, "On the Occurrence of Flint Implements," 286.

27. Evans, "On the Occurrence of Flint Implements," 289.

28. Evans, "On the Occurrence of Flint Implements," 292.

29. Evans, "On the Occurrence of Flint Implements," 292.

30. John Burrow, "Evolution and Anthropology in the 1860s: The Anthropological Society of London, 1863–71," *Victorian Studies* 7 (1963); Ronald Rainger, "Race, Politics, and Science: The Anthropological Society of London in the 1860s," *Victorian Studies* 2 (1978).

31. Sadiah Qureshi, "Robert Gordon Latham, Displayed Peoples, and the Natural History of Race, 1854–1866," *Historical Journal* 54, no. 1 (2011); Efram Sera-Shriar, "Observing Human Difference: James Hunt, Thomas Huxley and Competing Disciplinary Strategies in the 1860s," *Annals of Science* 70, no. 4 (2013); and Sera-Shriar, "The Scandalous Affair of the Anthropological Review: Hyde Clarke, James Hunt, and British Anthropology in the 1860s," in *Victorian Culture and the Origin of Disciplines*, ed. Bernard Lightman and Bennett Zon (New York: Routledge, 2019).

32. Peter Mandler, *The English National Character: The History of an Idea from Edmund Burke to Tony Blair* (New Haven, CT: Yale University Press, 2006).

33. Efram Sera-Shriar, "Race," in *Historicism and the Human Sciences in Victorian Britain*, ed. Mark Bevir (Cambridge: Cambridge University Press, 2017).

34. James Hunt, "Introductory Address on the Study of Anthropology," *Anthropological Review* 1, no. 1 (1863): 1–2.

35. Hunt, "Introductory Address," 2.

36. Hannah Franziska Augstein, *James Cowles Prichard's Anthropology: Remaking the Science of Man in Early Nineteenth-Century Britain* (Amsterdam: Rodopi, 1999).

37. James Hunt, "The President's Address," *Journal of the Anthropological Society of London* 2 (1864): lxxxi.

38. Hunt, "President's Address," lxxxiv.

39. James Hunt, "Address Delivered at the Third Anniversary Meeting of

the Anthropological Society of London," *Journal of the Anthropological Society of London* 4 (1866): lxv.

40. Hunt, "Address Delivered," lxv.

41. William Boyd Dawkins, *Early Man in Britain and His Place in the Tertiary Period* (London: Macmillan, 1880), 1.

42. Dawkins, *Early Man*, 1.

43. Dawkins, *Early Man*, 2.

44. Dawkins, *Early Man*, 3.

45. Dawkins, *Early Man*, 3.

46. Dawkins, *Early Man*, 2.

47. Dawkins, *Early Man*, 1–2.

48. Dawkins, *Early Man*, 163.

49. Dawkins, *Early Man*, 172.

50. Dawkins, *Early Man*, 239.

51. Alberti, *Nature and Culture*.

52. "Report on the Progress and Development of the Museum from October 1890, to March, 1894," [n.d.], Box 2: Offprints & Notes, William Boyd Dawkins Papers, Buxton Museum and Art Gallery, Buxton, UK [hereafter WBD Papers].

53. *General Guide to the Collections of the Manchester Museum*, 1915, Box 2, WBD Papers.

54. *General Guide*, 8.

55. *General Guide*, 16.

56. *General Guide*, 17.

57. Dawkins, "Memorandum on the History of the Museum, 31 May 1927," Box 61: Lectures, WBD Papers.

58. This breakdown is traced in Alberti, *Nature and Culture*, 38–53.

59. Dawkins, "Memorandum on the History of the Museum."

CHAPTER 8: DISCIPLINING THE FIELD

The work presented here is supported by the Carlsberg Foundation, grant CF20-0180.

1. The German name for the Association of German Natural Philosophers and Physicists is Versammlung Deutscher Naturforscher und Aerzte, located in Graz. The event was large and popular, with around four thousand visitors gathering in Graz in the Austrian Alps.

2. Susan Barr and Cornelia Lüdecke, eds., *The History of the International Polar Years (IPYs)* (Heidelberg: Springer Science & Business Media, 2010); Susan Barr, "The Expeditions of the First International Polar Year 1882–83," *Polar Research* 28, no. 2 (January 1, 2009): 311–12; Roger D. Launius, James Rodger Fleming, and David H. DeVorkin, eds., *Globalizing Polar Science: Reconsider-*

ing the *International Polar and Geophysical Years* (London: Palgrave Macmillan, 2010).

3. *Geographical Magazine*, April 1, 1876.

4. Julian Dowdeswell, "The Weddell Sea, Antarctica: Modern Science and the Search for Shackleton's Endurance," Geography Departmental Seminar Series talk, University of Cambridge, 2019, http://www.talks.cam.ac.uk/talk/index/128389.

5. Michael H. G. Hoffmann, Jan C. Schmidt, and Nancy J. Nersessian, "Philosophy of and as Interdisciplinarity," *Synthese* 190, no. 11 (2013): 1857.

6. Richard C. Powell, "Becoming a Geographical Scientist: Oral Histories of Arctic Fieldwork," *Transactions of the Institute of British Geographers*, n.s., 33, no. 4 (October 1, 2008); Klaus Dodds and Richard Powell, "Polar Geopolitics: New Researchers on the Polar Regions," *Polar Journal: Polar Geopolitics: New Researchers on the Polar Regions* 3, no. 1 (2013); Johanne M. Bruun and Anna Guasco, "Reimagining the 'Fields' of Fieldwork," *Dialogues in Human Geography*, 2023, first published online; Michael Bravo, *North Pole* (London: Reaktion Books, 2019); Robert G. David, *The Arctic in the British Imagination 1818–1914* (Manchester: Manchester University Press, 2000); Jen Hill, *White Horizon: The Arctic in the Nineteenth-Century British Imagination* (Albany: State University of New York Press, 2009); Janice Cavell, *Tracing the Connected Narrative: Arctic Exploration in British Print Culture, 1818–1860* (Toronto: University of Toronto Press, 2008).

7. E. Tammiksaar, N. G. Sukhova, and I. R. Stone, "Hypothesis versus Fact: August Petermann and Polar Research," *Arctic* 52, no. 3 (1999); John Woitkowitz, "Science, Networks, and Knowledge Communities: August Petermann and the Construction of the Open Polar Sea," unpublished conference paper, annual meeting of the Canadian Historical Association, Vancouver, BC, 2019.

8. Royal Society of Great Britain, *Manual and Instructions for the Arctic Expedition: Manual of the Natural History, Geology, and Physics of Greenland and the Neighbouring Regions; Prepared for the Use of the Arctic Expedition of 1875, under the Direction of the Arctic Committee of the Royal Society ... Together with Instructions Suggested by the Arctic Committee of the Royal Society for the Use of the Expedition* (London: Her Majesty's Stationary Office, 1875); Royal Geographical Society of Great Britain, *Arctic Geography and Ethnology: A Selection of Papers on Arctic Geography and Ethnology* (London: John Murray, 1875).

9. For my understanding of scientific naturalism, I draw on Gowan Dawson and Bernard V. Lightman, eds., *Victorian Scientific Naturalism: Community, Identity, Continuity* (Chicago: University of Chicago Press, 2014); Lightman, "The 'History' of Victorian Scientific Naturalism: Huxley, Spencer and the 'End' of Natural History," *Studies in History and Philosophy of Biology & Biomedical Science* 58 (2016); Lightman and Michael S. Reidy, eds., *The Age of Scientific Naturalism: Tyndall and His Contemporaries* (Abingdon, UK: Routledge, 2016).

10. Bernard Lightman, *Victorian Popularizers of Science: Designing Nature for*

New Audiences (Chicago: University of Chicago Press, 2009); Lightman, "The Story of Nature: Victorian Popularizers and Scientific Narrative," *Victorian Review* 25, no. 2 (2000); Ruth Barton, *The X Club: Power and Authority in Victorian Science* (Chicago: University of Chicago Press, 2018).

11. Mary Louise Pratt, *Imperial Eyes: Travel Writing and Transculturation*, 2nd ed. (London: Routledge, 2008), 8. See also Mia M. Bennett, "Unfreezing the Arctic: Science, Colonialism, and the Transformation of Inuit Lands," *Polar Geography* 41, no. 2 (April 3, 2018); Deborah Neill, *Networks in Tropical Medicine: Internationalism, Colonialism, and the Rise of a Medical Specialty, 1890–1930* (Stanford, CA: Stanford University Press, 2012); Michael Bravo, "The Postcolonial Arctic," *Moving Worlds: A Journal of Transcultural Writings* 15 (2015); Felix Driver, *Geography Militant: Cultures of Exploration and Empire* (Oxford: Blackwell Publishers, 2000).

12. At the time of departure, the crew consisted of 120 naval men.

13. "Notes", *Nature* (November 9, 1876): 50.

14. George S. Nares, *Narrative of a Voyage to the Polar Sea: During 1875–6 in H. M. Ships 'Alert' and 'Discovery,'* 2 vols. (London: Slow, Marston, Searle, & Rivington, 1878), 1:xiii.

15. Nares, *Narrative of a Voyage*, 1:xiii, xvii.

16. It is worth noting here that when the geographical goals of expeditions failed, as they so often did, the scientific research could be highlighted to show that the expedition had been worth the expenditure. For more on the relationship between science and travel in the Arctic, see Trevor H. Levere, *Science and the Canadian Arctic: A Century of Exploration, 1818–1918* (Cambridge: Cambridge University Press, 2004); Nanna Katrine Lüders Kaalund, *Explorations in the Icy North: How Travel Narratives Shaped Arctic Science in the Nineteenth Century*, Science and Culture in the Nineteenth Century (Pittsburgh: University of Pittsburgh Press, 2021).

17. John Ross, *A Voyage of Discovery, Made under the Orders of the Admiralty, in His Majesty's Ships Isabella and Alexander, for the Purpose of Exploring Baffin's Bay, and Inquiring into the Probability of a North-West Passage* (London: John Murray, 1819), 9.

18. Ross, *Voyage of Discovery*, 116–17.

19. Trevor Levere, "Henry Wemyss Feilden, Naturalist on HMS *Alert* 1875–1876," *Polar Record* 24, no. 151 (1988): 307.

20. "The Polar Expedition," *York Herald*, February 1, 1875, 3.

21. "The Arctic Expedition," *Hampshire Telegraph and Sussex Chronicle*, May 29, 1875, 3.

22. Thomas Henry Huxley to Charles Darwin, letter number DCP-LETT-9827, January 22, 1875, in *The Correspondence of Charles Darwin*, vol. 23, *1875*, ed. Frederick Burkhardt and James A. Secord (Cambridge: Cambridge University Press, 2015), 115.

23. Joseph Dalton Hooker to Darwin, letter number DCP-LETT-9932, April 15, 1875, in *Correspondence of Charles Darwin*, 23:198.

24. Frank M. Turner, *Between Science and Religion: The Reaction to Scientific Natural in Late Victorian England* (New Haven, CT: Yale University Press, 1974); Lightman, *Victorian Popularizers of Science*, 6 (quote).

25. British Association for the Advancement of Science, *Queries Respecting the Human Race, to Be Addressed to Travellers and Others. Drawn up by a Committee of the British Association for the Advancement of Science, Appointed in 1839* (London: printed by Richard and John E. Taylor for British Association for the Advancement of Science, 1841). In my analysis I draw in particular on these works: James Urry, *Before Social Anthropology: Essays on the History of British Anthropology* (London: Routledge, 2012); Efram Sera-Shriar, *The Making of British Anthropology, 1813–1871* (London: Pickering & Chatto, 2013); Sadiah Qureshi, "Science, Empire and Globalization in the Nineteenth Century," in *The Routledge Research Companion to Nineteenth-Century British Literature and Science*, ed. John Holmes and Sharon Ruston (Abingdon, UK: Routledge, 2017).

26. Janet Browne, "Biogeography and Empire," in *Cultures of Natural History*, ed. Nicholas Jardine, James A. Secord, and E. C. Spary (Cambridge: Cambridge University Press, 1996).

27. Joseph Dalton Hooker, "Instructions in Botany," in *Manual and Instructions for the Arctic Expedition*.

28. Thomas Henry Huxley, "Supplementary Instructions," in *Manual and Instructions for the Arctic Expedition*, 60–61.

29. John Tyndall, "Hints towards Observations in the Arctic Regions," in *Manual and Instructions for the Arctic Expedition*, 34–35.

30. The issue of glacial motion had formed part of Tyndall's wider debates with another group of scientific reformers, the so-called North British physicists. John Tyndall, *The Glaciers of the Alps: Being a Narrative of Excursions and Ascents, an Account of the Origin and Phenomena of Glaciers and an Exposition of the Physical Principles to Which They Are Related* (London: John Murray, 1860); Nanna Katrine Lüders Kaalund, "A Frosty Disagreement: John Tyndall, James David Forbes, and the Early Formation of the X-Club," *Annals of Science* 74, no. 4 (October 2, 2017); Bruce Hevly, "The Heroic Science of Glacier Motion," *Osiris* 11 (January 1, 1996).

31. George Stocking, *Victorian Anthropology* (New York: Simon and Schuster, 1991); Marc Flandreau, *Anthropologists in the Stock Exchange: A Financial History of Victorian Science* (Chicago: University of Chicago Press, 2016); Sera-Shriar, *Making of British Anthropology*.

32. Edward Burnett Tylor, "Religion, Sociology &c," in *Arctic Geography and Ethnology*, 282–83.

33. For more on the specific scientific naturalism and ethnography of Tylor, see Efram Sera-Shriar, "Historicizing Belief: E. B. Tylor, Primitive Culture,

and the Evolution of Religion," in *Historicizing Humans: Deep Time, Evolution, and Race in Nineteenth-Century British Sciences*, ed. Sera-Shriar (Pittsburgh: University of Pittsburgh Press, 2018). See also, Paul-François Tremlett, Graham Harvey, and Liam T. Sutherland, eds., *Edward Burnett Tylor, Religion, and Culture* (New York: Bloomsbury Academic, 2017).

34. Augustus Pitt Rivers, "Enquiries into Customs Relating to War"; Pitt Rivers, "Enquiries Relating to Certain Arrow-Marks and Other Signs in Use amongst the Eskimos"; Pitt Rivers, "Enquiries Relating to Drawing, Carving, and Ornamentation," all in *Arctic Geography and Ethnology*.

35. Pitt Rivers, "Enquiries Relating to Certain Arrow-Marks," 286–87.

36. Stocking, *Victorian Anthropology*, 181; Sera-Shriar, *Making of British Anthropology*, 155–57.

37. Joseph Barnard Davis, "Appendix: Questions for Explorers, General," in *Arctic Geography and Ethnology*, 281–82.

38. J. Lubbock, "On the Development of Relationships," *Journal of the Anthropological Institute* 1 (1871), as referenced by Davis in "Appendix: Questions for Explorers, General," in *Arctic Geography and Ethnology*, 281–82.

39. John Tyndall, "Hints towards Observations in the Arctic Regions," 34.

40. Tyndall, "Hints towards Observations in the Arctic Regions," 35.

41. For my understanding of the discipline formation of physics, I am particularly informed by Joseph Simon, "Writing the Discipline: Ganot's Textbook Science and the 'Invention' of Physics," *Historical Studies in the Natural Sciences* 46, no. 3 (2016); Graeme Gooday, *The Morals of Measurement: Accuracy, Irony, and Trust in Late Victorian Electrical Practice* (Cambridge: Cambridge University Press, 2004); Richard Noakes, *Physics and Psychics: The Occult and the Sciences in Modern Britain* (Cambridge: Cambridge University Press, 2019).

42. Kaalund, "Frosty Disagreement"; Hevly, "Heroic Science of Glacier Motion"; Michael S. Reidy, "John Tyndall's Vertical Physics: From Rock Quarries to Icy Peaks," *Physics in Perspective* 12, no. 2 (June 15, 2010).

43. I draw in particular on Bernard Lightman, *Evolutionary Naturalism in Victorian Britain: The "Darwinians" and Their Critics*, Variorum Collected Studies Series (Aldershot, UK: Ashgate, 2009); Lightman and Dawson, *Victorian Scientific Naturalism*.

44. Joseph D. Hooker, "Botany," in Nares, *Narrative of a Voyage*, 2:301.

45. George Busk, "Polyzoa, Arctic Expedition, 1875–76," in Nares, *Narrative of a Voyage*, 2:283.

46. For my understanding of the performative construction of veracity in the context of exploration, I am particularly informed by Innes M. Keighren, Charles W. J. Withers, and Bill Bell, *Travels into Print: Exploration, Writing, and Publishing with John Murray, 1773–1859* (Chicago: University of Chicago Press, 2015), 17.

47. For more on the evolutionary landscape, see Bernard Lightman and

Bennett Zon, *Evolution and Victorian Culture* (Cambridge: Cambridge University Press, 2014); Ian Hesketh, "A Good Darwinian? Winwood Reade and the Making of a Late Victorian Evolutionary Epic," *Studies in History and Philosophy of Science Part C: Studies in History and Philosophy of Biological and Biomedical Sciences* 51 (June 1, 2015); Lightman, *Global Spencerism: The Communication and Appropriation of a British Evolutionist* (Leiden: Brill, 2015); Janet E. Browne, "Charles Darwin and Ideology: Rethinking the Darwinian Revolution," *Mètode Science Studies Journal* 7 (2016).

48. Henry Wemyss Feilden, "Ethnology" in Nares, *Narrative of a Voyage*, 187.

49. Feilden, "Ethnology," 191.

50. Henry Wemyss Feilden, "Ornithology," in Nares, *Narrative of a Voyage*.

51. Lorraine Daston and Peter Galison, *Objectivity* (New York: Zone Books, 2007); H. G. Jones, "Teaching the Explorers: Some Inuit Contributions to Arctic Discoveries," *Polar Geography* 26, no. 1 (January 1, 2002); Penny Petrone, *Northern Voices: Inuit Writing in English* (Toronto: University of Toronto Press, 1992).

52. The main biography of Suersaq is: Jan Løve, *Hans Hendrik og Hans Ø: Beretningen om Hans Hendrik og de to Hans Øer* (Copenhagen: Det Grønlandske Selskab, 2016). Sursaq published a narrative from his Arctic travels as: *Memoirs of Hans Hendrik: The Arctic Traveller, Serving under Kane, Hayes, Hall and Nares, 1853–1876*, ed. George Stephens, trans. Hinrich [Henry] Rink (London: Trübner, 1878).

53. No known vital dates.

54. For more on nineteenth-century European and Euro-American expeditions and their narratives of exploration, see Cavell, *Tracing the Connected Narrative*; Kaalund, *Explorations in the Icy North*; Keighren Withers, and Bell, *Travels into Print*.

55. Feilden, "Ethnology," 188.

56. Pratt, *Imperial Eyes*. See also Stuart B. Schwartz, ed., *Implicit Understandings: Observing, Reporting and Reflecting on the Encounters between Europeans and Other Peoples in the Early Modern Era* (Cambridge: Cambridge University Press, 1994); Tiffany Shellam, "Miago and the 'Great Northern Men': Indigenous Histories from In-Between," in *Indigenous Mobilities*, ed. Rachel Standfield, Across and beyond the Antipodes (N.p [online]: ANU Press, 2018).

57. Fa-ti Fan, *British Naturalists in Qing China Science, Empire, and Cultural Encounter* (Cambridge, MA: Harvard University Press, 2004). For other key writings on the plurality of encounters and how we can conceptualize this in the history of science, see Felix Driver, "Exploration as Knowledge Transfer: Exhibiting Hidden Histories," in *Mobilities of Knowledge*, ed. Heike Jöns, Peter Meusburger, and Michael Heffernan, Knowledge and Space (Cham, Switzerland: Springer International Publishing, 2017); Felix Driver, "Intermediaries

and the Archive of Exploration," in *Indigenous Intermediaries*, ed. Shino Konishi, Maria Nugent, and Tiffany Shellam, New Perspectives on Exploration Archives (N.p. [online]: ANU Press, 2015).

58. I refer in particular to: Julie Cruikshank, *Do Glaciers Listen? Local Knowledge, Colonial Encounters, and Social Imagination* (Vancouver: UBC Press, 2010); Cruikshank, "Glaciers and Climate Change: Perspectives from Oral Tradition," *Arctic* 54, no. 4 (December 1, 2001); Mary Caroline Rowan, "Co-constructing Early Childhood Programs Nourished by Inuit Worldviews," *Études/ Inuit/Studies* 38, no. 1 (2014); Claudio Aporta, "The Trail as Home: Inuit and Their Pan-Arctic Network of Routes," *Human Ecology* 37, no. 2 (April 1, 2009); Peter Bates, "Inuit and Scientific Philosophies about Planning, Prediction, and Uncertainty," *Arctic Anthropology* 44, no. 2 (2007); Valerie Henitiuk, "'Memory Is So Different Now': The Translation and Circulation of Inuit-Canadian Literature in English and French," *Perspectives* 25, no. 2 (April 3, 2017); Lawrence F. Felt and David Natcher, "Ethical Foundations and Principles for Collaborative Research with Inuit and Their Governments," *Études/Inuit/Studies* 35, no. 1/2 (2011).

59. "The British Association," *Standard*, August 25, 1882.

60. Sarah Burton, "Becoming Sociological: Disciplinarity and a Sense of 'Home,'" *Sociology* 50, no. 5 (October 4, 2016).

61. Theodore M. Porter, *Trust in Numbers: The Pursuit of Objectivity in Science and Public Life* (Princeton, NJ: Princeton University Press, 1995).

CHAPTER 9: STRETCHING THE BOUNDARIES OF KNOWLEDGE

1. For the purpose of this chapter, I use the *Cambridge English Dictionary*'s definition of *interdisciplinary*: "involving two or more different subjects or areas of knowledge." *Cambridge English Dictionary*, s.v. "Interdisciplinary," accessed October 21, 2019, https://dictionary.cambridge.org/dictionary/english/ interdisciplinary.

2. Mark Bevir, "Historicism and the Human Sciences in Victorian Britain," in *Historicism and the Human Sciences in Victorian Britain*, ed. Mark Bevir (Cambridge: Cambridge University Press, 2017), 2.

3. Slade was previously accused of cheating in 1872. After sitting through two unsuccessful séances with Slade during his tour of New York, a self-proclaimed pickpocket and fake medium named John W. Truesdell alleged that Slade had confessed to him that his whole performance was an act of deception. Truesdell argued that Slade used various methods employed by professional conjurors to stage the phenomena. His allegations, however, did not gain much mileage. See John W. Truesdell, *The Bottom Facts Concerning the Science of Spiritualism: Derived from Careful Investigation Covering a Period of Twenty-Five Years* (New York: G. W. Carleton, 1883), 143–61.

4. Janet Oppenheim, *The Other World: Spiritualism and Psychical Research*

in England, 1850–1914 (Cambridge: Cambridge University Press, 1985); Logie Barrow, *Independent Spirits: Spiritualism and English Plebeians, 1850–1910* (London: Routledge and Kegan Paul, 1986); and Alex Owen, *The Darkened Room: Women, Power and Spiritualism in Late Victorian England* (London: Virago Press, 1989).

5. There is a wealth of literature exploring both psychology's and physics' engagement with modern spiritualism. Key examples include Oppenheim, *Other World*; Roger Luckhurst, *The Invention of Telepathy* (Oxford: Oxford University Press, 2002); Richard Noakes, "Haunted Thoughts of the Careful Experimentalist: Psychical Research and the Troubles of Experimental Physics," *Studies in History and Philosophy of Biological and Biomedical Sciences* 48 (2014); Shane McCorristine, *Spectres of the Self: Thinking about Ghosts and Ghost-Seeing in England, 1750–1920* (Cambridge: Cambridge University Press, 2010); Peter Lamont, *Extraordinary Beliefs: A Historical Approach to a Psychological Problem* (Cambridge: Cambridge University Press, 2013); Richard Noakes, *Physics and Psychics: The Occult and the Sciences in Modern Britain* (Cambridge: Cambridge University Press, 2019). See also Seymour Mauskopf and Micharl Roger McVaugh, *The Elusive Science: Origins of Experimental Psychical Research* (Baltimore: Johns Hopkins University Press, 1980); H. M. Collins and T. J. Pinch, *Frames of Meaning: The Social Construction of Extraordinary Science* (London: Routledge, 1982); Mary Roach, *Spook: Science Tackles the Afterlife*, (New York: W. W. Norton, 2005); Deborah Blum, *Ghost Hunters: William James and the Search for Scientific Proof of Life after Death* (London: Penguin Books, 2006); and Egil Asprem, *The Problem of Disenchantment: Scientific Naturalism and Esoteric Discourse 1900–1939* (Leiden: Brill, 2014).

6. For more secondary literature on the history of scientific observation within the natural and social sciences, see Deborah Coon, "Testing the Limits of Sense and Science: American Experimental Psychologists Combat Spiritualism, 1800–1920," *American Psychologist* 47 (1992); Anne Secord, "Artisan Naturalists: Science as Popular Culture in Nineteenth-Century England" (PhD diss., University of London, 2002), 135–206; Anna Grimshaw, *The Ethnographer's Eye: Ways of Seeing in Modern Anthropology* (Cambridge: Cambridge University Press, 2001); Cristina Grasseni, ed., *Skilled Visions: Between Apprenticeship and Standards* (Oxford: Berghahn Books, 2007); Lorraine Daston and Peter Galison, *Objectivity* (New York: Zone Books, 2007); Daston and Elizabeth Lunbeck, eds., *Histories of Scientific Observation* (Chicago: University of Chicago Press, 2011); Efram Sera-Shriar, *The Making of British Anthropology, 1813–1871* (London: Pickering & Chatto, 2013), 1–20; and Sera-Shriar, "What Is Armchair Anthropology? Observational Practices in Nineteenth-Century British Human Sciences," *History of the Human Sciences* 27 (2014). For more on the establishment of credible witnessing in the human sciences, see Sera-Shriar, "Arctic Observers: Richard King, Monogenism and the Historicization of Inuit

through Travel Narratives," *Studies in History and Philosophy of Biological and Biomedical Sciences* 51 (2015); Sera-Shriar, "Tales from Patagonia: Phillip Parker King and Early Ethnographic Observation in British Ethnology, 1826–1830," *Studies in Travel Writing* 19 (2015); and Sera-Shriar, *Psychic Investigators: Anthropology, Modern Spiritualism, & Credible Witnessing in the Late Victorian Age* (Pittsburgh: University of Pittsburgh Press, 2022).

7. Barbara Weisberg, *Talking to the Dead: Kate and Maggie Fox and the Rise of Spiritualism* (San Francisco: HarperCollins, 2004), 12–13.

8. For more on the Burned-Over District, see Whitney R. Cross, *The Burned-Over District: The Social and Intellectual History of Enthusiastic Religion in Western New York, 1800–1850* (Ithaca, NY: Cornell University Press, 1950); Linda K. Pritchard, "The Burned-Over District Reconsidered: A Portent of Evolving Religious Pluralism in the United States," *Social Science History* 8 (1984); and Judith Wellman, *Grassroots Reform in the Burned-Over District of Upstate New York: Religion, Abolitionism, and Democracy* (Boca Raton, FL: CRC Press, 2000).

9. Edward Clodd, *The Question: A Brief History and Examination of Modern Spiritualism* (London: Grant Richards, 1917), 34.

10. Clodd, *Question*, 40; Oppenheim, *Other* World, 11, 273, and 290; Owen, *Darkened Room*, 19; Barry H. Wiley, *The Thought Reader Craze: Victorian Science at the Enchanted Boundary* (London: McFarland, 2012), 9–17.

11. For more on Faraday's and Carpenter's experiments, see Michael Faraday, "Experimental Investigation of Table-Turning," *Journal of the Franklin Institute* 56 (1853); "Michael Faraday's Researches in Spiritualism," *Scientific Monthly* 83 (1956); and William Benjamin Carpenter, "Electro-biology and Mesmerism," *Quarterly Review* 93 (1853). See also Oppenheim, *Other World*, 327–28, 336–37.

12. The most comprehensive study of Home's career is Peter Lamont, *The First Psychic: The Extraordinary Mystery of a Notorious Victorian Wizard* (London: Abacus Books, 2005).

13. Peter Lamont, "Spiritualism and a Mid-Victorian Crisis of Evidence," *Historical Journal*, 47 (2004).

14. Frank Podmore, *Modern Spiritualism: A History and a Criticism*, 2 vols. (London: Methuen, 1902), 2:14.

15. For more on the rise of spiritualism in the nineteenth century, see Judith Walkowitz, "Science and the Séance: Transgressions of Gender and Genre in Late Victorian London," *Representations* 22 (1988); Julian Holloway, "Enchanted Spaces: The Séance, Affect, and Geographies of Religion," *Annals of the Association of American Geographies* 96 (2006); Owen Davies, *The Haunted: A Social History of Ghosts* (Basingstoke, UK: Palgrave Macmillan, 2007); Jennifer Bann, "Ghostly Hands and Ghostly Agency: The Changing Figure of the Nineteenth-Century Spectre," *Victorian Studies* 51 (2009); McCorristine, *Spectres of*

the Self, 12–13; David Nartonis, "The Rise of Nineteenth-Century American Spiritualism, 1854–1873," *Journal for the Scientific Study of Religion* 49 (2010); and Andreas Sommer, "Psychical Research in the History and Philosophy of Science: An Introduction and Review," *Studies in History and Philosophy of Biological and Biomedical Sciences* 48 (2014).

16. [James Burns], "Henry Slade: Spirit Medium," *Medium and Daybreak* 7 (1876): 626.

17. Peter J. Bowler, "Lankester, Sir (Edwin) Ray (1847–1929), Zoologist." *Oxford Dictionary of National Biography*, September 23, 2004, https://www.oxforddnb.com/view/10.1093/ref:odnb/9780198614128.001.0001/odnb-9780198614128-e-34406. The most detailed biography of Edwin Ray Lankester is Joseph Lester and Peter J. Bowler, eds., *E. Ray Lankester and the Making of Modern Biology* (Farringdon: British Society for the History of Science, 1995).

18. For more on the contest for cultural authority in the nineteenth century, see Frank Turner, "The Victorian Conflict between Science and Religion: A Professional Dimension," *Isis* 69 (1978); and Frank Turner, *Contesting Cultural Authority: Essays in Victorian Intellectual Life* (Cambridge: Cambridge University Press, 1993). For a more nuanced understanding of the boundaries between science and religion, see Bernard Lightman, *The Origins of Agnosticism: Victorian Unbelief and the Limits of Knowledge* (Baltimore: Johns Hopkins University Press, 1987); John Headley Brooke and Geoffrey Cantor, "Whose Science? Whose Religion?," in *Reconstructing Nature: The Engagement of Science and Religion*, ed. John Headley Brooke and Geoffrey Cantor (Oxford: Oxford University Press, 1998); Peter J. Bowler, *Reconciling Science and Religion: The Debate in Early Twentieth Century Britain* (Chicago: University of Chicago Press, 2001); Peter Harrison, "'Science' and 'Religion': Constructing the Boundaries," *Journal of Religion* 86 (2006); Lightman, "Victorian Sciences and Religion: Discordant Harmonies," *Osiris* 16 (2001); Lightman, "Does the History of Science and Religion Change Depending on the Narrator? Some Atheist and Agnostic Perspectives," *Science and Christian Beliefs* 24 (2012); John Headley Brooke, *Science and Religion: Some Historical Perspectives* (Cambridge: Cambridge University Press, 2014); Matthew Stanley, *Huxley's Church and Maxwell's Demon: From Theistic Science to Naturalistic Science* (Chicago: University of Chicago Press, 2014); and Peter Harrison, *The Territories of Science and Religion* (Chicago: University of Chicago Press, 2015).

19. *Encyclopaedia Britannica*, s.v. "The Times," last updated October 21, 2023, https://www.britannica.com/topic/The-Times.

20. For example, Cox published the following seminal work in psychology; Edward William Cox, *What Am I? A Popular Introduction to the Study of Psychology*, 2 vols. (London: Longman, 1873).

21. Trevor Hall, *The Spiritualists: The Story of Florence Cook and William Crookes* (London: Duckworth, 1962), 79–84.

22. William H. Brock, *William Crooks (1832–1919) and the Commercialization of Science* (Abingdon, UK: Routledge, 2008), 126.

23. Edward William Cox, *Spiritualism Answered by Science*, (New York: H. L. Hinton, 1871).

24. E. Ray Lankester, Horatio B. Donkin, and Alice Lane Fox, "A Spirit Medium," letter to the editor, *The Times*, September 16, 1876, 7.

25. Steven Shapin and Simon Schaffer, *Leviathan and the Air-Pump: Hobbes, Boyle, and the Experimental Life* (Princeton, NJ: Princeton University Press, 1985), 60–65.

26. Daniela Bleichmar, *Visible Empire: Botanical Expeditions and Visual Culture in the Hispanic Enlightenment* (Chicago: University of Chicago Press, 2012), 6–10.

27. Lankester, "Spirit Medium," 7.

28. Lankester, "Spirit Medium," 7.

29. Lankester, "Spirit Medium," 7.

30. Daston and Galison, *Objectivity*, 19–27. See also Bleichmar's discussion on long-distance observation; Bleichmar, *Visible Empire*, 66–72.

31. Lankester, "Spirit Medium," 7.

32. Lankester, "Spirit Medium," 7.

33. Lankester, "Spirit Medium," 7.

34. Charles Carter Blake, "A Spirit Medium," letter to the editor, *The Times*, September 18, 1876, 6.

35. James Hunt, "Introductory Address on the Study of Anthropology," *Anthropological Review* 1 (1863): 2.

36. J. F. Collingwood, "The Anthropological Society and the Davenports," *Human Nature: A Monthly Journal of Zoistic Science* 2 (August 1868).

37. For more on the Davenport Brothers and the spirit cabinet trick, see Thomas Low Nichols, *A Biography of the Brothers Davenport: With Some Account of the Physical and Psychical Phenomena Which Have Occurred in Their Presence, in America and Europe* (London: Sounders, Otley, 1864); Podmore, *Modern Spiritualism*, 2:55–61; and Harry Houdini, *A Magician among the Spirits* (London: Harper & Brothers, 1924), 17–35; Lamont, *First Psychic*, 149–67; and Lamont, *Extraordinary Beliefs*, 131–41, 158–62.

38. Blake, "Spirit Medium," 6.

39. For more on observational methods in Victorian anthropology, see Sera-Shriar, "What Is Armchair Anthropology?" For more on comparative observational methods in science more generally, see Daston and Galison, *Objectivity*, 19–27; and Bleichmar, *Visible Empire*, 66–72.

40. Blake, "Spirit Medium," 6.

41. C[harles] C[arleton] Massey, "A Spirit Medium," letter to the editor, *The Times*, September 18, 1876, 6.

42. Massey, "Spirit Medium," 6.

43. John Algernon Clarke, "A Spirit Medium," letter to the editor, *The Times*, September 18, 1876, 6.

44. Clarke, "Spirit Medium," 6.

45. Alfred Russel Wallace, *Miracles and Modern Spiritualism: Three Essays* (London: James Burns, 1875). For more on Wallace's spiritualism, see William Keezer, "Alfred Russel Wallace: Naturalist, Zoogeographer, Spiritualist, and Evolutionist," *Bios* 36 (1965); Malcolm Jay Kottler, "Alfred Russel Wallace, the Origin of Man, and Spiritualism," *Isis* 65 (1974); John Durant, "Scientific Naturalism and Social Reform in the Thought of Alfred Russel Wallace," *British Journal for the History of Science* 12 (1979); Peter Peels, "Spiritual Facts and Super-Visions: The Conversion of Alfred Russel Wallace," *Religion and Modernity* 8 (1995); Martin Fichman, "Science in Theistic Contexts: A Case Study of Alfred Russel Wallace on Human Evolution," *Osiris* 16 (2001); and Benjamin David Mitchell, "Capturing the Will: Imposture, Delusion, and Exposure in Alfred Russel Wallace's Defence of Spirit Photography," *Studies in History and Philosophy of Biological and Biomedical Sciences* 46 (2014). Wallace's spiritualism has also been discussed in the following biographies: Michael Shermer, *In Darwin's Shadow: The Life and Science of Alfred Russel Wallace* (Oxford: Oxford University Press, 2002); Ross Slotten, *The Heretic in Darwin's Court: The Life of Alfred Russel Wallace* (New York: Columbia University Press, 2004), 326–51; and Martin Fichman, *An Elusive Victorian: The Evolution of Alfred Russel Wallace* (Chicago: University of Chicago Press, 2004), 139–208.

46. Wallace refers to spirit investigations as a "new branch of anthropology" in his letter to the biologist and anatomist Thomas Henry Huxley in 1866; Alfred Russel Wallace to Thomas Henry Huxley, November 22, 1866, BL Add. 46439 f. 5, The British Library, London.

47. Wallace, *Miracles and Modern Spiritualism*, 100–104. See also Efram Sera-Shriar, "Credible Witnessing: A. R. Wallace, Spiritualism, and a 'New Branch of Anthropology,'" *Modern Intellectual History* 17 (2020).

48. Alfred Russel Wallace, "A Spirit Medium," letter to the editor, *The Times*, September 19, 1876, 4.

49. Edward William Cox, "A Spirit Medium," letter to the editor, *The Times*, September 20, 1876, 8. For more on the Psychological Society of Great Britain and Cox, see Luckhurst, *Invention of Telepathy*, 47–51. See also Cox, *Spiritualism Answered by Science*; and Cox, *What Am I?* For more on psychical research and its connections to psychology see: McCorristine, *Spectres of the Self*; Lamont, *Extraordinary Beliefs*.

50. Cox, "Spirit Medium," 8.

51. Cox, "Spirit Medium," 8.

52. Cox, "Spirit Medium," 8.

53. Henry Slade, "A Spirit Medium," letter to the editor, *The Times*, September 21, 1876, 3.

54. Slade, "Spirit Medium," 3

55. Slade, "Spirit Medium," 3.

56. "Conviction of Dr. Slade," *Hull Packet and East Riding Times*, November 3, 1876, 3.

57. Issue 340 of *The Medium and Daybreak* was completely devoted to Henry Slade. For more, see [James Burns], ed., "The Slade Number," *Medium and Daybreak* 7, no. 340 (1876).

58. For the details of the Slade trial, see "Spiritualism: The Charge against Dr. Slade," *Leeds Mercury*, October 3, 1876, 5; "Spiritualism on Trial: The Charge against Dr. Slade," *Cheshire Observer*, October 7, 1876, 3; "A Spirit Medium at Bow-Street," *Times*, October 11, 1876, 12; "The Prosecution of Spiritualists: The Charge against 'Dr.' Slade," *Birmingham Daily Post*, October 23, 1876, 6; "Prosecutions of Spiritualists: The Slade Case," *Times*, October 28, 1876, 11; "Conviction of Dr. Slade," *Hull Packet and East Riding Times*, November 23, 1876, 3.

59. For more on Slade's life and career, see Truesdell, *Bottom Facts Concerning the Science of Spiritualism*, 143–61; Henry Ridgely Evans, *Hours with the Ghosts or, Nineteenth-Century Witchcraft* (Chicago: Laird and Lee, 1897), 36; Podmore, *Modern Spiritualism*, 1:89, 204; Clodd, *Question*, 49–51; Joseph McCabe, *Spiritualism: A Popular History from 1847* (New York: Dodd, Mead, 1920), 160–61; Houdini, *Magician among the Spirits*, 79–116; Paul Kutz, *A Skeptic's Handbook of Parapsychology* (Amherst, NY: Prometheus Books, 1985), 253; and Lester and Bowler, *E. Ray Lankester*, 93–104.

CHAPTER 10: "ANIMALS CANNOT SUBSIST ON AIR"

1. "Professor Huxley on the Relation of Physical Science to Medical Science and Medical Education," *British Medical Journal* (May 12, 1866): 503.

2. "Professor Huxley," 503.

3. "Bantingism," *British Medical Journal* (October 22, 1864): 470.

4. Simon Schaffer, "Scientific Discoveries and the End of Natural Philosophy," *Social Studies of Science* 16, no. 3 (1986).

5. Jonathan R. Topham, "Scientific Publishing and the Reading of Science in Nineteenth-Century Britain: A Historiographical Survey and Guide to Sources," *Studies in History and Philosophy of Science Part A* 31, no. 4 (2000); James Secord, "Knowledge in Transit," *Isis* 95, no. 4 (2004).

6. Manuscript recipes have received a significant amount of attention in recent history of science scholarship. Our claim here is concerning nineteenth-century manuscript recipes and their relation to nutrition specifically.

7. Elizabeth Neswald, David F. Smith, and Ulrike Thoms, "Introduction," in *Setting Nutritional Standards: Theory, Policies, Practices*, ed. Neswald, Smith, and Thoms (Rochester, NY: University of Rochester Press, 2017), 2.

8. Neswald, Smith, and Thoms, "Introduction," 5.

9. Dariush Mozaffarian, Irwin Rosenberg, and Ricardo Uauy, "History of Modern Nutrition Science—Implications for Current Research, Dietary Guidelines, and Food Policy," *British Medical Journal* (2018), https://doi.org/10.1136/bmj.k2392. Typically, the formation of the American Institute for Nutrition and its corresponding *Journal of Nutrition* is mobilized as a rationale for locating the emergence of nutrition science as a concrete discipline in the early twentieth century. See "History of ASN," American Society for Nutrition, accessed August 9, 2019, https://nutrition.org/about-asn/asn-history/.

10. Neswald, Smith, and Thoms, "Introduction," 5.

11. Henry E. Sigerist, "The History of Dietetics," *Gesnerus* 46 (1989): 253, in "Articles on Dietetics," ed. Dr. Donald S. McLaren, PPDSM/3/2, Wellcome Library, London.

12. *Cookery Book*, 1800[?], images 004, 006, and 132, TD/006/001, Special Collections, University of Leeds (hereafter UoL).

13. *Cookery Book*, image 031.

14. *An Anonymous Collection of Culinary and Medical Receipts*, image 039, 1800[?], TD/005/001, UoL.

15. *Cookery Book MS 62*, image 029, 1850, TD/007/001, UoL.

16. *An Anonymous Recipe Book, Including Some Recipes from Surrey and Yorkshire MS 462*, image 058, 1800, TD/008/001, UoL.

17. *Anonymous Recipe Book*, image 056.

18. *Cookery Book MS 62*, images 032, 033, and 165.

19. *Cookery Book*, image 085; *Anonymous Recipe Book*, image 070.

20. *Cookery Book MS 59*, image 023, ca.1851, TD/011/001, UoL.

21. *Cookery Book*, image 117; *Cookery Book MS 62*, image 364.

22. *Cookery Book MS 62*, image 366.

23. *Cookery Book MS 62*, image 073.

24. *Anonymous Collection of Culinary and Medical Receipts*, image 137.

25. *Cookery Book MS 62*, image 367.

26. *Cookery Book MS 59*, image 021.

27. *Cookery Book MS 59*, image 020.

28. *Anonymous Collection of Culinary and Medical Receipts*, images 011, 113–14; *Cookery Book*, image 067.

29. *Cookery Book MS 59*, image 017.

30. *Anonymous Collection of Culinary and Medical Receipts*, images 136–37.

31. *Anonymous Recipe Book*, image 047.

32. *Anonymous Recipe Book*, image 072.

33. *Cookery Book MS 59*, image 019; *Cookery Book MS 62*, image 104. On Baillie's life, see John Jones, "Baillie, Matthew (1761–1823)," *Oxford Dictionary of National Biography* (Oxford: Oxford University Press, 2004), http://www.oxforddnb.com/view/article/1066; Franco Crainz, "The Editions and Translations of Dr Matthew Baillie's *Morbid Anatomy*," *Medical History* 26 (1982).

34. *Cookery Book MS 59*, image 050; *Cookery Book MS 62*, image 140.

35. *Cookery Book MS 59*, images 016–020.

36. Charles Lane, *Dietetics: An Endeavour to Ascertain the Law of Human Nutriment* (London: Whittaker, 1849), 3. See also W. H. G. Armytage, *Heavens Below: Utopian Experiments in England, 1560–1960* (London: Routledge, 2007), 173.

37. He also edited a food magazine, *The Healthian*, which similarly advocated a vegetarian diet, and went on to found a proto-vegan community, Fruitlands, in Massachusetts. Karen Iacobbo and Michael Iacobbo, *Vegetarian America: A History* (Westport, CT: Praeger, 2004), 79.

38. Lane, *Dietetics*, 20.

39. Lane, *Dietetics*, 5.

40. Animal nutrition was also, as one might expect, a significant object of study in the Victorian period, with dedicated treatises on the subject appearing from traditions of husbandry, veterinary science, and physiology. Indeed, the cross-fertilization of agriculture and scientific disciplines in the service of nutrition was visible in a large number of European institutions, such as the Agricultural Institute of Wiesbaden, which employed C. Remigius Fresenius (1818–1897) as professor of chemistry in the mid-nineteenth century. Fresenius, "XVI. Practical Application of the Law Pointed Out by Dr. R. D. Thomson, of the Proper Balance of the Food in Nutrition," *London, Edinburgh, and Dublin Philosophical Magazine and Journal of Science*, ser. 3, 35, no. 234 (1849).

41. Michael Kiernan, ed., *The Oxford Francis Bacon*, vol. 15, *The Essayes or Counsels, Civill and Morall* (Oxford: Oxford University Press, 2019), https://doi.org/10.1093/actrade/9780198186731.book.1.

42. James Gregory, "Vegetable Fictions in the Kingdom of Roast Beef: Representing the Vegetarian in Victorian Literature," in *Consuming Culture in the Long Nineteenth Century: Narratives of Consumption, 1700–1900*, ed. Tamara S. Wagner and Narin Hussain (Lanham, MD: Lexington Books, 2007), 25.

43. Robert Hunt, ed., *A Supplement to Ure's Dictionary of Arts, Manufactures and Mines* (New York: Appleton, 1866), 804.

44. Donald Cardwell, "Ure, Andrew (1778–1857)," in *Oxford Dictionary of National Biography* (Oxford: Oxford University Press, 2019), https://doi.org/10.1093/ref:odnb/28013; W. S. C. Copeman, "Andrew Ure, M.D., F.R.S. (1778–1857)," *Proceedings of the Royal Society of Medicine* 44, no. 8 (1951). John Hedley Brooke and Geoffrey Cantor describe Ure's geology as "scriptural" and subject to extensive critique from the likes of Adam Sedgwick; Brooke and Cantor, *Reconstructing Nature: The Engagement of Science and Religion* (Edinburgh: T&T Clark, 2000), 62, 71.

45. John Hughes Bennett, "Treatise on the Oleum Jecoris Aselli, or Cod Liver Oil, as a Therapeutic Agent in Certain Forms of Gout, Rheumatism, and Scrofula; with Cases," *British and Foreign Medical Review* 13, no. 25 (1842);

later reprinted as Bennett, *Treatise on the Oleum Jecoris Aselli, or Cod-Liver Oil, as a Therapeutic Agent in Certain Forms of Gout, Rheumatism, and Scrofula* (Edinburgh: MacLauchlan, Stewart, 1848), which drew on his experiences of practice in both Paris and Heidelberg.

46. John Hughes Bennett, *On Abnormal Nutrition, as Observed in Softening, Suppuration, Granulation, Reorganization of Tissue, Morbid Growths, &c.* (Edinburgh: Balfour and Jack, Printers, 1842), https://archive.org/details/b21475635.

47. Andrew Combe, *The Physiology of Digestion, Considered with Relation to the Principles of Dietetics*, 9th ed. (Edinburgh: MacLauchlan and Stewart, 1849), 64. Combe was a high-profile figure in early nineteenth-century medicine. His son George published an extensive series of papers and correspondence, including biographical reflections: George Combe, *The Life and Correspondence of Andrew Combe, M.D.* (Edinburgh: MacLauchlan and Stewart, 1850).

48. Andrew Combe, *The Physiology of Digestion, Considered with Relation to the Principles of Dietetics*, ed. James Coxe, 10th ed. (Edinburgh: MacLauchlan and Stewart, 1860), 53.

49. Ernst Feuchtersleben, *The Dietetics of the Soul* (London: John Churchill, 1852), 7. This was translated from the original German text: Ernst Feuchtersleben, *Zur Diätetik der Seele* (Wien, 1838).

50. Feuchtersleben, *Dietetics*, 3.

51. L. C. Burns, "A Forgotten Psychiatrist—Baron Ernst von Feuchtersleben," *Proceedings of the Royal Society of Medicine* 47, no. 3 (1954).

52. Feuchtersleben, *Dietetics*, 3.

53. Alfred Smee, *General Debility and Defective Nutrition: Their Causes, Consequences, and Treatment* (London: John Churchill, 1859). Smee's earlier works covered a wide range of topics, including the physical sciences and botany. See, for example, his *Elements of Electro-Metallurgy*, 2nd ed. (London: E. Palmer, 1843); and *The Potato Plant, Its Uses and Properties: Together with the Cause of the Present Malady. The Extension of That Disease to Other Plants, the Question of Famine Arising Therefrom, and the Best Means of Averting That Calamity* (New York: Wiley and Putnam, 1847).

54. Smee, *General Debility*, 2.

55. Smee, *General Debility*, 3 and 5. The elements in question, which according to Smee constituted all of the "food of man," were hydrogen, carbon, nitrogen, oxygen, sulfur, phosphorus, iron, potassium, sodium, chlorine, magnesium, and calcium.

56. Smee, *General Debility*, 10.

57. Smee, *General Debility*, 11.

58. Smee, *General Debility*, 16–17.

59. Smee, *General Debility*, 56.

60. Smee, *General Debility*, 56.

61. James Henry Bennet, *Nutrition in Health and Disease* (London: John Churchill, 1858), ix.

62. Bennet, *Nutrition in Health*, 1.

63. Bennet, *Nutrition in Health*, 3.

64. Bennet, *Nutrition in Health*, 14.

65. T[homas] Grainger Stewart, *On the Position and Prospects of Therapeutics: A Lecture Introductory to a Course on Materia Medica and Dietetics* (Edinburgh: Adam and Charles Black, 1862), 4.

66. William Brinton, *On Food and Its Digestion: Being an Introduction to Dietetics* (London: Longman, Green, Longman, and Roberts, 1861), v.

67. Brinton, *On Food*, 13–14. Liebig's contributions to organic chemistry are widely recognized; Neubauer and Vogel also published widely, both together and individually, in this area: Carl Neubauer and Julius Vogel, *A Guide to the Qualitative and Quantitative Analysis of the Urine, Designed for Physicians, Chemists and Pharmacists*, trans. Elbridge G. Cutler and Edward S. Wood (New York: William Wood, 1879).

68. Brinton, *On Food*, ix–xiv.

69. Brinton, *On Food*, 400.

70. Brinton, *On Food*, 400–401.

71. Brinton, *On Food*, 413.

72. Brinton, *On Food*, 428, 430.

73. Biographical dates for Alexander Murphy are unknown. Alexander Murray, *The Domestic Oracle: or, A Complete System of Modern Cookery and Family Economy Containing Directions for Purchasing, Keeping, and Dressing All Kinds of Butcher's Meat, Fish, Poultry, and Game* (London, 1850), 505.

74. Joel Pinney, *An Exposure of the Causes of the Present Deteriorated Condition of Health, and Diminished Duration of Longer Human Life, Compared with That Which Is Attainable by Nature* (London: Longman, Rees, Orme, Brown, and Green, 1830), title page, 144. Pinney went on to author at least two other texts exploring the merits of various lifestyle factors in determining health and longevity: *The Alternative: Disease or Premature Death, or Health and Long Life: Being an Exposure of the Prevailing Misconception of Their Respective Sources* (London: S. Highley, 1838); and *The Antidote for the Causes That Abridge the Natural Term of Human Existence; and an Outline of the Organs and Functions Subservient to Life* (London: S. Highley, 1847).

75. Pinney, *Exposure*, iv.

76. Pinney's last text explored connections between occupation and health, but was still rooted in personal experience, morality, and intemperance: *The Influence of Occupation on Health and Life, with a Remedy for Attaining the Utmost Length of Life Compatible with the Present Constitution of Man* (London: Longman, Brown, Green, and Longmans, 1856).

77. Michael Donovan, *Domestic Economy* (London: Longman, Orme,

Brown, Green, & Longman, 1837); Morse Peckham, "Dr Lardner's 'Cabinet Cyclopaedia,'" *Papers of the Bibliographical Society of America* 45, no. 1 (1951): 41.

78. Donovan, *Domestic Economy*, 2:251.

79. P[eter] Redfern, *On Anormal Nutrition in Articular Cartilages* (Edinburgh: Sutherland and Knox, 1849), 4. For more on Redfern, see "Peter Redfern, M.D. Lond., F.R.C.S.Eng., D.Sc," *British Medical Journal* 1, no. 2714 (January 4, 1913).

80. Richard Baron Howard, *An Inquiry into the Morbid Effects of Deficiency of Food, Chiefly with Reference to Their Occurrence amongst the Destitute Poor: Also Practical Observations on the Treatment of Such Cases* (London: Simpkin, Marshall; Manchester: George Sims, 1839), 26.

81. Biographical dates for William Grisenthwaite are unknown. William Grisenthwaite, *An Essay on Food* (London: William Crofts, 1838), 2.

82. "Art. XIII: An Essay on Food, in Which the Received Doctrine of Modern Physiologists Respecting the Waste of the Body Is Exploded," *British and Foreign Medical Review* 7 (1839): 538. In this text, it is clear that Grisenthwaite mobilized arguments that were straightforwardly natural theological in character, extolling the benefits of dietetics in harmony with natural laws, mobilizing Lavoisier, Davy, Priestley, and others in the process, according to a review of his book: "Grisenthwaite on Food," *Monthly Magazine*, March 1839. He also published on a broad range of other topics, including agriculture and the nature of genius. See, for example, William Grisenthwaite, *On Genius: In Which It Is Attempted to Be Proved, That There Is No Mental Distinction among Mankind* (London: Hamilton and Adams, 1830).

83. Charles T. Wolfe and Motoichi Terada, "The Animal Economy as Object and Program in Montpellier Vitalism," *Science in Context* 21, no. 4 (2008).

84. James Morison, *Morisoniana, or, Family Adviser of the British College of Health* (London: College of Health, 1831), 435.

85. James Morison, *The Hygeian Treatment of the Most Prevalent Diseases of India and of Warm Climates Generally* (London: George Taylor, 1836). Morison drew explicitly on the Greek figure of Hygeia, one of five deities who were associated with health and illness; he was far from alone in invoking classical traditions as a means of lending credibility to his system.

86. The preface was arranged as a dialogue between "The Doctor"—clearly Wilson himself—and the imagined "Reader," the latter remarking that "it's a *big* book." James Wilson, *The Principles and Practice of the Water Cure: And Household Medical Sciences: In Conversations on Physiology, on Pathology, or the Nature of Disease, and on Digestion, Nutrition, Regimen, and Diet* (London: John Churchill, 1854), xiii. The institution itself was far from outside the bounds of mainstream therapeutics; among the many high-profile visitors to Wilson and Gully's clinic were Alfred, Lord Tennyson, and Charles Darwin (who was a regular visitor and wrote to William Henry Benson on December 7, 1855 that "Dr Gully did

me *much* good"): Charles Darwin to William Henry Benson, December 7, 1855, DCP-LETT-4354, Darwin Correspondence Project, https://www.darwinproj ect.ac.uk/letter/?docId=letters/DCP-LETT-4354.xml.

87. Wilson, *Principles*, vii–viii.

88. Wilson, *Principles*, 163.

89. Wilson, *Principles*, 163–64. Wilson's concept of the animal economy likely stemmed from John Hunter's work, which had similarities with, but also some key differences from, the Montpellier school. See François Duchesneau, "Vitalism in Late Eighteenth-Century Physiology: The Cases of Barthez, Blumenbach and John Hunter," in *William Hunter and the Eighteenth-Century Medical World*, ed. W. F. Bynum and Roy Porter (Cambridge: Cambridge University Press, 1985).

90. Wilson, *Principles*, 171.

91. *Vital Nutrition: A Popular Application of the Principles of Modern Science to the Promotion of Health and Vital Energy* (London, 1859), 31.

92. *Vital Nutrition*, 32.

93. Jonah Horner, *Instruction to the Invalid on the Nature of the Water Cure: In Connection with the Anatomy and Physiology of the Organs of Digestion and Nutrition* (London: Simpkin, Marshall, 1855), 2.

94. Horner, *Instruction to the Invalid*, 3.

95. Horner, *Instruction to the Invalid*, 10.

96. Thomas Parry, *On Diet, with Its Influence on Man; Being an Address to Parents, &c., Or, How to Obtain Health, Strength, Sweetness, Beauty, Development of Intellect, and Long Life* (London: Samuel Highley, 1844), iii.

97. Biographical dates for Thomas Parry are unknown. Having addressed diets suitable for infants, children, and adolescents, Parry dedicated individual chapters to diets appropriate for the laboring man, mechanic, professional man, sinecurist, gentleman, and females, before considering diets calculated to promote "intellectual attainment," the latter of which "should be made upon corn food" on the grounds that "Luther wrote his great work upon bread and water, Newton his work upon light upon bread and water, and Byron his best productions upon biscuit and water"; Parry, *On Diet*, 118.

98. Henry C. Sherman, *The Nutritional Improvement of Life* (New York: Columbia University Press, 1950), quoted in E. N. Todhunter, "Some Aspects of the History of Dietetics," *World Review of Nutrition and Dietetics* 18 (1973): 3.

99. Benjamin, Count of Rumford, *Essays, Political, Economical, and Philosophical* (London: T. Cadell Junior and W. Davies, 1796), 1:191, original emphasis. Rumford's approach to diet has been described as "a Foucauldian disciplinary system of coercive, 'improving' institutions to hour the indigent poor, and a barley and legume-based gruel to feed them." Neswald, Smith, and Thoms, "Introduction," 6.

100. Dariush Mozaffarian, Irwin Rosenberg, and Ricardo Uauy, "His-

tory of Modern Nutrition Science—Implications for Current Research, Dietary Guidelines, and Food Policy," *British Medical Journal* (2018): 361:k2392, doi:https://doi.org/10.1136/bmj.k2392.

101. Bernard Lightman, "The Evolution of the Scientific Disciplines," in *Victorian Culture and the Origin of Disciplines*, ed. Lightman and Bennett Zon (New York: Routledge, 2020).

102. Neswald, Smith, and Thoms, "Introduction," 5.

Afterword

1. *Oxford English Dictionary*, s.v. "afterword," accessed October 30, 2020, https://doi.org/10.1093/OED/7753594635.

2. [William Whewell], *"On the Connexion of the Physical Sciences*, by Mrs. Somerville," *Quarterly Review* 51, no. 101 (March 1834): 59.

3. See Sydney Ross, "Scientist: The Story of a Word," *Annals of Science* 18, no. 2 (1962); see also Gowan Dawson and Bernard Lightman, eds., *Victorian Scientific Naturalism: Community, Identity, Continuity* (Chicago: University of Chicago Press, 2014), 3–10.

4. "On Leonardo da Vinci and Coreggio," *Blackwood's Edinburgh Magazine* 48, no. 398 (August 1840): 273.

5. Whewell, review, 58–59.

6. Whewell, review, 60.

7. Elizabeth S. Goodstein, *Georg Simmel and the Disciplinary Imaginary* (Stanford, CA: Stanford University Press, 2017).

8. Bernard Lightman and Bennett Zon, eds., *Evolution and Victorian Culture* (Cambridge: Cambridge University Press, 2014); and Lightman and Zon, eds., *Victorian Culture and the Origin of Disciplines* (New York: Routledge, 2019).

9. Bernard Lightman, chap. 1 in this volume.

10. Bernard Lightman, chap. 1 in this volume.

11. James F. Stark and Richard T. Bellis, chap. 10 in this volume.

12. Ian Hesketh, chap. 4 in this volume.

13. Helen Nicholson, editorial, *Research in Drama Education: The Journal of Applied Theatre and Performance* 11, no. 3 (2006): 172.

14. Christina Nadler, "Deterritorializing Disciplinarity: Towards an Immanent Pedagogy," *Cultural Studies ↔ Critical Methodologies* 15, no. 2 (2015): 145.

15. Lightman and Zon, *Victorian Culture*, 2–3.

16. Henri Bergson, *Creative Evolution*, trans. Arthur Mitchell (London: Macmillan, 1911), 362.

17. Bergson, *Creative Evolution*, 322.

18. Raymond Williams, "Structures of Feeling," in Williams, *Marxism and Literature* (Oxford: Oxford University Press, 1977), 128.

19. John Shotter, "More than Cool Reason: 'Withness-Thinking' or 'Sys-

temic Thinking' and 'Thinking about Systems,'" *International Journal of Collaborative Practices* 3, no. 1 (2012): 3.

20. Shotter, "More than Cool Reason," 4–5.

21. John Shotter, "Inside Processes: Transitory Understandings, Action Guiding Anticipations, and Withness Thinking," *International Journal of Action Research* 1, no. 2 (2005): 158.

22. Joe Moran, *Interdisciplinarity*, 2nd ed. (London: Routledge, 2010), 13, 71.

23. Steve Fuller, "Deviant Interdisciplinarity," in *Oxford Handbook of Interdisciplinarity*, ed. Robert Frodeman, Julie Thompson Klein, and Roberto C. S. Pacheco (Oxford: Oxford University Press, 2017), 54; Peter Osborne, "Problematizing Disciplinarity, Transdisciplinary Problematics," *Theory, Culture and Society* 32, no. 5–6 (2015); Marta B. Calas and Linda Smircich, "The Journey to Neo-disciplinarity," *Organization* 3, no. 2 (1996): 168.

24. Veronica Strang and Tom McLeish, *Evaluating Interdisciplinary Research: A Practical Guide*, Institute of Advanced Study, Durham University, UK, 2015.

25. Interdisciplinary Advisory Panel (IDAP), *Criteria Phase Report*, 2019, https://www.ref.ac.uk/panels/interdisciplinary-research-advisory-panel.

26. Arthur W. Frank, "What Is Dialogical Research, and Why Should We Do It?" *Qualitative Health Research* 15, no. 7 (2005): 968.

27. Scherto Gill, "'Holding Oneself Open in a Conversation'—Gadamer's Philosophical Hermeneutics and the Ethics of Dialogue," *Journal of Dialogue Studies* 3, no. 1 (2015): 20.

28. Hans-Herbert Kögler, "The Crisis of a Hermeneutic Ethic," *Philosophy Today* 58, no. 1 (2014).

29. Rebecca N. Mitchell, *Victorian Lessons in Empathy and Difference* (Columbus: Ohio State University Press, 2011), 22.

30. See Lightman and Zon, *Evolution and Victorian Culture*.

31. Audrey Jaffe, *Scenes of Sympathy: Identity and Representation in Victorian Fiction* (Ithaca, NY: Cornell University Press, 2000).

32. Nanna Katrine Lüders Kaalund, chap. 8 in this volume.

33. Efram Sera-Shria, chap. 9 in this volume.

34. Susan Blackmore, *The Meme Machine* (Oxford: Oxford University Press, 1999), 8.

35. Elsa Richardson, chap. 5 in this volume.

36. John Shotter, *Getting It: Withness-Thinking and the Dialogical . . . in Practice* (New York: Hampton Press, 2011), 125.

37. Adi Kidron and Yael Kali, "Boundary Breaking for Interdisciplinary Learning," *Research in Learning Technology* 23 (October 20, 2015), https://journal.alt.ac.uk/index.php/rlt/article/view/1646/xml_18.

38. Peter Reason and Brian Goodwin, "Towards a Science of Qualities

in Organizations: Lessons from Complexity Theory and Postmodern Biology," *Concepts and Transformation* 4, no. 3 (January 1999): 293, 281.

39. Shotter, "More Cool than Reason," 4.

40. Michelle Montague, "The Sense/Cognition Distinction," *Inquiry: An Interdisciplinary Journal of Philosophy* 66, no. 2 (December 30, 2018), https://www-tandfonline-com.ezphost.dur.ac.uk/doi/full/10.1080/0020174X.2018.1562371?scroll=top&needAccess=true.

41. See, for example, "Thought and Sense: On the Interface between Perception and Cognition," 2015–2018, University of Glasgow Centre for the Study of Perceptual Experience, https://www.gla.ac.uk/research/az/cspe/projects/thought-and-sense/.

42. Maggi Savin-Baden and Katherine Wimpenny, *A Practical Guide to Arts-Related Research* (Rotterdam: Sense Publishers, 2014), 143–44.

43. Raffaele De Luca Picione and Jaan Valsiner, "Psychological Functions of Semiotic Borders in Sense-Making: Liminality of Narrative Processes," *Europe's Journal of Psychology* 13, no. 3 (2017): 532.

44. Charles Darwin, "Notebook B: Transmutation of Species," 1837–1838, 36, DAR 121, Cambridge University.

45. Janet Browne, chap. 3 in this volume..

46. Shotter, *Getting It*, 100.

47. Shotter, *Getting It*, 100.

48. Istvan Kecskes, "From Pragmatics to Dialogue," in *The Routledge Handbook of Language and Dialogue*, ed. Edda Weigand (London: Routledge, 2017), 80.

49. Shotter, "More than Cool Reason," 4–5.

50. Iwan Rhys Morus, chap. 6 in this volume.

51. Shotter, "More than Cool Reason," 7.

52. See Jerry A. Jacobs, *In Defense of Disciplines: Interdisciplinarity and Specialization in the Research University* (Chicago: University of Chicago Press, 2013).

53. Tom McLeish, *The Poetry and Music of Science: Comparing Creativity in Science and Art* (Oxford: Oxford University Press, 2019), 27.

54. Whewell, review, 59.

55. James F. Stark and Richard T. Bellis, chap. 10 in this volume.

56. McLeish, *Poetry and Music of Science*, 28.

57. Hans-Georg Gadamer, *Truth and Method*, trans. Donald G. Marshal, rev. 2nd ed. (London: Bloomsbury, 2004), 113.

58. Gadamer, *Truth and Method*, 371.

59. Bruce Ellis Benson, *The Improvisation of Musical Dialogue: A Phenomenology of Music* (Cambridge: Cambridge University Press, 2003), 15.

60. Jeremy S. Begbie, *Theology, Music and Time* (Cambridge: Cambridge University Press, 2012), 204.

SELECTED BIBLIOGRAPHY OF SCHOLARLY SOURCES

This bibliography contains the key scholarly sources that have examined the relationship between interdisciplinarity and science in the Victorian period. Each contributor was asked to provide a list of the five most important scholarly sources they had come across as they conducted research for their chapter.

Alberti, Samuel. *Nature and Culture: Objects, Disciplines and the Manchester Museum.* Manchester: Manchester University Press, 2009.

Anderson, Amanda, and Joseph Valente, eds. *Disciplinarity at the Fin de Siècle.* Princeton, NJ: Princeton University Press, 2002.

Barton, Ruth. *The X Club: Power and Authority in Victorian Science.* Chicago: University of Chicago Press, 2018.

Beer, Gillian, M. Bowie, and B. Perrey, eds., *In(ter)discipline: New Languages for Criticism.* London: Routledge, 2007.

Bevir, Mark, ed. *Historicism and the Human Sciences in Victorian Britain.* Cambridge: Cambridge University Press, 2017.

Brooke, John Hedley. *Science and Religion: Some Historical Perspectives.* Cambridge: Cambridge University Press, 2014.

Buckland, Adelene, and Sadiah Qureshi. *Time Travelers: Victorian Encounters with Time and History.* Chicago: University of Chicago Press, 2020.

Bulstrode, Jenny. "The Industrial Archaeology of Deep Time." *British Journal for the History of Science* 49, no. 1 (2016): 1–25.

Chatterjee, Ronjaunee, Alicia Mireles Christoff, and Amy R. Wong. "Introduction: Undisciplining Victorian Studies." *Victorian Studies* 62, no. 3 (2020): 369–91.

Daston, Lorraine, and Peter Galison. *Objectivity.* New York: Zone Books, 2007.

Daston, Lorraine, and Elizabeth Lunbeck, eds. *Histories of Scientific Observation.* Chicago: University of Chicago Press, 2011.

Daunton, Martin, ed. *The Organisation of Knowledge in Victorian Britain.* Oxford: Oxford University Press, 2005.

Dawson, Gowan. *Darwin, Literature and Victorian Respectability.* Cambridge: Cambridge University Press, 2007.

Dawson, Gowan, and Bernard Lightman, eds. *Victorian Scientific Naturalism: Community, Identity, Continuity.* Chicago: University of Chicago Press, 2014.

Endersby, Jim. *Imperial Nature: Joseph Hooker and the Practices of Victorian Science.* Chicago: University of Chicago Press, 2008.

Finnegan, Diarmid. *The Voice of Science: British Scientists on the Lecture Circuit in Gilded Age America.* Pittsburgh: University of Pittsburgh Press, 2021.

Frodeman, Robert, Julie Thompson Klein, and Roberto C. S. Pacheco, eds. *Oxford Handbook of Interdisciplinarity.* Oxford: Oxford University Press, 2017.

Gadamer, Hans-Georg. *Truth and Method.* Translated by Donald G. Marshall. Rev. 2nd ed. London: Bloomsbury, 2004.

Gieryn, Thomas F. "Boundary-Work and the Demarcation of Science from Non-science: Strains and Interests in Profession Ideologies of Scientists." *American Sociological Review* 48, no. 6 (December 1983): 781–95.

Gooday, Graeme. *The Morals of Measurement: Accuracy, Irony and Trust in Late Victorian Electrical Practice.* Cambridge: Cambridge University Press, 2005.

Gooding, David. "In Nature's School: Faraday as an Experimentalist." In *Faraday Rediscovered: Essays on the Life and Work of Michael Faraday, 1791–1867,* edited by David Gooding and Frank James, 105–35. London: Macmillan, 1985.

Graff, Harvey J. *Undisciplining Knowledge: Interdisciplinarity in the Twentieth Century.* Baltimore: Johns Hopkins University Press, 2015.

Grasseni, Cristina, ed. *Skilled Visions: Between Apprenticeship and Standards.* Oxford: Berghahn Books, 2007.

Grayson, Donald K. *The Establishment of Human Antiquity.* New York: Academic Press, 1983.

Gregory, James. *Of Victorians and Vegetarians: The Vegetarian Movement in Victorian Britain.* London: Tauris, 2007.

Harrison, Peter. *The Territories of Science and Religion.* Chicago: University of Chicago Press, 2015.

Haushofer, Lisa. *Wonder Foods: The Science and Commerce of Nutrition*. Berkeley: University of California Press, 2022.

Hesketh, Ian. *The Science of History in Victorian Britain: Making the Past Speak*. Pittsburgh: University of Pittsburgh Press, 2011.

Holmes, John, and Sharon Ruston, eds. *The Routledge Research Companion to Nineteenth-Century British Literature and Science*. London: Routledge, 2017.

Hunt, Bruce. *The Maxwellians*. Ithaca, NY: Cornell University Press, 1991.

Hunt, Verity. "Narrativizing 'The World's Show': The Great Exhibition, Panoramic Views and Print Supplements." In *Popular Exhibitions, Science and Showmanship, 1840–1910*, edited by Joe Kember, John Plunkett, and Jill Sullivan, 115–32. New York: Pickering & Chatto, 2012.

Hyland, Ken. *Disciplinary Identities: Individuality and Community in Academic Discourse*. Cambridge: Cambridge University Press, 2012.

Jacobs, Jerry A. *In Defense of Disciplines: Interdisciplinarity and Specialization in the Research University*. Chicago: University of Chicago Press, 2013.

Klancher, Jon. *Transfiguring the Arts and Sciences: Knowledge and Cultural Institutions in the Romantic Age*. Cambridge: University of Cambridge Press, 2013.

Kohler, Robert E. *Landscapes and Labscapes: Exploring the Lab-Field Border in Biology*. Chicago: University of Chicago Press, 2002.

Leong, Elaine, and Sara Pennell. "Recipe Collections and the Currency of Medical Knowledge in the Early Modern 'Medical Marketplace.'" In *Medicine and the Market in England and Its Colonies, c. 1450–c.1850*, edited by Mark Jenner and Patrick Wallis, 133–52. Basingstoke, UK: Macmillan, 2007.

Levine, Philippa. *The Amateur and the Professional: Antiquarians, Historians and Archaeologists in Victorian England 1838–1886*. Cambridge: Cambridge University Press, 2008.

Lightman, Bernard, and Bennett Zon, eds. *Victorian Culture and the Origin of Disciplines*. New York: Routledge, 2019.

Mandler, Peter. *The English National Character: The History of an Idea from Edmund Burke to Tony Blair*. New Haven, CT: Yale University Press, 2006.

Moran, Joe. *Interdisciplinarity*. 2nd ed. London: Routledge, 2010.

Morrell, Jack, and Arnold Thackray. *Gentlemen of Science: Early Years of the British Association for the Advancement of Science*. Oxford: Clarendon Press, 1981.

Morus, Iwan. "More the Aspect of Magic than Anything Natural: The Philosophy of Demonstrations in Victorian Popular Science." In *Science in the Marketplace: Nineteenth-Century Sites and Experiences*, ed. Bernard Lightman and Aileen Fyfe, 336–70. Chicago: University of Chicago Press, 2007.

Neswald, Elizabeth, David F. Smith, and Ulrike Thoms, eds. *Setting Nutritional Standards: Theory, Policies, Practices*. Rochester, NY: University of Rochester Press, 2017.

Noakes, Richard. *Physics and Psychics: The Occult and the Sciences in Modern Britain*. Cambridge: Cambridge University Press, 2019.

O'Hara, J. May. "Foods or Medicine? A Study in the Relationship between Foodstuffs and Materia Medica from the Sixteenth to Nineteenth Century." *Transactions of the British Society for the History of Pharmacy* 1 (1971): 61–97.

Otter, Chris. "The British Nutrition Transition and Its Histories." *History Compass* 10, no. 11 (2012): 812–25.

Pandora, Katherine. "The Permissive Precincts of Barnum's and Goodrich's Museums of Micellaneity: Lessons in Knowing Nature for New Learners." In *Science Museums in Transition: Anglo-American Cultures of Display in the Nineteenth Century*, edited by Carin Berkowitz and Bernard Lightman, 36–64. Pittsburgh: University of Pittsburgh Press, 2017.

Qureshi, Sadiah. "Robert Gordon Latham, Displayed Peoples, and the Natural History of Race, 1854–1866." *Historical Journal* 54, no. 1 (2011): 143–66.

Rich, Rachel. *Bourgeois Consumption: Food, Space and Identity in London and Paris, 1850–1914*. Manchester: Manchester University Press, 2011.

Richards, Evelleen. *Darwin and the Making of Sexual Selection*. Chicago: University of Chicago Press, 2017.

Riper, A. Bowdoin Van. *Men among the Mammoths: Victorian Science and the Discovery of Human Prehistory*. Chicago: University of Chicago Press, 1993.

Rudwick, Martin. *Earth's Deep History: How It Was Discovered and Why It Matters*. Chicago: University of Chicago Press, 2014.

Schaffer, Simon. "Late Victorian Metrology and Its Instrumentation: A Manufactory of Ohms." In *Invisible Connections: Instruments, Institutions, and Science*, edited by Robert Bud and Susan E. Cozzens, 24–55. Bellingham, WA: SPIE Optical Engineering Press, 1992.

Secord, James. "Knowledge in Transit." *Isis* 95, no. 4 (2004): 654–72.

Sera-Shriar, Efram, ed. *Historicizing Humans: Deep Time, Evolution, and Race in Nineteenth-Century British Sciences*. Pittsburgh: University of Pittsburgh Press, 2018.

Sera-Shriar, Efram. *The Making of British Anthropology, 1813–1871*. London: Pickering & Chatto, 2013.

Shapin, Steven. *The Scientific Life: A Moral History of a Late Modern Vocation*. Chicago: University of Chicago Press, 2008.

Shotter, John. *Getting It: Withness-Thinking and the Dialogical . . . in Practice*. New York: Hampton Press, 2011.

Smith, Crosbie. *The Science of Energy: A Cultural History of Energy Physics in Victorian Britain*. London: Athlone Press, 1998.

Snow, C. P. *The Two Cultures and the Scientific Revolution*. Cambridge: Cambridge University Press, 1959.

Stocking, George W., Jr. *Victorian Anthropology*. New York: Free Press, 1987.

Strang, Veronica, and Tom McLeish. *Evaluating Interdisciplinary Research: A Practical Guide*. Institute of Advanced Study, Durham University, 2015.

Topham, Jonathan R. "Scientific Publishing and the Reading of Science in Nineteenth-Century Britain: A Historiographical Survey and Guide to Sources." *Studies in History and Philosophy of Science Part A* 31, no. 4 (2000): 559–612.

Treitel, Corinna. *Eating Nature in Modern Germany: Food, Agriculture and Environment c. 1870–2000*. Cambridge: Cambridge University Press, 2017.

Turner, Frank M. *Contesting Cultural Authority: Essays in Victorian Intellectual Life*. Cambridge: Cambridge University Press, 1993.

Wagner, Tamara S., and Narin Hussain, eds. *Consuming Culture in the Long Nineteenth Century: Narratives of Consumption, 1700–1900*. Lanham, MD: Lexington Books, 2007.

Warwick, Andrew. *Masters of Theory: Cambridge and the Rise of Mathematical Physics*. Chicago: University of Chicago Press, 2003.

Weigand, Edda, ed. *The Routledge Handbook of Language and Dialogue*. London: Routledge, 2017.

Wilson, Adrian. "Science's Imagined Pasts." *Isis* 108, no. 4 (2017): 814–26.

Wise, M. Norton, ed. *The Values of Precision*. Princeton, NJ: Princeton University Press, 1995.

Yeo, Richard. *Defining Science: William Whewell, Natural Knowledge and Public Debate in Early Victorian Cambridge*. Cambridge: Cambridge University Press, 1993.

NOTES ON CONTRIBUTORS

Richard Bellis is associate lecturer in medical humanities at the University of St. Andrews Medical School. He is a historian of medicine, specializing in the history of anatomy, disease, and science communication in the eighteenth and nineteenth centuries. He has published on subjects such as the use of the senses in eighteenth-century anatomical investigations, and the use of epistemic genre in medical case histories. He is currently working on a monograph on Matthew Baillie and the development of morbid anatomy in Britain.

Janet Browne is the Aramont Professor Emerita in the History of Science at Harvard University. She is now retired. Her interests range widely over the history of biology. She is best known for a two-volume biography of Darwin (Knopf, 1995, 2002) that was awarded the James Tait Black award for non-fiction, the W. H. Heinemann Prize from the Royal Literary Society, and the Pfizer Prize from the History of Science Society. Before moving to Harvard, she was based at the Wellcome Trust Centre for the History of Medicine at University College London. She has an honorary degree from Trinity College Dublin, 2009.

Geoffrey Cantor is professor emeritus of the history of science at the University of Leeds. His publications include *Michael Faraday: Scientist and Sandemanian*

(Macmillan, 1991), *Quakers, Jews, and Science* (Oxford University Press, 2005), *Religion and the Great Exhibition of 1851* (Oxford University Press, 2011), and with John Hedley Brooke, *Reconstructing Nature: The Engagement of Science and Religion* (the 1995–1996 Gifford Lectures at Glasgow; T&T Clark, 1998; Oxford University Press, 2000). He is also an accredited mental health mentor working principally with PhD students.

Ian Hesketh is associate professor of history at the University of Queensland. He is the author of *A History of Big History* (Cambridge University Press, 2023), *Victorian Jesus: J. R. Seeley, Religion, and the Cultural Significance of Anonymity* (University of Toronto Press, 2017), *The Science of History in Victorian Britain* (University of Pittsburgh Press, 2020), *Of Apes and Ancestors: Evolution, Christianity and the Oxford Debate* (University of Toronto Press, 2009), and editor of *Imagining the Darwinian Revolution: Historical Narratives of Evolution from the Nineteenth Century to the Present* (University of Pittsburgh Press, 2022).

Nanna Katrine Lüders Kaalund is a postdoctoral research associate at Aarhus University, Denmark, where she holds a Carlsberg Fellowship. The work presented here is supported by the Carlsberg Foundation, grant CF20-0180. She previously worked at the University of Cambridge and the University of Leeds, and is the author of the book *Explorations in the Icy North: How Travel Narratives Shaped Arctic Science in the Nineteenth Century* (University of Pittsburgh Press, 2021). Her research examines the intersection of Arctic exploration, race, print culture, science, religion, and medicine in the modern period with a focus on the British, North American, and Danish imperial worlds.

Bernard Lightman is distinguished research professor in the Humanities Department at York University, and past president of the History of Science Society. Lightman's research focuses on the cultural history of Victorian science. Among his most recent publications are the edited collections *Rethinking History, Science and Religion, Science Periodicals in Nineteenth Century Britain,* and *Identity in a Secular Age* (coedited with Fern Elsdon-Baker). He is one of the general editors of the John Tyndall Correspondence Project, an international collaborative effort to obtain, digitalize, transcribe, and publish all surviving letters to and from Tyndall.

Chris Manias is senior lecturer in the history of science and technology at King's College London. His research focuses on the development and cultural implications of the human, evolutionary, and deep-time sciences. He studied for his PhD at Birkbeck and has previously held positions at the German Historical Institute London and the universities of Bristol, Exeter, and Manchester. He is the author of *Race, Science and the Nation: Reconstructing the Ancient Past in Britain, France and Germany, 1800–1914* (Routledge, 2013) and *The Age of Mammals: Nature, Development and Paleontology in the Long Nineteenth Century* (University of Pittsburgh Press, 2023).

Iwan Rhys Morus is professor of history at Aberystwyth University. He has published extensively on Victorian science and culture, most recently *Nikola Tesla and the Electrical Century* (Icon Books, 2019) and *How The Victorians Took Us to the Moon* (Icon Books, 2022). He is a fellow of the Royal Historical Society and of the Learned Society of Wales.

Elsa Richardson is a chancellor's fellow in the history of health and well-being at the University of Strathclyde, Glasgow. Her research examines British health cultures in the nineteenth and twentieth centuries, touching on topics as diverse as Christianity, neurology, folklore, surgery, spiritualism, nutrition, literature, and science. She has published on histories of the supernatural, psychoanalysis, vegetarianism, dietetics, and health food, and is the author of *Second Sight in the Nineteenth Century: Prophecy, Imagination and Nationhood* (Palgrave Macmillan, 2017) and the forthcoming *Rumbles: A Curious History of the Gut* (Palgrave Macmillan, 2024).

Efram Sera-Shriar is a Copenhagen-based historian and writer. He received his PhD from the University of Leeds and has worked in higher education and the museum sector for nearly twenty years. As an associate professor in English studies at the University of Copenhagen, he teaches about the history and culture of the English-speaking world. He is also associate director of research for the Centre for Nineteenth-Century Studies International at Durham University. Prior to taking on these roles, he was a senior researcher and research grants manager for the Science Museum Group in the UK and lecturer of modern history at Leeds Trinity University. His major works include *The Making of British Anthropology, 1813–1871* (Pickering & Chatto, 2013) and *Psychic Investigators: Anthropology, Modern Spiritualism, and Credible Witnessing in the Late Victorian Age* (University of Pittsburgh Press, 2022).

James Stark is professor of medical humanities at the University of Leeds. He is a historian of modern medicine and science with wide-ranging interests across the field, including the histories of advertising, food, ageing, and microbes. His publications include two monographs, *The Making of Modern Anthrax* (Pickering & Chatto, 2013; University of Pittsburgh Press, 2020) and *The Cult of Youth* (Cambridge University Press, 2020). In addition to his own scholarly work, he has recently completed a three-year term as director of the Leeds Arts and Humanities Research Institute (2020–2023) and is a founding Advisory

Board member of The Humanities in the World book series, published by the University of Rochester Press.

Bennett Zon is professor of music at Durham University and director of its Centre for Nineteenth-Century Studies. He is general editor of *Nineteenth-Century Music Review* and the Music in Nineteenth-Century Britain book series, and an editor of the *Yale Journal of Music and Religion*, the Congregational Music Studies book series, and *Nineteenth-Century Contexts*. He was recently elected inaugural president of the International Nineteenth-Century Studies Association. Zon researches the relationship of music, religion, and science in the long nineteenth century. Recent books include *Evolution and Victorian Musical Culture* (2017) and *Victorian Culture and the Origin of Disciplines* (Cambridge University Press, 2020), coedited with Bernard Lightman.

INDEX

Moran, Joe, 229
Morgan, Augustus de, 186
Morison, James, 220
Morley, John, 93
Moses, William Stainton, 197
Murray, Alexander, 217
Murray, James, 5
Murray, John 75–77
Myers, Frederic W. H., 203

Nares, George Strong, 13, 159,
    162–64, 167, 173, 174, 177, 180
natural history, 4, 6, 7, 20–21, 222
natural philosophy, 6, 7
Nature, 23, 25, 34
Neubauer, Carl, 216
Nevill, Dorothy, 73
Newth, George, 56
Newton, Isaac, 85
Nicholson, William, 43, 46
North British Physicists, 31
nutrition, 13–14, 204–23

Oliver, Daniel, 174
Owen, Richard, 68, 78
Owen, Robert, 186
Owens College, 155
Paget, James, 116, 117
Palaeontographical Society, 77
paleontology, 12, 139–57
Paley, William, 66, 69
Panopticon of Science and Art, 108
Parry, Thomas, 222
Pasteur, Louis, 108, 110, 172
Pattison, Mark, 86
Pavy, Frederick William, 112
Pepper, John Henry, 133
Perthes, Jacques Boucher de Crève-
    coeur, 142
Petermann, August, 160
Pharmaceutical Society, 8
Philosophical Club, 123

Physical Society, 135
physics, 12, 118–36
Picione, Raffaele De Luca, 233
Pickering, Edward Charles, 130
Pinney, Joel, 217–18
Podmore, Frank, 187
Polar Committee of the Royal Soci-
    ety, 166
Powell, Baden, 65
prehistory, 139–57
Prestwich, Joseph, 141, 143, 145–47
Prichard, James Cowles, 3, 149, 171
professionalization, 9–10, 20, 37, 80
psychical research, 13

Queen Victoria, 215

Ramsay, William 51–52, 56
Rance, C. E. De, 175
Rayleigh, Lord, *See* Strutt, John
    William
Ray, John, 77
Ray Society, 77
Reason, Peter, 232
Redfern, Peter, 218
Richardson, John, 164
Rink, Hinrich, 169
Rivers, Augustus Pitt, 170, 171, 176
Roberts, Mrs., 186
Roscoe, H. E., 11, 22, 25–27, 29–32,
    50–51, 56
Ross, John, 164
Rousseau, E., 53
Royal College of Chemistry, 52
Royal Geographical Society, 161,
    167, 169, 170
Royal Institution, 88, 123–25, 132,
    143
Royal Society of London, 43, 62,
    122–23, 143, 146, 161, 165–67,
    169
Rue, Warren de la, 133